DECISION MAKING IN SYSTEMS ENGINEERING AND MANAGEMENT

DECISION MAKING IN SYSTEMS ENGINEERING AND MANAGEMENT

Second Edition

Edited by

GREGORY S. PARNELL
PATRICK J. DRISCOLL
DALE L. HENDERSON

A JOHN WILEY & SONS, INC., PUBLICATION

Published by John Wiley & Sons, Inc., Hoboken, New Jersey
Published simultaneously in Canada

For general information on our other products and services or for technical support, please contact our
Customer Care Department within the United States at 877-762-2974, outside the United States at
317-572-3993 or fax 317-572-4002.

Wiley also publishes its books in a variety of electronic formats. Some content that appears in print
may not be available in electronic formats. For more information about Wiley products, visit our web
site at www.wiley.com.

Library of Congress Cataloging-in-Publication Data:

Decision making in systems engineering and management / [edited by] Gregory S. Parnell, Patrick J.
Driscoll, Dale L. Henderson.—2nd ed.
    p. cm.—(Wiley series in systems engineering and management ; 79)
    ISBN 978-0-470-90042-0 (hardback)
    1. Systems engineering–Management. 2. Systems engineering–Decision making. I. Parnell,
Gregory S. II. Driscoll, Patrick J. III. Henderson, Dale L.
    TA168.D43 2010
    620.001'171–dc22

                                                                                        2010025497

Printed in the United States of America

oBook ISBN: 978-0-470-92695-6
ePDF ISBN: 978-0-470-92695-6
ePub ISBN: 978-0-470-93471-5

10   9   8   7

Systems engineers apply their knowledge, creativity, and energy to making things better. Rarely do we assume grave personal risk to do so.

We dedicate this book to our colleagues from the Department of Systems Engineering at The United States Military Academy who have sacrificed their lives to make the world a place where systems engineers are free to make things better.

Contents

13 Solution Implementation   447
Kenneth W. McDonald and Daniel J. McCarthy

14 Summary   477
Gregory S. Parnell

Foreword to the Second Edition

The first edition of this book was developed by the faculty of the Department of Systems Engineering at the United States Military Academy and two colleagues at the University of Arkansas. We used the book in draft and final form for four years as a text for undergraduate courses and professional continuing education courses for systems engineers and engineering managers, and the book has been used as a text for undergraduate and graduate courses at other universities. In addition, we used the foundational material on systems thinking, systems engineering, and systems decision making on very diverse and important research and consulting projects by our students and faculty. The development and use of this text resulted in restructuring part of our curriculum and has significantly improved our academic programs and the research of our faculty and our students.

However, we have continued to develop new material and refine the techniques that we use to present the material. The second edition keeps the problem-solving focus on systems thinking, systems engineering, and systems decision making but incorporates our learning based on teaching students and helping senior leaders solve significant challenges in many important problem domains.

The major changes include an increased focus on risk analysis as a key tool for systems thinking and decision making; explicit inclusion of cost analysis in our solution design phase; additional techniques for the analysis of uncertainty and risk in the decision making phase; and a revised solution implementation chapter more aligned with project management literature.

With the new material, this second edition can be used as an undergraduate or a graduate text in systems engineering, industrial engineering, engineering management, and systems management programs. In addition, the book is an excellent resource for engineers and managers whose professional education is not in systems engineering or engineering management.

We hope that the material in this book will improve your problem solving skills by expanding your system thinking ability, increasing your understanding of the roles of systems engineers, and improving the systems decision making processes required to solve the complex challenges in your organization.

BRIGADIER GENERAL TIM TRAINOR, PH.D.
Dean of the Academic Board

United States Military Academy
West Point, New York
September 2010

Foreword to the First Edition

The Department of Systems Engineering is the youngest academic department at the United States Military Academy. Established in 1989, the department has developed into an entrepreneurial, forward-looking organization characterized by its unique blend of talented military and civilian faculty. This book is our effort to leverage that talent and experience to produce a useful undergraduate textbook focusing on the practical application of systems engineering techniques to solving complex problems. Collectively, the authors bring nearly two centuries of experience in both teaching and practicing systems engineering and engineering management. Their work on behalf of clients at the highest levels of government, military service, and industry spans two generations and a remarkably broad range of important, challenging, and complex problems. They have led thousands of systems engineering, engineering management, information engineering, and systems management students through a demanding curriculum focused on problem solving.

Teaching systems engineering at the undergraduate level presents a unique set of challenges to both faculty and students. During the seven years I served as the department head, we searched for a comprehensive source on systems engineering for undergraduates to no avail. What we found was either too narrowly focused on specific areas of the systems engineering process or more intended for practitioners or students in masters or doctoral programs.

While conceived to fill the need for an undergraduate textbook supporting the faculty and cadets of the United States Military Academy, it is designed to be used by faculty in any discipline at the undergraduate level and as a supplement to graduate level studies for students who do not have a formal education or practical experience in systems engineering.

The book is organized around the principles we teach and apply in our research efforts. It goes beyond exposing a problem-solving procedure, offering students the opportunity to grow into true systems thinkers who can apply their knowledge across the full spectrum of challenges facing our nation.

BRIGADIER GENERAL (Ret.) MICHAEL MCGINNIS, PH.D.

Formerly
Professor and Head,
Department of Systems Engineering, 1999–2006
United States Military Academy

Executive Director
Peter Kiewit Institute
University of Nebraska

Preface to the Second Edition

WHAT IS THE PURPOSE OF THE BOOK?

The purpose of this book is to contribute to the education of systems engineers by providing them with the concepts and tools to successfully deal with systems engineering challenges of the twenty-first century. The book seeks to communicate to the reader a philosophical foundation through a systems thinking world view, a knowledge of the role of systems engineers, and a systems decision process (SDP) using techniques that have proven successful over the past 20 years in helping to solve tough problems presenting significant challenges to decision makers. This SDP applies to major systems decisions at any stage of their system life cycle. The second edition makes several important refinements to the SDP based on our teaching and practice since the first edition was published in 2008. A sound understanding of this approach provides a foundation for future courses in systems engineering, engineering management, industrial engineering, systems management, and operations research.

WHAT IS THIS BOOK?

This book provides a multidisciplinary framework for problem solving that uses accepted principles and practices of systems engineering and decision analysis. It has been constructed in a way that aligns with a structure moving from the broad to the specific, using illustrative examples that integrate the framework and demonstrate the principles and processes for systems engineering. The book is not a detailed engineering design book nor a guide to system architecting. It is a complement to engineering design and system architecting. It introduces tools and techniques sufficient for a complete treatment of systems decision making with references for future learning. The text blends the mathematics of multiple objective decision analysis with select elements of stakeholder theory, multi-attribute value theory, risk analysis, and life cycle cost analysis as a foundation for trade studies and the analysis of design solutions.

WHO IS THIS BOOK FOR?

The first edition of this book was intended primarily to be a textbook for an undergraduate course that provides an introduction to systems engineering or systems management. Based on the recommendations and requests from a host of academic and professional practitioners, this second edition extends much of the existing material and adds new material to enable the book to be comfortably adopted as a graduate text or a text in support of professional continuing education while remaining a valuable resource for systems engineering professionals. The book retains all of the features that readers identified as useful for any individual who is leading or participating in a large, complex systems engineering or engineering management process. Not surprisingly, readers of the first edition have highlighted the usefulness of the approach we present to other disciplines as well, such as human factors engineering, law, history, behavioral sciences, and management, in which the object of focus can be conceptualized as a system.

WHY DID WE WRITE THIS BOOK?

We authored the first edition of this book to fill a critical gap in available resources that we (and others) needed to support systems engineering projects that our faculty, and hence our students as future systems engineers, were being asked to engage with concerning high-visibility, high-impact systems in both government and corporate settings. Moreover, it was nearly always the case in these projects that key stakeholders vested in the potential solutions demanded-large amounts of decision support throughout the engagement horizon. Thus, systems engineering with a systems decision-making emphasis had evolved to be our primary professional practice with clients and yet the field was lacking a single source that students and practitioners could turn to for guidance.

Specifically, there were three immediate needs driving us to the task. First, we needed a textbook for our lead-in systems engineering courses offered by the Department of Systems Engineering at the United States Military Academy at West Point. Second, we needed to more fully describe the problem solving process that we developed and successfully applied since the Systems Engineering Department was formed in 1989. The process introduced in this book, called the systems decision process (SDP), is the refined version of this process we currently use. Lastly, we wanted to document the problem solving lessons we have learned by hard knocks, happenstance, and good fortune as leaders, military officers, engineering managers, systems engineers, teachers, and researchers.

We teach two foundational systems engineering undergraduate courses at West Point that serve a broad clientele. SE301, Foundations of Engineering Design and System Management, is the first course we offer to our approximately 100 academic majors each year. These majors include systems engineering, engineering management, and systems management. The first two of these are programs accredited by ABET Inc.

This is the course where our faculty make "first contact" with each new class of talented students. Based on a host of discussions with students, faculty, and external stakeholders to our curriculum, we concluded that this needed to be the flagship course of the department, taught by our most experienced faculty; to communicate a fundamentally different thought process than that emphasized by other engineering fields; and to change the way our students thought about problem solving and their role in the process. Moreover, the course needed to set the professional standards required to put our students in front of real-world clients with real-world systems decision problems at the start of their senior year, to support the requirement of their year-long senior capstone experience.

The other course, SE300, Introduction to Systems Engineering, is the first course in a three-course Systems Engineering sequence taken by 300–400 nonengineering majors each year. Rather than simply providing an introduction to a field that was not their academic major, we structure this course to deliver value to the students both in their chosen majors and as future decision makers in their role as military officers. These design considerations became part of our plan for the first edition of the textbook, and we retained these for the second edition as well.

HOW DID WE WRITE THE BOOK?

We wrote the book in the manner that we advocate good systems engineering be applied in practice. The editors led a team effort that leveraged the expertise of each of the authors, several of whom were personally responsible for the structure of the downstream courses for each of our academic majors. In this manner, each author could craft critical material in direct support of later courses so that the book retained value as a reference beyond the initial program course.

A host of regularly scheduled collaboration and communication sessions were used to develop and refine the terminology, content, and voice used throughout the book. The concept maps in each chapter serve two purposes. First, they define the key concepts of the chapter. Second, they help us identify a common lexicon for the book. Since the book includes a systems decision process, we tried to incorporate several illustrative examples as an integrating tool that would carry the reader through the various systems decision process chapters. Our faculty and students read and evaluated each of the chapters for clarity, consistency, and ease of use.

As with most iterative processes, we learned a great deal about our own programs in the process. The writing of this book became a wonderful means of cross-leveling knowledge and understanding among the faculty as to the emphasis and content that was being taught across our curriculum. This book and the approach contained within have significantly contributed to our curriculum assessment process, enabling us to more clearly articulate program and course outcomes and objectives in a manner that communicates value return while aligning with accepted professional standards. Valuable feedback from faculty and students using the initial three preliminary printings and the first edition has been incorporated into this edition.

HOW IS THIS BOOK ORGANIZED?

The book is organized in three parts. Part I provides an introduction to systems thinking, system life cycles, risk management, systems modeling and analysis, and life cycle costing. Part II provides an introduction to systems engineering, the practice of systems engineering, and systems effectiveness. Part III introduces the systems decision process (SDP) and describes the four phases of our systems decision process: problem definition, solution design, decision making, and solution implementation, in addition to the primary environmental factors that house important stakeholders and their vested interests. The systems decision process can be used in all stages of a system life cycle. The final chapter provides a summary of the book.

Gregory S. Parnell and Patrick J. Driscoll

West Point, New York
July 2010

Acknowledgments

We would like to acknowledge several individuals for their contributions and support for this second edition. Our design editor, Dale Henderson, again did a superb job on many design details that add quality to this work. The department leadership under COL Robert Kewley continues to provide great support and encouragement for the project. Thanks also go to many of the U.S. Military Academy Department of Systems Engineering faculty contributed to what was to become the Systems Decision Process (SDP).

The editors would like to thank the chapter authors for their hard work and flexibility as we defined and refined many of the concepts included in the book. Crafting a text such as this is a challenging undertaking. Having a tight production schedule adds to this challenge in a significant way. Their continuing level of patience, professionalism, and commitment to the project is acknowledged with our heartfelt gratitude.

A great example of this flexibility was how the Rocket Problem, developed for the first edition by Dr. Paul West, was quickly accepted and used as the example to present the concepts in Chapters 10–13. It continues to prove its usefulness for many of the extended concepts and new material of this second edition. We would also like to acknowledge COL Kewley's development of the Curriculum Management System example, along with the real system that has been implemented at our institution as a result. We also thank COL Donna Korycinski for a very careful read of the initial manuscript and many helpful suggestions for clarification.

We continue to extend thanks to the many, many cadets who have taken courses in the Department of Systems Engineering. We honor their commitment to service with our best efforts to inspire and lead them. Their enthusiasm and high standards make us all better teachers and better leaders. Finally, the entire project team would like to thank their families for their selfless support and encouragement during this demanding book project.

G. S. P.
P. J. D.

Thoughts for Instructors

COURSE DESIGN USING THE BOOK

This book has been designed as a systems engineering and management textbook and as a reference book for systems engineers and managers. There are lots of ways to use this material for undergraduate and graduate courses. Chapter 1 is always a good place to start! Part I (Chapters 2 through 5) present systems thinking. Most courses would probably want to start with at least Chapters 2 and 3 to set a good foundation in systems thinking and the system life cycle. Chapters 4 and 5 can be introduced next or during presentation of the systems decision process in Part III. Part III is designed to be presented sequentially but is based on knowledge provided in Chapter 1 through Chapter 5. Chapters 6 and 7 introduce systems engineering and describe systems engineering practice. They can be presented before or after Part III. The most advanced mathematics of the book is in Chapter 8, and Chapter 11, Section 11.4. These can be omitted in an introductory course since they may be covered in other courses in your student's academic program. Instructors will want to supplement the course with additional material.

AN EXAMPLE UNDERGRADUATE COURSE DESIGN

We use the text for our undergraduate systems engineering and management fundamentals course, our introduction to systems engineering course for nonengineering majors, and our year long capstone design course for academic majors. The fundamentals course is taken by our systems engineering, engineering management, and systems management majors, whereas the introductory course is the first of a three course systems engineering sequence taken annually by about

350–400 students. The capstone design course is the final, integrative experience for our students. We have found it useful to have the students learn the systems decision process from three perspectives: a personal systems decision with known or relatively easy to determine alternatives (e.g., buying a car); a complex systems integration problem involving multiple decision makers and stakeholders (e.g., adding new components to perform new missions with an existing unmanned aircraft system); and a complex systems design involving multiple stakeholders with challenging implementation issues (e.g., the IT illustrative example presented at the end of each chapter in Part III of the text).

Figure 0.1 provides the flow of the course material using this approach. We begin with Chapters 1 through 3 to provide an introduction to the course material

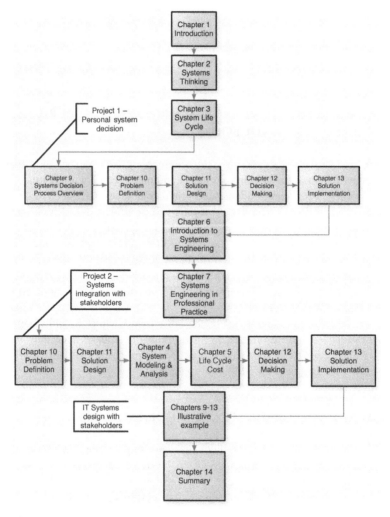

Figure 0.1   Course design with two projects and one illustrative example.

and a good understanding of systems thinking and the system life cycle. Next, we introduce Project 1, a system decision problem that the students may encounter in which, as the assumed primary decision maker, they can easily determine their values, solution alternatives, measure scores, and basic life cycle costs. Example problems might be buying a car or selecting a graduate program of study. The students read Chapter 9 and the introductory parts of the four chapters describing the four phases in the systems decision process (Chapters 10–13). They then apply these concepts to their system decision problem. The effort culminates with a presentation and a paper that demonstrate the degree to which each student understands the multiple objective decision analysis (MODA) mathematics used to evaluate the value of the alternatives. Following this, we present the fundamentals and the practice of systems engineering using Chapters 6 and 7. This is also a good time to give the first exam.

Next, we introduce Project 2. For this project, we look to a systems integration and/or systems design project that has one or more decision makers and multiple stakeholders influencing the system requirements and subsequent trade space. We require the students to perform more extensive research, stakeholder interviews and surveys to develop the data and modeling components required by the MODA approach. Proceeding to Chapters 10 to 13, we introduce additional material to help the students address design and analysis issues associated with more complex systems decision problems. Modeling and simulation techniques introduced in Chapter 4 are used for solution design and evaluation. Time permitting, we include material from Chapter 5 addressing life cycle cost estimating.

While the students are completing their analysis of Project 2, we discuss the design of a system from system need to implementation. The IT illustrative example presented at the end of Chapters 9–13 was included in the book to provide an example of a complete application of the systems decision process. We conclude the Project 2 effort with student presentations and a summary of the course.

EXAMPLE GRADUATE PROGRAM SUPPORT

As mentioned previously, we received a significant number of suggestions for enhancements to the book from academicians and practitioners since the publication of the first edition. A number of these expressed a desire to use the book in support of their graduate programs or for workshops they were offering as continuing professional education. Figure 0.2 shows one perspective that might be helpful in this regard. It describes how that each chapter might support program and course objectives for a select number of graduate programs listed. It is intended purely as illustrative course topic coverage based on the editors' experience teaching courses in these types of programs. Any specific curriculum design would and should obviously be driven by the academic program specifics and course objectives. In addition, several of the chapters include material and associated mathematical content that may be appropriate for advanced undergraduate or graduate courses. These are predominantly:

Program / Chapter	Industrial & Systems Engineering	Engineering Management	Traditional Engineering	Acquisition Management/ Professional Continuing Education
1. Introduction	Introduces key systems concepts and role of stakeholders	Introduces key systems concepts and role of stakeholders	Introduces key systems concepts and role of stakeholders	Introduces key systems concepts and role of stakeholders
2. Systems Thinking	Foundational systems thinking principles and techniques	Foundational systems thinking principles and techniques	Foundational systems thinking principles and techniques	Foundational systems thinking principles and techniques
3. Systems Life Cycle	Importance of life cycle and risk management for system solutions	Importance of life cycle and risk management for system management	Importance of life cycle and risk management for system engineered solutions	Optional
4. Systems Modeling & Analysis	Introduces roles and techniques of M&S in systems analysis	Introduces roles and techniques of M&S in systems analysis	Introduces roles of M&S in systems analysis	Introduces roles of M&S in systems analysis
5. Life Cycle Costing	Supplements engineering economy course	Supplements engineering economy course	Foundational cost analysis principles and techniques	Foundational cost analysis principles and techniques
6. Introduction to SE	Provides setting and context for SE and system complexity	Provides setting and context for SE and system complexity	Provides setting and context for SE and system complexity	Provides setting and context for SE and system complexity
7. SE in Professional Practice	Help system engineers understand the roles and activities of SEs	Helps engineering manager understand the roles of SE	Help traditional engineers understand the role of SE	Helps acquisition manager understand the role of SE
8. Systems Reliability	Overview of systems effectiveness principles and techniques	Optional	Overview of systems effectiveness principles and techniques	Optional
9. Systems Decision Process Overview	Useful to introduce and compare SDP with other DM processes	Useful to introduce and compare SDP with other DM processes	Useful to introduce and compare SDP with other DM processes	Useful to introduce and compare SDP with other DM processes
10. Problem Definition	Emphasizes importance of problem definition and demonstrates key qualitative techniques	Emphasizes importance of problem definition and demonstrates key qualitative techniques	Emphasizes importance of problem definition and demonstrates key qualitative techniques	Emphasizes importance of problem definition and demonstrates key qualitative techniques
11. Solution Design	Teaches fundamental system design principles and techniques	Teaches fundamental system design principles and techniques	Shows relationship of traditional engineering design in systems design	Teaches fundamental system design principles and techniques
12. Decision Making	Demonstrates sound mathematical techniques for trade studies and analysis of alternatives and presentations to DMs	Demonstrates sound mathematical techniques for trade studies and analysis of alternatives and presentations to DMs	Demonstrates sound mathematical techniques for trade studies and analysis of alternatives and presentations to DMs	Demonstrates sound mathematical techniques for trade studies and analysis of alternatives and presentations to DMs
13. Solution Implementation	Introduces key project management techniques	Reinforces project management techniques	Introduces key project management techniques	Reinforces project management techniques
14. Summary	Summarizes key messages of the book	Summarizes key messages of the book	Optional	Optional

Figure 0.2   Example graduate program topical support.

- Chapter 5: Life Cycle Costing, CER
- Chapter 8: System Reliability
- Chapter 11: Solution Design (section on experimental design and response surface methodology)
- Chapter 12: Decision Making (the sections on Decision-Focused Transformation, Monte Carlo simulation, and decision trees)

Presentation and decision quality	Criteria and grade			
	Worst	Adequate	Very good	Ideal
Bottom line up front	0	7	9	10
	Not used	One bullet chart that summarizes the major results.	Bullet and graphs or pictures that illustrate the major results.	Briefing could be presented in one chart.
1. Appropriate frame	0	14	18	20
	No problem definition	Problem definition based on stakeholder analysis performed using interviews and surveys. Findings, conclusions, and recommendations.	Insightful problem definition clearly supported by stakeholder analysis.	New insights provided to the client.
2. Creative doable alternatives	0	14	18	20
	No alternatives	Alternative generation table used to generate feasible alternatives.	Alternatives identified that have potential to provide high value.	Create alternatives client has not considered.
3. Meaningful, reliable information	0	14	18	20
	No documentation on presentation.	Adequate documentation for values and scores. Assumptions identified.	Appropriate sources.	Very credible sources.
4. Clear values and tradeoffs	0	28	36	40
	No value modal	Value model (measures aligned with system functions), value functions, and swing weight model implemented without errors.	Clear, meaningful objectives and credible measures.	Insightful objectives and direct measures.
5. Logically correct reasoning	0	14	18	20
	No rotational for recommendations	Value versus cost plot presented and interpretated correctly.	Logic for recommendation explained in three sentences.	Logic for recommendation explained in one sentences.
6. Commitment to action	0	14	18	20
	No discussion of implementation	Implementation plan presented using WBS and performance measures. Key risks identified.	Clean plan to reduce implementation risk.	Identified stakeholders who may not support and develop plan to obtain their support.
Total	0	105	135	150

Figure 0.3 A systems-based, project evaluation rubric.

DECISION ANALYSIS SOFTWARE

The text is designed to be independent of software. All of multiple objective decision analysis in Chapter 12 can be performed in a spreadsheet environment. For the case of Microsoft® Excel, macros that perform a linear interpolation useful for converting measure scores to units of value via value functions exist (Kirkwood,

1997).[1] In several of the illustrative examples, we call upon the Excel Solver to support component optimization. Any similar utility within other spreadsheet software would serve this purpose just as well. Certainly, one alternative to using a spreadsheet would be to employ decision analysis software, a number of which we highlight in this text where appropriate. Any Excel templates we use are available upon request from the editors.

STUDENT EVALUATION

Systems engineers face a continuing challenge of balancing robust processes with quality content. Creative ideas without a solid systems decision process will seldom be defended and successfully implemented. However, a wonderful, logical process is of little value without creativity and innovation. We believe we must impart to our students the importance of both process and creativity without sacrificing the benefits of either. Consequently, we used the concepts introduced in this book—the systems decision process, quality decisions, and presentation guidance—to develop a project grading mechanism that rewards both process and content. Figure 0.3 shows our Project 2 grading sheet. The decision quality terms in the first column are explained in Chapter 9. Insofar as grades are concerned, a student able able to perform the process correctly will earn a "C." Performing the process and having very good context will earn a "B." Demonstrating a mastery of the process, appropriate creativity, and producing outstanding insights will typically result in a grade of "A." We have found this grading approach helpful for recognizing student performance and for conveying course expectations.

FINAL THOUGHT

While we have attempted to incorporate all the suggestions and great ideas we have received from readers of the first edition, we wholeheartedly recognize the value of continuous improvement. Thus, while we are certainly limited in the degree to which the outstanding publication staff at Wiley allow us to alter content between printings without engaging in a third edition, we welcome feedback and suggestions whenever they occur.

GREGORY S. PARNELL AND PATRICK J. DRISCOLL

West Point, New York
July 2010

---

[1] Kirkwood, CW. *Strategic Decision Making: Multiple Objective Decision Analysis with Spreadsheets.* Pacific Grove, CA: Duxbury Press, 1997.

Contributors

Roger C. Burk, Ph.D. Dr. Burk is an Associate Professor in the Department of Systems Engineering at the United States Military Academy (USMA) at West Point. He retired from the (U.S.) Air Force after a career in space operations, space systems analysis, and graduate-level instruction; afterwards he worked in industry as a systems engineer supporting national space programs before joining the USMA faculty. He teaches courses in statistics, decision analysis, mathematical modeling, systems engineering, and systems acquisition and advises senior research projects. He also consults in the areas of decision analysis and mathematical modeling in the space and national defense domains. Dr. Burk has a bachlor in Liberal Arts from St. John's College, Annapolis; an M.S. in Space Operations from the Air Force Institute of Technology; and a Ph.D. in Operations Research from the University of North Carolina at Chapel Hill. He is a member of the Institute for Operations Research and the Management Sciences, the Military Operations Research Society, the American Society for Engineering Education, and Alpha Pi Mu.

Robert A. Dees, M.S. Major Robert Dees is an instructor and senior analyst with the Department of Systems Engineering at United States Military Academy. MAJ Dees has degrees in Engineering Management (United States Military Academy) and Industrial and Systems Engineering (M.S. Texas A&M University). MAJ Dees conducts applications research in the fields of decision analysis, systems engineering, and simulation for the U.S. Department of Defense and is an integral part of the teaching faculty at USMA. He is a member of the Military Applications Society of the Institute for Operations Research and the Management Sciences, the Decision Analysis Society of the Institute for Operations

Research and the Management Sciences, and the Military Operations Research Society.

Patrick J. Driscoll, Ph.D. Dr. Pat Driscoll is a Professor of Operations Research at the United States Military Academy at West Point. He has systems experience modeling and improving a wide range of systems including university academic timetabling, information quality in supply chains, vulnerability and risk propagation in maritime transportation, infrastructure modeling and analysis, and value structuring in personnel systems. He also serves on the boards of several nonprofit organizations. Dr. Driscoll has degrees in Engineering (U.S. Military Academy, West Point), Operations Research (Stanford University), Engineering-Economic Systems (Stanford University), and Industrial and Systems Engineering (OR) (Ph.D, Virginia Tech). He is a member of the Institute for Operations Research and the Management Sciences, the Institute of Industrial Engineers, the IEEE, the Military Operations Research Society, the Operational Research Society, and is President of the Military Applications Society of INFORMS.

Bobbie L. Foote, Ph.D. Dr. Bobbie Leon Foote served as a senior member of the faculty in Systems Engineering at the United States Military Academy. He has created systems redesign plans for Compaq, American Pine Products, the United States Navy, and Tinker Air Force Base. He was a finalist for the Edelman prize for his work with Tinker Air Force Base. He is the author of four sections on systems, forecasting, scheduling and plant layout for the 2006 Industrial and Systems Engineering Handbook and the 2007 Handbook of Operations Research. He jointly holds a patent on a new statistical test process granted in 2006 for work done on the Air Warrior project. He is a fellow of the Institute of Industrial Engineers.

Simon R. Goerger, Ph.D. Colonel Simon R. Goerger is the Director of the DRRS Implementation Office and Senior Readiness Analyst for the U.S. Office of the Secretary of Defense. Col. Goerger is a former Assistant Professor in the Department of Systems Engineering at the United States Military Academy. He has taught both systems simulations and senior capstone courses at the undergraduate level. He holds a Bachelor of Science from the United States Military Academy and a Masters in Computer Science and a Doctorate in Modeling and Simulations from the Naval Postgraduate School. His research interests include combat models, agent-based modeling, human factors, training in virtual environments, and verification, validation, and accreditation of human behavior representations. LTC Goerger has served as an infantry and cavalry officer for the U.S. Army as well as a software engineer for COMBAT XXI, the U.S. Army's future brigade and below analytical model for the twenty-first century. He is a member of the Institute for Operations Research and the Management Sciences, the Military Operations Research Society, and the Simulation Interoperability Standards Organization.

Dale L. Henderson, Ph.D., Design Editor   LTC Dale Henderson is a senior military analyst for the TRADOC Analysis Center (Ft. Lee) and a former Assistant Professor of Systems Engineering at the United States Military Academy at West Point. He has systems engineering and modeling experience in support of large-scale human resources systems and aviation systems. He graduated from West Point with a B.S. in Engineering Physics and holds an M.S. in Operations Research from the Naval Postgraduate School and a Ph.D. in Systems and Industrial Engineering from the University of Arizona. He is a member of the Institute for Operations Research and the Management Sciences, the Military Operations Research Society, and Omega Rho.

Robert Kewley, Ph.D.   COL Robert Kewley is the Professor and Head of the Department of Systems Engineering at the United States Military Academy Department of Systems Engineering. He has systems analysis experience in the areas of battle command, combat identification, logistics, and sensor systems. He has also conducted research in the areas of data mining and agent-based modeling. He has taught courses in decision support systems, system simulation, linear optimization, and computer-aided systems engineering. COL Kewley has a bachelor's degree in mathematics from the United States Military Academy and has both a master's degree in Industrial and Managerial Engineering and a Ph.D. in Decision Science and Engineering Systems from Rensselaer Polytechnic Institute. He is a member of the Military Operations Research Society.

John E. Kobza, Ph.D.   Dr. John E. Kobza is a Professor of Industrial Engineering and Senior Associate Dean of Engineering at Texas Tech University in Lubbock, Texas. He has experience modeling communication, manufacturing, and security systems. He has taught courses in statistics, applied probability, optimization, simulation, and quality. Dr. Kobza has a B.S. in Electrical Engineering from Washington State University, a Master's in Electrical Engineering from Clemson University, and a Ph.D. in Industrial and Systems Engineering from Virginia Tech. He is a member of Omega Rho, Sigma Xi, Alpha Pi Mu, the Institute for Operations Research and the Management Sciences, the Institute of Industrial Engineers, and the Institute of Electrical and Electronics Engineers and is a registered professional engineer in the state of Texas.

Paul D. Kucik III, Ph.D.   LTC Paul Kucik is an Academy Professor and Director of the Operations Research Center at the United States Military Academy at West Point. He has extensive systems experience in the operations and maintenance of military aviation assets. He has taught a variety of economics, engineering management, and systems engineering courses. LTC Kucik conducts research in decision analysis, systems engineering, optimization, cost analysis, and management and incentive systems. LTC Kucik has degrees in Economics (United States Military Academy), Business Administration (MBA, Sloan School of Management, Massachusetts Institute of Technology), and Management Science and Engineering (Ph.D., Stanford University). He is a member of the Military

Applications Society of the Institute for Operations Research and the Management Sciences, the Military Operations Research Society, the American Society for Engineering Management, and the Society for Risk Analysis.

Michael J. Kwinn, Jr., Ph.D. Dr. Michael J. Kwinn, Jr. is the Deputy Director for the System-of-Systems Engineering organization for the Assistant Secretary of the U.S. Army for Acquisition, Logistics and Technology and is a former Professor of Systems Engineering at the United States Military Academy at West Point. He has worked on systems engineering projects for over 15 years. Some of his recent work is in the areas of acquisition simulation analysis, military recruiting process management, and condition-based maintenance implementation. He has also applied systems engineering techniques while deployed in support of Operation Iraqi Freedom (OIF) and Operation Enduring Freedom (OEF). He teaches systems engineering and operations research courses and has served as an advisory member for the Army Science Board. Dr. Kwinn has degrees in General Engineering (U.S. Military Academy), Systems Engineering (MSe, University of Arizona), National Security and Strategic Studies (MA, US Naval War College), Management Science (Ph.D., University of Texas at Austin). He is the past-President of the Military Operations Research Society and is a member of the International Council on Systems Engineering, the American Society for Engineering Education, and the Institute for Operations Research and the Management Sciences.

LTC Daniel J. McCarthy is an Academy Professor and the Director of the Systems Engineering and Operations Research Programs in the Department of Systems Engineering at the United States Military Academy. He has systems analysis experience in the areas of personnel management, logistics, battle command, and unmanned systems. He has also conducted research and has experience in the areas of system dynamics, project management, product development, strategic partnership and strategic assessment. He has taught courses in systems engineering design, system dynamics, production operations management, mathematical modeling, decision analysis, and engineering statistics. LTC McCarthy has degrees in Organizational Leadership (U.S. Military Academy), Systems Engineering (University of Virginia), and Management Science (Ph.D., Massachusetts Institute of Technology). He is a member of the Military Operations Research Society (MORS), the International Council of Systems Engineering (INCOSE), the System Dynamics Society, the American Society of Engineering Education (ASEE), and the Institute of Industrial Engineers (IIE).

Kenneth W. McDonald, Ph.D. LTC Kenneth W. McDonald is an Associate Professor and Engineering Management Program Director in the Department of Systems Engineering at the United States Military Academy at West Point. He has extensive engineering management experience throughout a 25-year career of service in the U.S. Army Corps of Engineers and teaching. He teaches engineering management and systems engineering courses while overseeing the

Engineering Management program. LTC McDonald has degrees in Civil Engineering, Environmental Engineering, Geography, City and Regional Planning, Business and Information Systems. He is also a registered Professional Engineer (PE), a Project Management Professional (PMP), and a certified professional planner (AICP). He is a member of the Institute of Industrial Engineering, the American Society of Engineering Management, the American Institute of Certified Planners and the Society of American Military Engineers. He is also an ABET evaluator for Engineering Management programs.

Heather Nachtmann, Ph.D. Dr. Heather Nachtmann is an Associate Professor of Industrial Engineering and Director of the Mack-Blackwell Rural Transportation Center at the University of Arkansas. Her research interests include modeling of transportation, logistics, and economic systems. She teaches in the areas of operations research, engineering economy, cost and financial engineering, and decision analysis. Dr. Nachtmann received her Ph.D. in Industrial Engineering from the University of Pittsburgh. She is a member of Alpha Pi Mu, the American Society for Engineering Education, the American Society for Engineering Management, the Institute for Operations Research and the Management Sciences, the Institute of Industrial Engineers, and the Society of Women Engineers.

Gregory S. Parnell, Ph.D. Dr. Gregory S. Parnell is a Professor of Systems Engineering at the United States Military Academy at West Point. He has systems experience operating space systems, managing aircraft and missile R&D programs, and leading a missile systems engineering office during his 25 years in the U.S. Air Force. He teaches decision analysis, operations research, systems engineering, and engineering management courses. He also serves as a senior principal with Innovative Decisions Inc., a leading decision analysis consulting company. He serves on the Technology Panel of the National Security Agency Advisory Board. Dr. Parnell has degrees in Aerospace Engineering (State University of New York at Buffalo), Industrial and Systems Engineering (University of Florida), Systems Management (University of Southern California) and Engineering-Economic Systems (Ph.D., Stanford University). Dr. Parnell is a member of the American Society for Engineering Education, the International Council on Systems Engineering, the Institute for Operations Research and the Management Sciences, and the Military Operations Research Society.

Edward Pohl, Ph.D. Dr. Edward A. Pohl is an Associate Professor and John L. Imhoff Chair of Industrial Engineering at the University of Arkansas. Prior to joining the faculty at Arkansas, Dr. Pohl served as an Associate Professor of Systems Engineering at the United States Military Academy, and as an Assistant Professor of Systems Engineering at the Air Force Institute of Technology. During his 21 years of service in the United States Air Force, Dr. Pohl held a variety of systems engineering and analysis positions. He worked as a systems engineer on the B-2 Weapon Systems Trainer and worked as a reliability, maintainability, and availability engineer on a variety of strategic and tactical missile systems. Finally, he worked as a systems analyst on the staff of the Secretary of Defense, Programs Analysis and Evaluation on a variety of space and missile defense systems. Dr. Pohl has degrees in Electrical Engineering (Boston University), Engineering Management (University of Dayton), Systems Engineering (Air Force Institute of Technology), Reliability Engineering (University of Arizona), and Systems and Industrial Engineering (Ph.D., University of Arizona). He is a member of the International Council on Systems Engineering, the Institute for Operations Research and the Management Sciences, the Institute of Industrial Engineers, the Institute of Electrical and Electronics Engineers, and the Military Operations Research Society.

Robert Powell, Ph.D. COL Robert A. Powell was a former Academy Professor and Director of the Systems Management program at the United States Military Academy at West Point. Prior to his death in 2008, he had a vast and varied background of academic, research, and government experience in the engineering profession that spanned more than 21 years. He conducted research in decision analysis, systems engineering, battlefield imagery, optimization, and project management, as well as on the value of integrating practice into engineering curriculums. While on the faculty at USMA, COL. Powell taught courses in production operations management, engineering economics, and project management. COL Powell held a Ph.D. in Systems Engineering from Stevens Institute of Technology, a Master of Military Art and Science from the U.S. Army Command and General Staff College, an M.S. in Operations Research/Management Science from George Mason University, and a B.S. in Industrial Engineering from Texas A&M University. COL. Powell was also a member of the American Society for Engineering Education, the International Council on Systems Engineering, the Military Operations Research Society, and the National Society of Black Engineers.

Timothy Trainor, Ph.D. Brigadier General Timothy E. Trainor is the Dean of Academics and former Professor and Head of the Department of Systems Engineering at the United States Military Academy at West Point. He has systems experience in the operations of military engineering organizations. He teaches engineering management, systems engineering, and decision analysis courses. BG Trainor has degrees in Engineering Mechanics (United States Military Academy), Business Administration (MBA, Fuqua School of Business, Duke University), and Industrial Engineering (Ph.D., North Carolina State University). He is a member of the Military Applications Society of the Institute for Operations Research and the Management Sciences, the Military Operations Research Society, the American Society for Engineering Education, and the American Society of Engineering Management. Colonel Trainor is a member of the Board of Fellows for the David Crawford School of Engineering at Norwich University.

Paul D. West, Ph.D. Dr. Paul D. West is an Assistant Professor in the Department of Systems Engineering at the United States Military Academy at West Point. His systems engineering experience ranges from weapon system to state and local emergency management system design. He has taught courses in combat modeling and simulation, system design, and engineering economics. He designed and implemented an immersive 3D virtual test bed for West Point and chaired the Academy's Emerging Computing Committee. Other research interests include the design and operation of network-centric systems and human behavior modeling. He holds a bachelor's degree in Liberal Studies from the State University of New York at Albany, an M.B.A. degree from Long Island University, a Master of Technology Management degree from Stevens Institute of Technology, and a Ph.D. in Systems Engineering and Technology Management, also from Stevens. He is a member of the Military Operations Research Society, the American Society of Engineering Management, and the Institute for Operations Research and the Management Sciences.

Acronyms

ABET	Formerly Accreditation Board for Engineering and Technology, now ABET, Inc.
AFT	Alternative-Focused Thinking
AoA	Analysis of Alternatives
ASEM	American Society for Engineering Management
ASI	American Shield Initiative
BRAC	Base Realignment and Closure Commission
CAS	Complex Adaptive System
CCB	Configuration Control Board
CER	Cost Estimating Relationship
CFR	Constant Failure Rate
CIO	Chief Information Officer
CM	Configuration Manager (or Management)
CPS	Creative Problem Solving
CTO	Chief Technology Officer
DDDC	Dynamic, Deterministic, Descriptive, Continuous
DDDD	Dynamic, Deterministic, Descriptive, Discrete
DDPC	Dynamic, Deterministic, Prescriptive, Continuous
DDPD	Dynamic, Deterministic, Prescriptive, Discrete
DFR	Decreasing Failure Rate
DoD	Department of Defense
DOE	Design of Experiments

DPDC	Dynamic, Probabilistic, Descriptive, Continuous
DPDD	Dynamic, Probabilistic, Descriptive, Discrete
DPPC	Dynamic, Probabilistic, Prescriptive, Continuous
DPPD	Dynamic, Probabilistic, Prescriptive, Discrete
EM	Engineering Manager (or Management)
ESS	Environmental Stress Screening
FIPS	Federal Information Processing Standards
FRP	Full Rate Production
GIG	Global Information Grid
ICD	Interface Control Document
ICOM	Inputs, Controls, Outputs, and Mechanisms
IDEF	Integrated Definition for Function Modeling
IE	Industrial Engineer (or Engineering)
IED	Improvised Explosive Device
IFR	Increasing Failure Rate
IIE	Institute of Industrial Engineers
ILS	Integrated Logistic Support
INCOSE	International Council on Systems Engineering
INFORMS	Institute for Operations Research and the Management Sciences
IV& V	Independent Verification and Validation
LCC	Life Cycle Costing
LRIP	Low Rate Initial Production
M& S	Modeling and Simulation
MA	Morphological Analysis
MAS	Multi-Agent System
MTF	Mean Time to Failure
MODA	Multiple Objective Decision Analysis
MOE	Measure of Effectiveness
MOP	Measures of Performance
MORS	Military Operations Research Society
NCW	Network Centric Warfare
NPV	Net Present Value
NSTS	National Space Transportation System
OR/MS	Operations Research & Management Science
ORS	Operational Research Society
PM	Program Manager (or Management)
QA	Quality Assurance
R/C	Radio Controlled
RFP	Request for Proposal
RMA	Reliability, Maintainability, and Availability
RPS	Revised Problem Statement
RSM	Response Surface Method
SADT	Structured Analysis and Design Technique
SCEA	Society of Cost Estimating and Analysis

SDDC	Static, Deterministic, Descriptive, Continuous
SDDD	Static, Deterministic, Descriptive, Discrete
SDP	Systems Decision Process
SDPC	Static, Deterministic, Probabilistic, Continuous
SDPD	Static, Deterministic, Prescriptive, Discrete
SE	System(s) Engineer (or Engineering)
SEC	Securities and Exchange Commission
SEMP	Systems Engineering Master Plan
SME	Subject Matter Expert
SPDC	Static, Probabilistic, Descriptive, Continuous
SPDD	Static, Probabilistic, Descriptive, Discrete
SPPC	Static, Probabilistic, Prescriptive, Continuous
SPPD	Static, Probabilistic, Prescriptive, Discrete
TEMP	Test and Evaluation Master Plan
U.S.	United States
VFT	Value-Focused Thinking
WBS	Work Breakdown Structure

Chapter **1**

Introduction

GREGORY S. PARNELL, Ph.D.
PATRICK J. DRISCOLL, Ph.D.

> To be consistent, you have to have systems. You want systems, and not rules. Rules
> create robots. Systems are predetermined ways to achieve a result. The emphasis is
> on achieving the results, not the system for the system's sake... Systems give you
> a floor, not a ceiling.
>
> —Ken Blanchard and Sheldon Bowles

1.1 PURPOSE

This is the first chapter in a foundational book on a technical field. It serves two
purposes. First, it introduces the key terms and concepts of the discipline and
describes their relationships with one another. Second, it provides an overview
of the major topics of the book. All technical fields have precisely defined terms
that provide a foundation for clear thinking about the discipline. Throughout this
book we will use the terms and definitions recognized by the primary professional
societies informing the practice of contemporary systems engineering:

The International Council on Systems Engineering (INCOSE) [1] is a not-for-
profit membership organization founded in 1990. INCOSE was founded to
develop and disseminate the interdisciplinary principles and practices that
enable the realization of successful systems. INCOSE organizes several meetings each year, including the annual INCOSE international symposium.

Decision Making in Systems Engineering and Management, Second Edition
Edited by Gregory S. Parnell, Patrick J. Driscoll, Dale L. Henderson
Copyright © 2011 John Wiley & Sons, Inc.

The American Society for Engineering Management (ASEM) [2] was founded in 1979 to assist its members in developing and improving their skills as practicing managers of engineering and technology and to promote the profession of engineering management. ASEM has an annual conference.

The Institute for Operations Research and the Management Sciences (INFORMS) [3] is the largest professional society in the world for professionals in the fields of operations research and the management sciences. The INFORMS annual conference is one of the major forums where systems engineers present their work.

The Operational Research Society (ORS) [4] is the oldest professional society of operations research professionals in the world with members in 53 countries. The ORS provides training, conferences, publications, and information to those working in operations research. Members of the ORS were among the first systems engineers to embrace systems thinking as a way of addressing complicated modeling and analysis challenges.

Figure 1.1 shows the concept map for this chapter. This concept map relates the major sections of the chapter, and of the book, to one another. The concepts shown in round-edge boxes are assigned as major sections of this chapter. The underlined items are introduced within appropriate sections. They represent ideas and objects that link major concepts. The verbs on the arcs are activities that we describe briefly in this chapter. We use a concept map diagram in each of the chapters to help identify the key chapter concepts and make explicit the relationships between

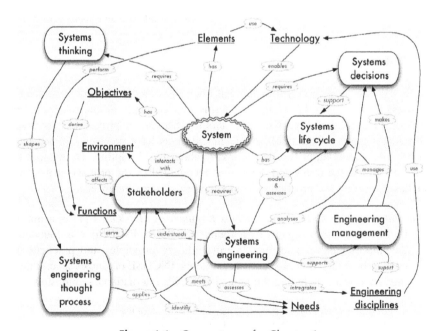

Figure 1.1   Concept map for Chapter 1.

key concepts we explore. This book addresses the concepts of systems, system life cycles, system engineering thought process, systems decisions, systems thinking, systems engineering, and engineering management.

1.2 SYSTEM

There are many ways to define the word *system*. The Webster Online Dictionary defines a system as "a regularly interacting or interdependent group of items [elements] forming a unified whole" [5]. We will use the INCOSE definition:

A system is "an integrated set of elements that accomplishes a defined objective. These elements include products (hardware, software, firmware), processes (policies, laws, procedures), people (managers, analysts, skilled workers), information (data, reports, media), techniques (algorithms, inspections, maintenance), facilities (hospitals, manufacturing plants, mail distribution centers), services (evacuation, telecommunications, quality assurance), and other support elements."[1]

As we see in Figure 1.1 a system has several important attributes:

- Systems have interconnected and interacting elements that perform systems functions to meet the needs of consumers for products and services.
- Systems have objectives that are achieved by system functions.
- Systems interact with their environment, thereby creating effects on stakeholders.
- Systems require systems thinking that uses a systems engineering thought process.
- Systems use technology that is developed by engineers from all engineering disciplines.
- Systems have a system life cycle containing elements of risk that are (a) identified and assessed by systems engineers and (b) managed throughout this life cycle by engineering managers.
- Systems require systems decisions, analysis by systems engineers, and decisions made by engineering managers.

Part I of this book discusses systems and systems thinking in detail.

1.3 STAKEHOLDERS

The primary focus of any systems engineering effort is on the stakeholders of the system, the definitions of which have a long chronology in the management sciences literature [6]. A stakeholder, in the context of systems engineering, is a person or organization that has a vested interest in a system or its outputs. When such a system is an organization, this definition aligns with Freeman's: "any group of individuals who can affect or is affected by the achievement of the organization's objectives" [7]. It is this vested interest that establishes stakeholder importance within any systems decision process. Sooner or later, for any systems decision problem, stakeholders will care about the decision reached because it will in one way or another affect them, their systems, or their success. Consequently, it is prudent and wise to identify and prioritize stakeholders in some organized fashion and to integrate their needs, wants, and desires in any possible candidate solution. In the systems decision process (SDP) that we introduce in Chapter 9, we do this by constructing value models based on stakeholder input. Their input as a group impacts system functions and establishes screening criteria which are minimum requirements that any potential solution must meet. Alternatives failing to meet such requirements are eliminated from further consideration.

It is important to recognize that all stakeholder input is conditionally valid based upon their individual perspectives and vested interests. In other words, from their experience with and relationship to the problem or opportunity being addressed, and within the environment of openness they have chosen to engage the systems engineering team, the information they provide is accurate. The same can be said of the client's information. What acts to fill any gaps in this information is independent research on the part of the team. Research never stops once it has begun, and it begins prior to the first meeting with any client. This triumvirate of input, so critical to accurately defining a problem, is illustrated in Figure 1.2.

Managing stakeholder expectations has become so intrinsic to project success that a number of other formalizations have been developed to understand the interrelationship between key individuals and organizations and the challenges that could arise as a project unfolds. Mitchell et al. [6] posit that stakeholders can be identified by their possessing or being attributed to possess one, two, or all three of the following attributes, which we generalize here to systems.

1. The stakeholder's *power* to influence the system.
2. The *legitimacy* of the stakeholder's relationship to the system.
3. The *urgency* of the stakeholder's claim on the system.

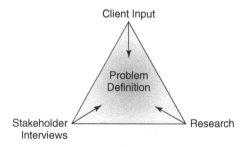

Figure 1.2   Three required ingredients for proper problem definition.

These attributes interact in a manner that defines *stakeholder salience*, the degree to which managers give priority to competing stakeholder claims. Salience then results in a classification of stakeholders by eight types shown in Figure 1.3. Throughout the systems decision process (SDP), there is a strong emphasis on identifying, engaging with, cultivating a trust relationship with, and crafting high value system solutions for a stakeholder called the *decision authority*. Mitchell's characterization clearly illustrates why this is so. The *decision maker* is a salience type 7 in Figure 1.3. The decision maker possesses an *urgency* to find a solution to the dilemma facing the system, the *power* to select and implement a value-based solution, and a recognized *legitimacy* by all stakeholders to make this selection.

Beyond understanding how stakeholders relate to one another and the system, these attributes are relevant to systems decision problems because Matty [8] has connected them to elements of value, a characteristic that comprises one-half of the tradeoff space advocated by the approach presented in this book (see Chapter 9). Stakeholder legitimacy strongly influences value identification; power strongly influences value positioning; and urgency strongly influences value execution.

Two other recent approaches have garnered broad interest in professional practice: the Stakeholder Circle™ and the Organizational Zoo [9].

The Stakeholder Circle™ is a commercially available software tool (www.stakeholder-management.com) which originated in a doctoral thesis [10, 23] motivated by several decades of project management experience in which "poor stakeholder engagement due to not seeing where some stakeholders were coming from led to project delivery failure." The software provides a visualization tool that measures and illustrates various stakeholders' power, influence, and positioning. It leverages a useful metaphor of stakeholders in a concentric circle surrounding the project itself. A five-step methodology is used to manage the stakeholder pool over the complete life cycle of a project: Identify the stakeholders and their needs, prioritize the stakeholders, visualize their relationship to the project, develop an engagement strategy, and monitor changes over time.

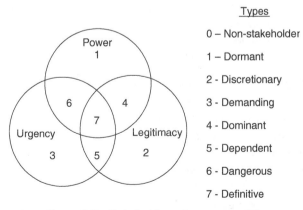

Figure 1.3   Stakeholder salience types [8].

The "Organizational Zoo" concept uses the metaphor of an animal kingdom and its familiar inhabitants to persuade stakeholders to see "how various situations and environments can facilitate or inhibit a knowledge-sharing culture." By associating key individuals with stereotypical behaviors expressed by lions, eagles, ants, mice, rattlesnakes, hyenas, unicorns, and other creatures, stakeholders gain an understanding of how and why they are likely to react to project-related situations. This approach is more stakeholder-centric in its application than the Stakeholder Circle™, though both methods possess similarities to the use of rich pictures in soft system methodology [11].

Notice that this notion of a stakeholder makes no distinction based on the motivation of stakeholder vested interest. We should allow the possibility that for any system of reasonable presence in its surrounding environment, there exists a subset of adversarial stakeholders who are not interested in the success and well-being of the system under study. On the contrary, they might have a vested interest in its demise, or at the very least the stagnation or reduction in the growth of the system, its outputs, and linkages. Market competitors, advocates of opposing political ideologies, members of hostile biological systems, and so on, are obvious examples of adversarial groups that might typify this malevolent category of stakeholders. Cleland [12] and Winch [13] introduce and elaborate upon several useful techniques for mitigating the risk to project success posed by hostile stakeholders.

More complex and challenging to identify are the less obvious stakeholders, namely, those persons and organizations that are once, twice, and further removed from direct interaction with the system under study but nonetheless have a vested interest that needs to be considered in a systems decision problem. A once removed stakeholder could be described as one whose direct vested interest lies in the output of a system that is dependent on output of the system under study. A similar relationship exists for further removed stakeholders. The environmental factors shown in the SDP of Figure 1.7 are very helpful in this regard. They are frequently used as memory cues during stakeholder identification.

For our purposes, the simplest complete taxonomy of stakeholders contains six types. In some systems decisions it may be useful to include additional types of stakeholders. For example, it may be helpful to divide the User group into two subgroups—operators and maintainers—to more clearly identify their role in interacting with the system and to better classify their individual perspectives.

Decision Authority. The stakeholder(s) with ultimate decision gate authority to approve and implement a system solution.

Client. The person or organization that solicited systems decision support for a project; the source of project compensation; and/or the stakeholder that principally defines system requirements.

Owner. The person or organization responsible for proper and purposeful system operation.

User. The person or organization accountable for proper and purposeful system operation.

Consumer. The person(s) or organization(s) that realize direct or indirect benefits from the products or services of the system.

Interconnected. The persons or organizations that will be virtually or physically connected to the system and have potential benefits, costs, and/or risks caused by the connection.

For any given systems decision problem, it is perhaps easiest to identify the Client first, then the Decision authority, followed by the others in any convenient order. For example, on a recent rental car system re-design, the Client solicited assistance in identifying creative alternatives for marketing nonrecreational vehicle rental in his region. When asked, the Client stated that while he would be making the intermediate gate decisions to move the project forward, any solutions would have to be approved by his regional manager prior to implementation. His regional manager is therefore the Decision authority.

An example will help to distinguish between a User and an Owner. A technology company purchases computer systems for its engineers to use for computer-aided design. The company owns the computers and is held responsible for maintaining proper accountability against loss. The engineers use the computers and typically sign a document acknowledging that they have taken possession of the computers. If, on a particularly bad Friday, one of the engineers (User) tosses her computer out the window and destroys it, she will be held accountable and have to pay for the damages or replacement. The managing supervisor of the engineer, as the company's representative (Owner), is held responsible that all proper steps were taken to protect and safeguard the system against its loss or damage.

This taxonomy can then be further divided into an *active* set and a *passive* set of stakeholders. The active set contains those stakeholders who currently place a high enough priority on the systems decision problem to return your call or participate in an interview, focus group, or survey in order to provide the design team with relevant information. The passive set contains those who do not. Membership in these two sets will most likely change throughout the duration of a systems decision project as awareness of the project and relevance of the impact of the decisions made increases in the pool of passive stakeholders.

1.4 SYSTEM LIFE CYCLE

Systems are dynamic in the sense that the passage of time affects their elements, functions, interactions, and value delivered to stakeholders. These observable effects are commonly referred to as system maturation effects. A system life cycle is a conceptual model that is used by system engineers and engineering managers to describe how a system matures over time. It includes each of the stages in the conceptualization, design, development, production, deployment, operation, and retirement of the system. For most systems decision challenges and all system design problems, when coupled with the uncertainties associated with cost, performance, and schedule, life cycle models become important tools to help these same engineers and managers understand, predict, and plan for how a system will evolve into the future.

A system's performance level, its supportability, and all associated costs are important considerations in any systems decision process. The process we introduce in Section 1.9 is fundamentally life cycle centered. In each stage of a system's useful life, systems owners make decisions that influence the well-being of their system and determine whether the system will continue to the next stage of its life cycle. The decision of whether or not to advance the system to the next stage is called a *decision gate*.

The performance of a system will degrade if it is not maintained properly. Maintaining a system consumes valuable resources. At some point, system owners are faced with critical decisions as to whether to continue to maintain the current system, modify the system to create new functionality with new objectives in mind, or retire the current system and replace it with a new system design. These decisions should be made taking into consideration the entire system life cycle and its associated costs, such as development, production, support, and "end of life" disposal costs, because it is in this context that some surprising costs, such as energy and environmental costs, become clearly visible.

Consider, for example, the life cycle costs associated with a washing machine [14] in terms of percentage of its overall contributions to energy and water consumption, air and water pollution, and solid waste. One might suspect that the largest solid waste costs to the environment would be in the two life cycle stages at the beginning of its life cycle (packaging material is removed and discarded) and at the end (the machine is disposed of). However, as can be seen in Figure 1.4, the operational stage dominates these two stages as a result of the many packets of washing detergent and other consumables that are discarded during the machine's life. It is just the opposite case with the environmental costs associated with nuclear power facilities. The disposal (long-term storage) costs of spent nuclear fuel have grown over time to equal the development and production costs of the facility [15].

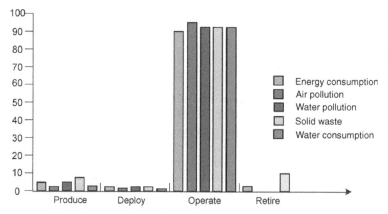

Figure 1.4   Life cycle assessment of environmental costs of a washing machine [14].

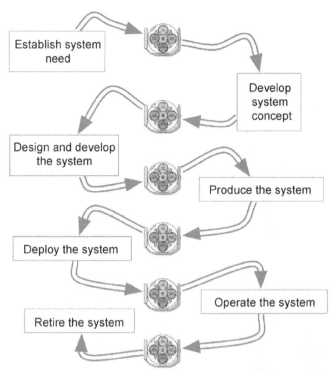

Figure 1.5 Systems decision process used throughout a system life cycle.

We use the system life cycle shown in Figure 1.5 throughout the book. Chapter 3 develops the life cycle in detail so that it can be used to assess any system in support of systems decisions. The stages of this life cycle are aligned with how a system matures during its lifetime. We assume in our approach that there also exist decision gates through which the system can only pass by satisfying some explicit requirements. These requirements are usually set by system owners. For example, a system typically will not be allowed to proceed from the design and development stage to the production stage without clearly demonstrating that the system design has a high likelihood of efficiently delivering the value to stakeholders that the design promises. Decision gates are used by engineering managers to assess system risk, both in terms of what it promises to deliver in future stages and threats to system survivability once deployed.

Throughout all of these considerations, uncertainties are present to varying degrees. While some cost components can be fixed through contractual agreements, others are dependent upon environmental factors well beyond the control and well outside of the knowledge base of systems engineering teams. Illness, labor strikes, late detected code errors, raw material shortages, weather-related losses, legal challenges, and so on, are all phenomena of the type that impose cost increases despite the best intentions and planning of the team. Important modeling parameters such as

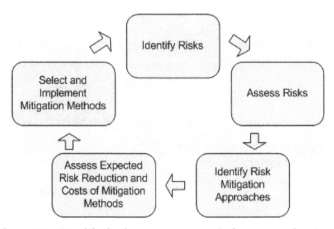

Figure 1.6   Simplified risk management cycle for systems decisions.

cost coefficients used in cost estimating relationships and component performance estimates are based on past data which, as all investment professionals will proclaim, are no guarantee of future performance. Performing the proper due diligence to identify, assess, and manage the potential downside impact of events driven by uncertainty such as these is the role of risk management.

As will be discussed in Chapter 3 in more detail, risk management involves a constant cycle of activities whose purpose is to leverage the most accurate information concerning uncertain events that could threaten system success to construct effective plans that eliminate, mitigate, relocate, or accept (and adapt to) the occurrence of these events [22]. Figure 1.6 shows a simplified risk management cycle whose elements are in common to all risk planning efforts.

Risk is a fundamental concept in systems decision making. Various forms of risk present themselves throughout the life cycle of a system: business risk (does it make sense for the project team to undertake the effort?), market risk (is there a viable and profitable market for the products and/or services the system is designed to deliver?), system program risk (can technical, schedule, and program risks be identified, mitigated, or resolved in a manner that satisfies system owners?), decision risk (is there a sufficient amount of accurate information to make critical decisions?), and implementation risk (can the system be put into action to deliver value?). Risk management, including risk forecasting and mitigation planning, starts early and continues throughout a system's life cycle.

1.5   SYSTEMS THINKING

Systems have become increasingly more complex, dynamic, interconnected, and automated. Both the number and diversity of stakeholders have increased, as global systems have become more prevalent. For example, software companies take advantage of time zone differences to apply continuous effort to new software systems by positioning development teams in the United States, Europe, India, and Japan. Financial systems previously operating as independent ventures now involve banks, businesses, customers, markets, financial institutions, exchange services, and national and international auditing agencies. Changes occurring in one system impact in a very short time those they are connected to. A change in the Tokyo market, for example, propagates quickly to the U.S. market because of strong relationships existing not only between these markets but also among the monetary exchange rates, trade balance levels, manufacturing production levels and inventory levels as well. In order to respond quickly to these market changes, buy and sell rules are automated so as to keep disrupting events from escalating out of control over time.

Military systems have dramatically increased in complexity as well. Currently, complex, interconnected systems use real-time satellite data to geo-locate themselves and find, identify, and classify potential targets using a worldwide network of sensor systems. These, in turn, are connected to a host of weapons platforms having the capacity to place precision guided munitions on targets. With systems such as these, a host of systems decisions arise. Is there a lower limit to human participation in a targeting process such as these? Are these limits defined by technological, cultural, moral, legal, or financial factors? Likewise, should there be an upper limit on the percentage of automated decision making? What measures of effectiveness (MOE) are appropriate for the integrated system behavior present only when all systems are operational?

In general then, for complex systems, how many systems interactions do we need to consider when we are faced with analyzing a single system? Answers to this question shape both the system boundaries and scope of our analysis effort. How can we ensure that critical interactions and relationships are represented in any model we build and that those that play only a minor role are discounted but not forgotten? For this and other important considerations to not be overlooked, we need a robust and consistent systems decision process driven by systems thinking that we can repeatedly apply in any life cycle stage of any system we are examining.

As is addressed in detail in Chapter 2, systems thinking is a holistic philosophy capable of uncovering critical system structure such as boundaries, inputs, outputs, spatial orientation, process structure, and complex interactions of systems with their environment [16]. This way of thinking considers the system as a whole, examining the behavior arising from the total system without assuming that it is necessary to decompose the system into its elements in order to improve or modify its performance. Understanding system structure enables system engineers to design, produce, deploy, and operate systems focused on delivering high value capabilities to customers. The focus on delivering value is what underscores every activity of modern systems engineering [17].

Systems thinking is a holistic philosophy capable of uncovering critical system structure such as boundaries, inputs, outputs, spatial orientation, process structure, and complex interactions of systems with their environment [16].

Systems thinking combined with engineering principles focused on creating value for stakeholders is a modern world view embedded in systems engineering capable of addressing many of the challenges posed by the growing complexity of systems. Systems engineers necessarily must consider both hard and soft systems analysis techniques [11].

In applying the SDP that we introduce in Section 1.9 and use throughout this book, a significant amount of time is consumed in the early steps of the process, carefully identifying the core issues from stakeholders' perspectives, determining critical functions that the system must perform as a whole in order to be considered successful, and clearly identifying and quantifying how these functions will deliver value to stakeholders. Many of the techniques used to accomplish these tasks are considered "soft" in the sense that they are largely subjective and qualitative, as opposed to "hard" techniques that are objective and quantitative. Techniques used in later steps of the SDP involving system modeling and analysis, which are introduced in Chapter 4, lean more toward the quantitative type. Together, they form an effective combination of approaches that makes systems engineering indispensable.

1.6 SYSTEMS ENGINEERING THOUGHT PROCESS

The philosophy of systems thinking is essentially what differentiates modern systems engineering from other engineering disciplines such as civil, mechanical, electrical, aerospace, and environmental. Table 1.1 presents some of the more significant differences [18]. While not exhaustive in its listings, the comparison clearly illustrates that there is something different about systems engineering that is fundamental to the discipline.

The engineering thought process underpinning these other engineering fields assumes that decomposing a structure into its smallest constituent parts, understanding these parts, and reassembling these parts will enable one to understand the structure. Not so with a systems engineering thought process. Many of these engineering fields are facing problems that are increasingly more interconnected and globally oriented. Consequently, interdisciplinary teams are being formed using professionals from a host of disciplines so that the team represents as many perspectives as possible.

> The systems engineering thought process is a holistic, logically structured sequence of cognitive activities that support systems design, systems analysis, and systems decision making to maximize the value delivered by a system to its stakeholders for the resources.

Systems decision problems occur in the context of their environment. Thus, while it is critical to identify the boundaries that set the system under study apart from its environment, the system is immediately placed back into its environment for all subsequent considerations. The diversity of environmental factors shown in the SDP of Figure 1.7 clearly illustrates the need for systems engineering teams to

TABLE 1.1 Comparison of Engineering Disciplines

Comparison Criteria	Systems Engineering	Traditional Engineering Discipline
Problem characteristics	Complex, multidisciplinary, incrementally defined	Primarily requiring expertise in no more than a couple of disciplines; problem relatively well-defined at the onset
Emphasis	Leadership in formulating and framing the right problem to solve; focus on methodology and process; finding parsimonious solutions; associative thinking	Finding the right technique to solve; focus on outcome or result; finding parsimonious explanations; vertical thinking
Basis	Aesthetics, envisioning, systems science, systems theory	Physical sciences and attendant laws
Key challenges	Architecting unprecedented systems; legacy migration; new/legacy system evolution; achieving multilevel interoperability between new and legacy software-intensive systems	Finding the most elegant or optimal solution; formulating hypothesis and using deductive reasoning methods to confirm or refute them; finding effective approximations to simplify problem solution or computational load
Complicating factors	SE has a cognitive component and oftentimes incorporates components arising from environmental factors (see SDP)	Nonlinear phenomena in various physical sciences
Key metric examples	Cost and ease of legacy migration; system complexity; system parsimony; ability to accommodate evolving requirements; ability to meet stakeholder expectations of value	Solution accuracy, product quality, and reliability; solution robustness

be multidisciplinary. Each of these factors represent potential systems, stakeholders, and vested interests that will affect any systems decision and must be considered in the design and implementation of any feasible system solutions.

1.7 SYSTEMS ENGINEERING

The definition used by the INCOSE, the world's leading systems engineering professional society, aligns with the philosophy of this book.

Systems engineering is "an interdisciplinary approach and means to enable the realization of successful systems. It focuses on defining customer needs and required functionality early in the development cycle, documenting requirements, then proceeding with design synthesis and system validation while considering the complete problem." [19]

This definition highlights several key functions of systems engineering as a professional practice:

- Understanding stakeholders (including clients, users, consumers) to identify system functions and objectives to meet their needs.
- Measuring how well system elements will perform functions to meet stakeholder needs.
- Integrating multiple disciplines into the systems engineering team and in consideration of systems alternatives: engineering (aerospace, bioengineering, chemical, civil, electrical, environmental, industrial, mechanical, and others), management, finance, manufacturing, services, logistics, marketing, sales, and so on.
- Remaining involved in many tasks throughout the system life cycle (defining client and user needs and required functionality; documenting requirements; design; identifying, assessing, and managing risks; and system validation).
- Participating in system cost analysis and resource management to ensure cost estimate credibility and system affordability.
- Performing system modeling and analysis to ensure that a sufficient and comprehensive system representation is being considered at each decision gate of the system life cycle.
- Supporting engineering managers' decision making as they manage the system throughout the system life cycle.

These functions, among others, serve to clarify an important point: systems engineering and engineering management are inextricably linked. They work in a complementary fashion to design, develop, deploy, operate, maintain, and eventually retire successful systems that deliver value to stakeholders. So, what is expected of a systems engineer?

Systems engineers are leaders of multidisciplinary technical teams. Azad Madni, an INCOSE Fellow, describes the expectations of systems engineers in the following way [18]: Systems engineers are required to be broad thinkers, capable of generating creative options and synthesizing solutions. They are lateral thinkers at heart, which underscores the natural multidisciplinary structure of systems engineering teams. They must be capable of formulating the right problem to solve and to challenge *every* assumption prior to accepting any. Systems engineers must have the necessary skills and knowledge to imbed aesthetics into systems (solutions), to create required abstractions and associations, to synthesize solutions using metaphors, analogies, and heuristics, and to know where and where not to infuse cognitive engineering in the system life cycle.

1.8 ENGINEERING MANAGEMENT

The American Society for Engineering Management (ASEM) developed a definition of engineering management that aligns with the philosophy of this book:

> Engineering management is "the art and science of planning, organizing, allocating resources, and directing and controlling activities which have a technological component." [2]

In the complex, global, competitive world of technology-driven products and services, there is a need for engineers who understand the essential principles of both engineering and management. Figure 1.7 shows the four dimensions of this

Figure 1.7 Engineering management.

engineering management discipline: entrepreneurship, engineering, management, and leadership.[1] Entrepreneurship is the term used to describe how engineering managers creatively use research and experimentation to develop new technologies to provide products and services that create value for customers. Engineering is used to describe the multidisciplinary teams of individuals from engineering disciplines that apply science and technology to develop these products and services for customers. Management includes the techniques used to plan, staff, organize, and control activities that effectively and efficiently use resources to deliver value to customers. Leadership includes the ability to develop a vision, motivate people, make decisions, and implement solutions while considering all the appropriate environmental factors and stakeholder concerns.

Figure 1.7 also identifies the four critical resources that engineering managers must effectively and efficiently manage: finances, technology, time, and people. All four of these resources are linked together in their effects, but a brief comment on each is appropriate here. Sufficient financing is a key to any engineering management project; it takes money to make money. Technology provides a means of providing products and services to support an engineering management project, whether as stand-alone or networked devices and applications. Time is the third key resource inextricably linked to money. Projects that are managed in such a way that they adhere to schedule have a greater opportunity to maintain the organizational support needed to successfully complete the project and satisfy stakeholder needs. People, the fourth resource, are the most critical resource that an engineering manager must control. Recruiting, motivating, developing, using, and retaining key human resources directly determines the success of any engineering management project.

1.9   SYSTEMS DECISION PROCESS

As a system operates and matures, it competes for resources necessary to maintain its ability to deliver value to stakeholders. Systems decisions involving the allocation of these resources are inevitably made during all phases of a system life cycle up to and including the point where system owners decide to retire the system from operation. As long as a system is operating successfully, other system owners will look to leverage its capabilities to increase the performance of their systems as well. There are many examples of this leveraging taking place, particularly in transportation, software systems, and telecommunications.

As a consequence, systems decisions have become more and more complicated as the number of dependencies on a system's elements or functions grows. Systems engineers need a logically consistent and proven process for helping a system owner (including all stakeholders) make major systems decisions, usually to continue to the next life cycle stage. The process we advocate is shown in Figure 1.8.

---

[1]Modified from original management diagram developed by our West Point colleague, Dr. John Farr.

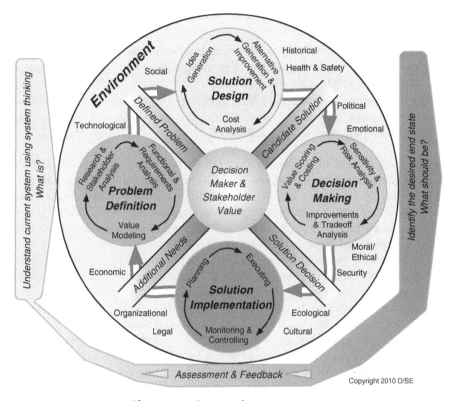

Figure 1.8   Systems decision process.

The systems decision process (SDP) is a collaborative, iterative, and value-based decision process that can be applied in any system life cycle stage.

Part III of this book develops a detailed understanding of the SDP. However, among its many advantages, five inherent characteristics are worth highlighting at this point:

- The SDP encapsulates the dynamic flow of system engineering activities and the evolution of the system state, starting with the current status (what is) and ending with a system that successfully delivers value to system stakeholders (what should be).
- It is a collaborative process that focuses on the needs and objectives of stakeholders and decision makers concerned with the value being delivered by the system.
- It has four major phases organized into a logical progression (problem definition, solution design, decision making, and solution implementation) that

embrace systems thinking and apply proven systems engineering approaches, yet are highly iterative.

- It explicitly considers the environment (its factors and interacting systems) that systems operate in as critical to systems decision making, and thus it highlights a requirement for multidisciplinary systems engineering teams.
- It emphasizes value creation (value modeling, idea generation and alternative improvement, and value-focused thinking) in addition to evaluation (scoring and sensitivity analysis) of alternatives.

The mathematical foundation of the SDP is found in multiobjective decision analysis [20]. This approach affords an ability to qualitatively and quantitatively define value by identifying requirements (solution screening criteria) and evaluation criteria that are essential to guide the development and evaluation of system solutions in all life cycle stages. Chapter 10 describes and illustrates the role of both qualitative and quantitative value models in the SDP. Chapter 11 describes and illustrates the process of using requirements to screen alternatives in order to develop feasible candidate solutions. Chapter 12 describes and illustrates the use of the quantitative value model to evaluate, analyze, and improve the candidate system solutions. Chapter 13 describes the use of planning, executing, and monitoring and control to ensure that value is delivered to the stakeholders.

The definition of a "systems decision" is very encompassing because a system can be defined in many ways. The SDP is a broadly applicable process that can be used to support a variety of enterprise and organizational decisions involving strategy, policy analysis, resource allocation, facility design and location, personnel hiring, event planning, college selection, and many others. The concepts and techniques arising from a systems thinking approach define systems, and the SDP provides the collaborative, dynamic, value-focused decision process that subsequently informs decision makers.

The SDP is a process, an organized way of thinking and taking action that maximizes the likelihood of success when supporting a systems decision. It captures the iterative, cyclical flow of activities that should be performed prior to passing through each of the critical decision gates shown. In practice and in educational settings, highlighting the modeling and analysis flow that typically accompanies the activities prescribed by the SDP greatly facilitates work breakdown and task assignments for team members. Figure 1.9 illustrates this product perspective of the decision support effort. While all of the elements shown are addressed in the chapters that follow, a few comments at this point will be helpful.

The diagram flows from top to bottom, aligning with the first three phases of the SDP: Problem Definition, Solution Design, and Decision Making. It culminates with a comprehensive trade space that supports the system solution decision gate immediately preceding the Solution Implementation phase. All of the analysis products developed in this flow carry over to the Solution Implementation phase once the solution decision has been made.

The top block contains the three primary products of the Problem Definition phase that must be developed before proceeding on: proper identification

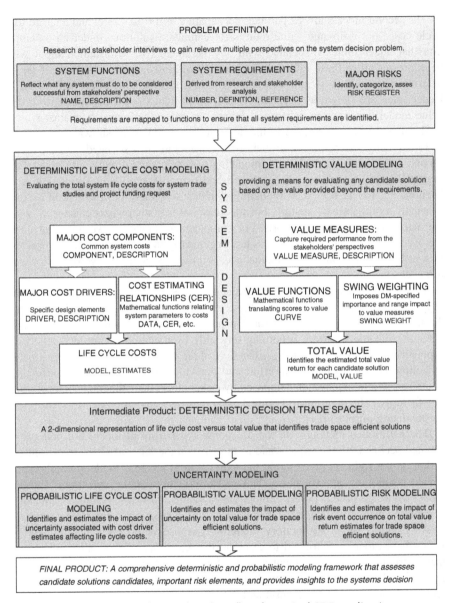

Figure 1.9 Modeling and analysis flow for typical SDP application.

and listing of systems functions, identifying and cataloging requirements, and identifying, categorizing, and assessing major risks. These represent what the system is expected to do, what every alternative must contain to be considered a feasible candidate solution, and the due diligence with respect to risk that every systems decision project must receive.

The second block shows a parallel, yet separate, effort to model and analyze life cycle costs and estimated value returns for candidate solutions under an assumption that uncertainties associated with any parameters or information input will be addressed after these efforts have been successfully concluded. Both of these deterministic analyses require candidate solutions against which they will be used. Hence, they are shown as intrinsic to solution design. Cost is separated from the value model construction because it defines the tradeoff dimension against which total value return is compared in the deterministic trade space shown.

Finally, any uncertainty or probabilistic considerations associated with the models and with their input, output, or modeling parameters are directly addressed. For most SDP applications, this is accomplished using Monte Carlo simulation (see Sections 5.5 and 12.6.1). Risk modeling, whether subjective or objective in nature, is usually a probabilistic venture. For this reason it is not shown as concurrent with the deterministic modeling efforts. For completeness however, we note that once the overall risk management process has begun early in a systems decision project, it is sustained throughout the systems decision process in each stage of the system life cycle.

The modeling and analysis flow ingrained in the SDP results in powerful decision support models. Teams developing these models need to keep in mind both who the models are being developed for and purpose they are intended to serve. The latter prevent models from becoming unwieldy by containing unneeded levels of sophistication and detail, or by exceeding their design scope. Adhering to the modeling purpose focuses team effort and prevents function creep from occurring as a result of late requirements imposed by stakeholders once the model is operating satisfactorily. The diagram in Figure 1.10 shows one such approach to identifying the modeling purpose [21].

The partitioned rectangle on top illustrates a spectrum of model use being distributed between 100% frequent and routine use on the left and 100% human interaction on the right. Arrayed along the axis below it are four modeling archetypes whose positioning approximates their characterization in the spectrum. Thus, a model whose purpose is purely exploratory in nature and whose results are intended to promote discussion among stakeholders would a position to the right extreme.

An example of an exploratory modeling purpose within the SDP framework would be a model constructed to examine the feasibility of futuristic, fully automated ground force engagement systems for the military. The interest in such a hypothetical case would not be in designing a system to accommodate stakeholder requirements, but rather to expose and discuss the implications with respect to the

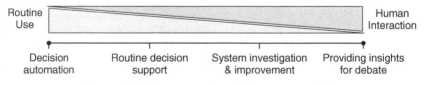

Figure 1.10   Spectrum of modeling purposes [21].

various environmental factors shown in Figure 1.8. Conversely, a decision support model built to aid a one-time systems decision might fall somewhere between the decision automation and routine decision support archetypes shown. While the cost, value, and risk models developed for a one-time decision require stakeholder interaction during their construction, they typically would not require intensive human interaction after their purpose has been served. Building sophisticated user interfaces to these models would not be a wise investment of the team's effort.

1.10 OVERVIEW

The organization of this book follows a logical and sequentially building pattern. Part I defines and describes system concepts. Chapter 2 introduces systems thinking as a discipline for thinking about complex, dynamic, and interacting systems, and it describes methods for representing systems that improve the clarity of our thinking about systems. Chapter 3 introduces the concept of a system life cycle and describes the system life cycle we use in this book. It also introduces the concept of risk, how risk affects systems decision making, and a technique for assessing the levels of various risk factors early in the system life cycle. Chapter 4 introduces system modeling and analysis techniques used to validate system functions and assess system performance. Chapter 5 introduces life cycle cost and other economic analysis considerations that are essential for systems engineering trade studies and ensuring system affordability.

Part II introduces the role of systems engineering in engineering management. Chapter 6 describes the fundamentals of systems engineering. Chapter 7 delineates the role of systems engineering in each phase of the system life cycle. Chapter 8 introduces the system effectiveness considerations and provides models of system suitability that enable a system to perform the function that it was designed for in the user environment.

Part III proposes, describes, and illustrates a systems decision process that can be used in all phases of the system life cycle. A rocket design problem and an academic information technology problem are used to explain the concepts and serve as illustrative examples. Chapter 9 introduces our recommended systems decision process and the illustrative problem. Chapter 10 describes and illustrates the problem definition phase, Chapter 11 the solution design phase, Chapter 12 the decision-making phase, and Chapter 13 the solution implementation phase. Finally, Chapter 14 summarizes the book and discusses future challenges of systems engineering.

1.11 EXERCISES

1.1. Do any of the four professional societies mentioned in this chapter have programs and resources specifically designed for students? If so, provide a brief summary of the services or products they provide that you might find valuable now.

1.2. Answer the following questions regarding a concept map.

    (a) How would you define a concept map? Is there a standard positioning of nouns and verbs on a concept map?

    (b) Is a concept map different from Checkland's [11] LUMAS model? Explain. Draw the LUMAS model that is associated with its definition.

    (c) Where on the spectrum of Figure 1.10 would you position a concept map as a model? Why?

1.3. Draw a concept map that illustrates the relationship between the objects within each of the following sets:

    (a) Space, planet, astronaut, satellite, space shuttle, NASA, missions, food, fuel, control center.

    (b) River, fish, water, insects, rocks, oxygen, riverbanks, trees, pollutants, EPA, boats.

    (c) Teachers, students, books, software applications, models, computers, graphics.

    (d) Facebook™, friends, hackers, pictures, personal information, lost classmates, jobs, services, movies.

1.4. Write a single sentence about each of the eight relationships of systems identified in the concept map in Figure 1.1.

1.5. Consider the automobile as a system.

    (a) Select a specific automobile introduced this year, and identify the major components of its system.

    (b) What new functions does this automobile do that set it apart from previous versions offered by the manufacturer?

    (c) Describe the life cycle this automobile.

    (d) Describe the major environmental factors that should have been considered when designing this new automobile. Do you think they were? Explain.

    (e) Using the environmental factors shown in the SDP, identify the major stakeholders whom you think have a vested interest this new automobile as a system.

    (f) For each of the stakeholders that you identified in part (e), conduct a sufficient amount of research to confirm the vested interest you suspected they held. List the source.

1.6. For each of the systems decision problems below, identify any possible stakeholders who could be classified into the six stakeholder taxonomy categories. Provide a brief justification for each choice.

    (a) The day manager of Todd French's up-scale dining restaurant called "Prunes" hires you to help "modernize" the restaurant's table reservation system.

(**b**) The Commissioner of the State of New York's Highway Department asks you to assist in selecting a new distributed computer simulation program for use in its Albany office.

(**c**) Danita Nolan, a London-based independent management consultant, asks you to help her with an organizational restructuring project involving the headquarters of DeWine Diamond Distributors.

(**d**) Fedek DeNut, one of the principals of a new high-technology company called GammaRaze, has hired you to help them design an internet firewall software application that automatically sends a computer disabling virus back to the "From" address on any spam email passing through the firewall.

(**e**) The musician Boi Rappa has reached such success with his last five DVD releases that he is planning on creating a new line of casual clothing for suburban teenagers. He hires you to help design a successful system to accomlish this.

1.7. Which future stages of a system life cycle are most important to be considered during the system concept stage? Explain.

1.8. Define "systems thinking." Does it have any utility outside of systems engineering? Explain.

1.9. What is systems engineering? List four of the major activities that systems engineers engage in.

1.10. What is engineering management and what do engineering managers do? List four of their major activities.

1.11. What is the relationship between systems engineers and engineering managers?

1.12. Describe the four phases of the SDP. Describe the relationships that exist between the SDP and a system life cycle.

1.13. Are there any environmental factors missing from those listed in the SDP? Why would you include these, if at all?

REFERENCES

1. International Council on Systems Engineering (INCOSE), http://www.incose.org.
2. American Society for Engineering Management (ASEM), http://www.asem.org.
3. The Institute for Operations Research and the Management Sciences (INFORMS), http://www.informs.org.
4. The Operational Research Society (ORS), http://www.orsoc.org.uk.
5. Merriam-Webster Online, http://www.m-w.com/dictionary/system.
6. Mitchell, RK, Agle, BR, Wood, DJ. Toward a theory of stakeholder identification and salience: Defining the principle of who or what really counts. *Academy of Management Review*, 1997;22(4):853–886.

7. Freeman, RE. *Strategic Management: A Stakeholder Approach*. Boston, MA: Pitman, 1984.

8. Matty, D. Stakeholder Salience Influence on Value Creation. Doctoral Research, Engineering Systems Division. Cambridge, MA: Massachusetts Institute of Technology, 2010.

9. Shelley, A. *The Organizational Zoo: A Survival Guide to Workplace Behavior*. Fairfield, CT: Aslan Publishing, 2007.

10. Bourne, L. Project relationship management and the stakeholder circle. Doctor of Project Management, Graduate School of Business, RMIT University, Melbourne, Australia, 2005.

11. Checkland, P. *Systems Thinking, Systems Practice*. New York: John Wiley & Sons, 1999.

12. Cleland, DI. *Project Management Strategic Design and Implementation*. Singapore: McGraw-Hill, 1999.

13. Winch, GM. Managing project stakeholders. In: Morris, PWG, and Pinto, JK, editors. *The Wiley Guide to Managing Projects*. New York: John Wiley & Sons, 2004, pp. 321–339.

14. The University of Bolton postgraduate course offerings. Available at http://www.ami.ac. uk/courses/topics/. Accessed 2006 July 18.

15. *The New Economics of Nuclear Power*. Available at http://www.world-nuclear.org/ Accessed 2006 July 18.

16. Systems Thinking Definition. Available at http://en.wikipedia.org/wiki/Systems_ thinking. Accessed 2006 January 26.

17. Keeney, RL. *Value-Focused Thinking: A Path to Creative Decisionmaking*. Boston, MA: Harvard University Press, 1992.

18. Madni, AM. The intellectual content of systems engineering: A definitional hurdle or something more? *INCOSE Insight*, 2006;9(1):21–23.

19. What is Systems Engineering? Available at http://www.incose.org/practice/whatissys temseng.aspx. Accessed 2006 January 26.

20. Keeney, R, Raiffa, H. *Decision Making with Multiple Objectives: Preferences and Value Tradeoffs*. New York: Cambridge University Press, 1976.

21. Pidd, M. Why modeling and model use matter. *Journal of Operational Research Society*, 2010;61(1):14–24.

22. Hubbard, DW. *The Failure of Risk Management*. Hoboken, NJ: John Wiley & Sons, 2009.

23. Walker, D, Shelley, A, Bourne, L. Influence, stakeholder mapping and visualization. *Construction Management and Economics*, 2008;26(6), 645–658.

Part I

Systems Thinking

Chapter 2

Systems Thinking

PATRICK J. DRISCOLL, Ph.D.

> No man is an island entire of itself; every man is a piece of the continent, a part of the main.
>
> —John Donne (1572–1631)

> Is the world actually composed of interacting sets of entities that combine to create system? Or is the notion of a system simply a convenient way of describing the world? In the final analysis, we simply do not know.
>
> —Micheal Pidd [1]

2.1 INTRODUCTION

As his quote from *Meditation XVII* clearly illustrates above, the Irish poet John Donne was very much a systems thinker. He envisioned the human race as being strongly interconnected to the point that "any man's death diminishes [him], because [he is] involved with mankind." Had he been living today, he no doubt would have been a strong advocate of social network theory [2].

As one of the current champions for using a holistic approach when developing models of real-world phenomena, Michael Pidd poses two interesting questions above that go right to the heart of the discipline of systems engineering (SE). In a sense, whether systems are actual entities or simply one by-product of human perception and reasoning is irrelevant. If systems are the natural means by which we cope with and understand the highly connected, information-intensive world we live in, it would seem illogical to not incorporate a strong consideration of the impact of this connectedness when making decisions about this same world. This is what systems thinking is all about.

Why system thinking matters: How you think is how you act is how you are. The way you think creates the results you get. The most powerful way to improve the quality of your results is to improve the way you think [3].

Systems thinking in support of decision making consistently leads to deep understanding of most, if not all, of the various factors affecting possible alternatives. This success is largely due to two distinguishing characteristics: (a) the manner in which it departs from analytic thinking and (b) its natural ability to reveal subtle but important intricacies hidden to myopic or pure decomposition approaches that fail to consider "the big picture."

Applied to a systems decision problem, analytic thinking starts with the current system, identifies problems and issues that require fixing, applies focused modeling techniques to understand these deficiencies and identify possible solutions, and concludes with recommending solutions for changing some controllable dimensions of system activities that improve the system end state. Although the specific steps used in various disciplines may differ, this is the prevalent style of thinking imbedded in modern education. This way of thinking is most successful in situations where fine tuning of some system performance measures is called for but the system structure itself is assumed to be acceptable. These decision problems are often referred to as "well-structured." Improving the efficiency of established logistic supply networks, increasing the system reliability of telecommunication networks, and reducing transportation costs in package pickup and delivery systems are good examples of where analytic thinking has successfully supported decision making. In all these cases (and others), the *operation* of the system lies at the heart of the decision to be made and not the system itself.

In contrast, systems thinking first and foremost centers *on the system itself* [28, 29]. Operational improvements such as the ones noted above, representing only one dimension of the overall system structure, are identified as part of alternative system solutions crafted with stakeholder ideals clearly in mind. This style of thinking drives the systems decision process (SDP) introduced earlier.

For any system decision problem, system thinking starts with the system output ("What should the system do? What is desired by the stakeholders?") and proceeds to work backwards to identify system functions, processes, objectives, structure, and elements necessary to achieve this desired output. It then assesses the current state of the system ("Where is the system currently?") and asks, "What actions need to be taken to move the system from where it is to where it needs to be in order to maximize the value it delivers to stakeholders?" This natural focus on output (i.e., results, effects) provided by systems thinking creates a goal-oriented frame of reference that produces long-term, effective *system-level solutions* rather than short-term, *symptom-level solutions*. This point bears emphasis.

Possessing the frame-of-reference afforded by systems thinking enables one to distinguish between symptom-level and system-level phenomena. Symptom-level phenomena are typically of short duration and easily observable. When they repeat, they tend to vary in character, intensity, and impact, thereby appearing as new phenomena to the untrained eye. Eliminating symptom-level problems provides short-term relief but will not prevent their recurrence in the future because the underlying system structure from which these symptoms arise is unchanged.

System-level phenomena are persistent, presenting themselves across a layer of commonality among all system components. These phenomena endure because they are an element or aspect of the underlying structure and organization of the system components, how these components interact, and the common ingredients that sustain their activity. System-level issues are identified using techniques that focus on identifying failure modes, such as root cause analysis [4] that attempt to trace collections of symptom-level effects back to shared sources of generation. System-level solutions provide long-term, fundamental system performance changes, some of which may not be predictable.

While symptom-level solutions can provide an immediate value return (e.g., stop crime in a neighborhood), they are not going to alter structural elements of the system that give rise to the observed symptom (e.g., cultural beliefs). Another way of saying this is that system-level solutions alter the fundamental system dynamics and relationships between system components; symptom-level solutions provide spot-fixes where these dynamics and relationships are failing. Risk-to-return on a specific investment instrument is a symptom-level phenomenon; elevated systemic risk shared across the entire derivatives market because of widespread use of collateralized debt obligations and credit default swaps is a system-level phenomenon [31]. A single company deciding to no longer participate in these financial products would provide localized risk-relief, but the underlying system-wide risk exposure still untreated will cause (possibly new) risk events to appear elsewhere.

It is possible in this framework that what are currently perceived as issues might not need fixing; they may actually be opportunities needing reinforcement. The perceived issues may very well be evidence of system functionality that is being imposed on the system by its users and is pushing against the constraints of the formally established structure and practices. In large organizations, "work-arounds" created by employees in order to properly accomplish tasks and the emergence of informal leaders assuming ad hoc roles and responsibilities outside of the established hierarchy are often indicators of just such a situation. The stakeholder analysis so critical to the successful application of the SDP properly frames these issues within a broader system perspective as they arise without assuming they must be eliminated.

Adopting and maintaining a value focus in this setting is essential because as a system evolves through its life cycle, changing in size and complexity, the value being delivered by the system likewise changes. Sometimes this occurs in an undesirable way that drives system redesign efforts. Other times a more subtle adaptation occurs in which the system shifts its value focus, obscuring from system owners and stakeholders exactly how and where it is delivering value. Systems thinking provides a frame through which to identify these subtle, but important, considerations.

> *Systems thinking* is a holistic mental framework and world view that recognizes a system as an entity first, as a whole, with its fit and relationship with its environment being primary concerns.

This philosophy underscores a systems engineering thought process predicated on the belief that the study of the whole should come before that of the parts, recognizing that there are system level behaviors, interactions, and structural characteristics that are not present when the system is decomposed into its elements. This sets apart systems engineering from classical engineering whose thought process is founded on the principle of decomposition as the basis of understanding. This philosophy has become indispensable when addressing modern systems whose size and complexity were not feasible less than a decade ago. Systems of systems engineering [5], model-oriented systems engineering [6], and techniques for designing complex systems [7] have emerged from the systems engineering community in response to this growing challenge. None of these approaches and their associated methods would exist in the absence of systems thinking.

The reason for departing from a pure decomposition principle at the onset of a systems decision problem is that decomposition is an activity that focuses on individual system element characteristics. It uses these individual characteristics to logically group or arrange elements so that the extent of shared characteristics becomes evident, thereby providing insights into how a more efficient or effective systems structure might be realized (by combining elements) or how a systems analysis might be more simply performed (because the analytical results associated with one element might apply to other elements possessing a high degree of shared characteristics with it).

Focusing on individual system elements tends to miss crucial interactions between the elements of a system or between composite groups of systems interacting as a whole. When these interactions or interdependencies are overlooked or insufficiently emphasized, the resulting modeling and analysis can suggest potential solutions that exhibit suboptimal characteristics. Such solution alternatives, while attractive to the performance of individual elements, can actually hinder or degrade some performance measure of the overall system. The risk (return volatility) for a portfolio of investments can easily increase because of an inappropriate over-investment in one particular asset, resulting in a loss of optimality for the portfolio [8]. In a similar fashion, installing high-intensity discharge (HID) headlamps into an older model vehicle may increase the lumens output (maximization effect for the headlamp element), but because the older model car is designed to take filament bulbs, doing so results in improperly focused beam patterns and excessive glare to other road users. A safety measure associated with the older vehicle as a transportation system would likely degrade as a result.

Systems have emergent properties that are not possessed by any of its individual elements. Bottlenecks and the "accordion effect" observed in dense highway traffic flows are examples of emergent properties not possessed by individual automobiles; they become evident only when the transportation elements are viewed as a whole system. These properties result from the relationships between system elements, commonly described as the system *structure* (Figure 2.1). In many cases, this structure can be described mathematically. The functions and expressions resulting from these mathematical descriptions directly support modeling the system as described in Chapter 4.

Systems thinking enables one to progress beyond simply seeing isolated events to recognizing patterns of interactions and the underlying structures that are responsible for them [9]. It reveals the structure in a way that enables us to specify the boundary of the system, which sets apart the system and its internal functions from the environment external to it. Knowing this boundary enables us to identify key system inputs and outputs and to visualize the *spatial arrangement* of the system within its environment. Critical systems thinking of the type required for systems engineering encourages creativity simply because of the strong interplay between conceptual visualization, detailed analysis, and unique measures of effectiveness produced by synthesizing ideas.

The combination of systems thinking with best engineering practices has produced a variation of systems engineering second to none. The particular application

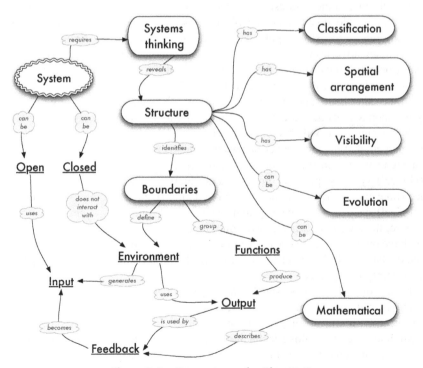

Figure 2.1 Concept map for Chapter 2.

of systems thinking advocated in this book is an adaption of general systems theory [10] with an emphasis on systems decision problems.

2.2 STRUCTURE

Understanding the workings of a system requires some conceptual thinking about its underlying *structure*. For systems engineering, this means several things: identifying the elements of the system, understanding how the elements are connected and how they interact with each other to achieve the system's purpose, and understanding where this system lies relative to other systems that can impact on its behavior or vice versa.

Different systems afford different levels of visibility into their inner workings. The relationships between elements can change dynamically over time, causing system evolution to become a concern. A portion of the output of a system can undergo *feedback* in which the environment uses the system output to create conditioned input that is fed back into the system. Interaction with the environment requires that a system be *open*, because a *closed* system has no interaction across its boundaries.

Consider for a moment the decision associated with purchasing an automobile. On the one hand, the decision might appear to be a simply structured one: go to a local dealership, examine their inventory, compare prices to budget, and purchase the vehicle that comes closest to satisfying desire without violating the working budget limits. No doubt many people exercise this strategy on a routine basis. Knowing this propensity of potential buyers enables automobile manufacturers to craft enticing advertisements and marketing campaigns that can effectively reshape what a potential buyer thinks they want into what the manufacturer wants to sell. This principle lies at the heart of effective marketing [11].

On the other hand, applying a small amount of systems thinking to the car buying decision reveals the purchase decision as highly connected in its structure. The car being purchased will become part of a system in which the automobile is a new element that interacts in various ways with other major systems: the transportation highway system (health & safety), banking (financial) systems, fuel logistic (technical) systems, personal prestige and entertainment (social) systems, insurance (financial) systems, motor vehicle regulation (legal) systems, communication (technical) systems, ecological systems, and so on. From this systems thinking perspective, the purchase decision takes on a much greater level of importance than it may have otherwise. It clearly has an impact on each of these other systems in some manner that should be either taken into consideration or intentionally disregarded, but certainly not ignored.

In fact, one of the most significant failings of the current U.S. transportation system is that the automobile was never thought of as being part of a system until recently. It was developed and introduced during a period that envisioned the automobile as a stand-alone technology largely replacing the horse and carriage. As long as it outperformed the previous equine technology, it was considered a success. This success is not nearly so apparent if the automobile is examined from a systems thinking perspective. In that guise, it has managed to fail miserably across a host of dimensions. Many of these can be observed in any major U.S. city today: oversized cars and trucks negotiating tight roads and streets, bridges and tunnels incapable of handling daily traffic density, insufficient parking, poor air quality induced in areas where regional air circulation geography restricts free flow of wind, a distribution of the working population to suburban locations necessitating automobile transportation, and so on. Had the automobile been developed as a multilateral system interconnected with urban (and rural) transportation networks and environmental systems, cities would be in a much different situation than they find themselves in today.

What is important here is not that the automobile could have been developed differently, but that in choosing to design, develop, and deploy the automobile as a stand-alone technology, a host of complementary transportation solutions to replace the horse and buggy were not considered. Systems thinking would have helped to identify these potentially feasible solutions. If they were subsequently rejected, it would have been for logically defendable reasons directly related to stakeholder requirements. In the business of supporting decision making, limiting the span of potential solutions tends to degrade the quality of the chosen solution, certainly against a criteria of robustness.

An example in a social setting can further illustrate this point. In the United States during the late 1960s, it was not uncommon to hear people taking positions on issues of behavior choices by saying, "Why should it matter to anyone else what I do? If I decide to do this, I am the only one affected. It's my life." While choice certainly is within an individual's control, this statement ignores any and all connections and interactions between the individual expressing this position and subsystems of the metasystem within which the individual lives. John Donne recognized the importance of these connections nearly four centuries earlier, as evidenced by his quote at the start of this chapter.

2.3 CLASSIFICATION

Expanding the way we think about challenging systems decisions requires a top-down classification scheme that starts with the "big picture" of the system and its observable behavior. As stated earlier, we formally define a system as follows:

> A *system* is an integrated set of elements that accomplish a defined objective.

Systems occur in many different forms. We generally use three classes for describing the various system types: physical, abstract, and unperceivable. A *physical* system, also referred to as a concrete system [10], exists in the reality of space–time and consists of at least two elements that interact in a meaningful manner. This is a system that is clearly evident in the real world and directly observable to the trained and perhaps untrained eye. An automobile is a good example of a physical system.

Physical systems further subdivide into four subclasses that can overlap in some cases:

- *Nonliving* has no genetic connections in the system and no processes that qualitatively transform the elements together with the whole and continuously renew these same elements.
- *Living*, typically referred to as an organic system, is a system subject to the principles of natural selection such as an ant colony, a predator–prey relationship between species, and human biophysiology.
- *Manmade* physical systems intentionally created to augment human life such as transportation road networks, logistic resupply systems, and communication systems.
- *Natural* are those systems coming into being by natural processes, such as waterway networks created by natural processes associated with glaciers, tectonic plate shifts, and weather.

An *abstract* system is a system of concepts that are linked together in some manner, generally in an attempt to convey an initial design, a strategic policy, or some other idea that has not been implemented in some other form. Abstract systems are organizations of ideas expressed in symbolic form that can take the form of words, numbers, images, figures, or other symbols. In a sense, this type of system is an intermediate system pinched between reality and the completely intangible as its elements may or may not be empirically observable, because they are relationships abstracted from a particular interest or theoretical point of view.

Figure 2.2 is an example of the field of engineering management (EM) expressed as an abstract system. In its organization, the diagram conveys the order and purpose of the professional field of EM as emerging from the unique interaction of four separate disciplines: leadership, management, economics, and engineering. Each of these four disciplines could in turn be represented as systems. Permeating all of these systems are the environmental resource considerations of people, technology, time, and finances. A similar illustration could be used to show how the car buying decision (substituted in place of "EM" in the picture) impacts the other systems mentioned earlier.

An interesting point to note with regard to this figure is that the EM system as illustrated does not exist apart from the four discipline systems shown. It exists solely because of the interaction of these four systems. It is, in fact, a professional field delivered only at a holistic level; it is not possible to decompose the abstract EM system representation into its multilateral system elements and still retain the complete character of EM. Bear this observation in mind when the concept of designing "system-level measures of performance" arises later in this book.

An *unperceivable* system is a classification that is largely based on the limitations of our ability to observe the system rather than some innate characteristics it may possess. These systems exist when an extreme number of elements and the complexity of their relationships mask the underlying system structure or organization. A system representation used to describe an unperceivable system is an

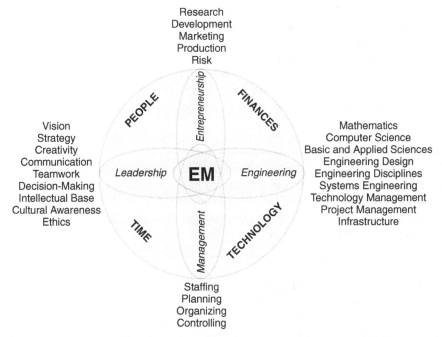

Research
Development
Marketing
Production
Risk

PEOPLE   Entrepreneurship   FINANCES

Vision   Mathematics
Strategy   Computer Science
Creativity   Basic and Applied Sciences
Communication   Engineering Design
Teamwork   Leadership   **EM**   *Engineering*   Engineering Disciplines
Decision-Making   Systems Engineering
Intellectual Base   Technology Management
Cultural Awareness   Project Management
Ethics   Infrastructure

TIME   Management   TECHNOLOGY

Staffing
Planning
Organizing
Controlling

Figure 2.2 Conceptualization of engineering management system. (Courtesy of Dr. John Farr, U.S. Military Academy.)

approximation of the system at best, typically containing only as much detail so as to enable its inputs and outputs to be identified.

An example of an unperceivable system is the U.S. economy. A complete description that includes all of its elements and interrelationships defies our comprehension. At best, we resort to surrogate measures such as gross domestic product (GDP), the Dow Jones composite index, unemployment estimates, inflation rate estimates, foreign trade balance, monetary exchange rates, and new housing starts as indicators of how well (or poorly) the U.S. economy is performing. The complex nature of this system due to the number of elements, number of interactions, and evolving states of both these make it unperceivable as a system. This goes a long way toward explaining why, despite technology advances and Nobel laureate awardees in economics, error-free future state forecasts of this system remain impossible to attain.

2.4 BOUNDARIES

The concept of establishing a *system boundary* is fundamental to working with systems. From a practical standpoint, a system boundary serves to delineate those elements, interactions, and subsystems we believe should be part of a system definition from those we see as separate. To a good extent, what is included within a system boundary is influenced by the systems decision motivating the analysis. Unfortunately, this means that there is no easy answer to setting a system boundary; every systems decision problem motivates its own appropriate system boundary and system definition. Systems engineers accomplish this using the information resulting from research and extensive interaction with stakeholders during the problem definition phase of the SDP. In this regard, the system boundary is one tool available to a systems engineer to help define the scope of a particular project.

A *system boundary* is a physical or conceptual boundary that encapsulates all the essential elements, subsystems, and interactions necessary to address a systems decision problem. The system boundary effectively and completely isolates the system under study from its external environment except for inputs and outputs that are allowed to move across the system boundary.

A system's boundary distinguishes the system from its environment. *Open* systems interact with their environment in a specific way in which they accept inputs from the environment in order to produce outputs that return to the environment. *Closed* systems are isolated and hermetic, accepting no inputs beyond those used to initialize the system and providing no outputs to its environment. Closed systems do not need to interact with their environment to maintain their existence. The clearest example of a closed system is one associated with physics and mechanical engineering: a perpetual motion device. Once initial energy is provided to put the device in motion, it stays in motion forever with absolutely no inputs from outside of its boundary.

Closed systems generally do not occur in nature. More often, human intervention in the form of controlled scientific experiments create closed systems. Some systems are considered closed systems because some particular exchange across its boundary is discounted or ignored. Atoms and molecules can be considered closed systems if their quantum effects are ignored. Greenhouses could be considered closed systems if energy (heat) exchange is ignored. In mathematics, an abstract system, vector spaces are closed systems: Valid operations performed on the elements of a vector space remain in the vector space forever. Edwin A. Abbott (1838–1926), an English schoolmaster and theologian, wrote an interesting book titled *Flatland: A Romance of Many Dimensions* [12] about what life would be like in a two-dimensional world. He had this closed system idea in mind.

Most, if not all, systems we encounter and operate in our daily lives are open systems. Consequently, if an initial inspection of a system makes it appear to be a closed system, chances are we are overlooking some multilateral or hierarchical systems that are affecting input to the system.

The boundary of a system is the actual limits of the major elements. Inputs and outputs cross boundaries of systems. The boundary of a system must be selected to include all of the important interacting elements of the system of interest and exclude all those that do not impact the system behavior that makes it a system. Isolating the core system is an important part of a systems engineering project because doing so allows lateral systems, subsystems, and metasystems to be identified as well.

Figure 2.3 illustrates a hypothetical open system with each of the major system structural elements shown. The system accepts input from its environment, acts on that input via its internal functions, and again interacts with its environment by producing output. Input can take the form of physical material, as in sand being used as a source of silicon dioxide for making silicon wafers for computer microchips. Input can also possess a less tangible form, as in the case of information. For decision support systems used to predict stock performance or assign airline ticket prices, information is one input flowing across the boundary of these open systems.

While it may be natural to think of these inputs and outputs in terms of matter, energy, and materials such as raw materials used in a manufacturing or production facility, they can just as well be services, or induced effects such as influences and other psychological entities. These can all be appropriately defined as crossing a particular system boundary. The strength of this conceptualization predominately lies in its broad application across a very wide spectrum of disciplines and applications.

Figure 2.3 also illustrates the two major versions of system *feedback*: internal and external [13].

Internal feedback is the feedback of a system that is modified and recycled with the system boundary to alter inputs delivered to a system.

Internal feedback is entirely controlled by the system and is typically not visible from outside the system boundary. A common example of internal feedback is a manufacturing quality control process that sends work-in-progress back through manufacturing stages for rework prior to releasing the product for

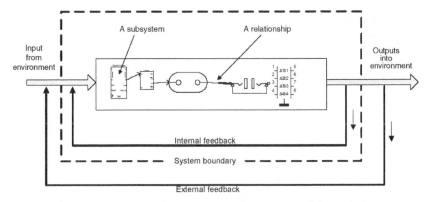

Figure 2.3 Structural organization of a system with boundaries.

distribution. A consumer only sees the end product, not being privy to the internal workings of the manufacturing company. In education, intradepartmental reviews of courses, syllabi, programs, and teaching philosophies are other examples of internal feedback if a system boundary for an academic major were drawn around a department's operations.

Identifying internal feedback depends on having access to the inner workings of a system. If the owner of a system allows access to its internal processes and structure, then it is possible to identify internal feedback. Otherwise, internal feedback is invisible to external observation. When a Securities and Exchange Commission (SEC) accountant performs an in-depth financial audit of a publicly traded company in the United States, full access is required so that the inner workings of the company (aka: systems internal processes) can be identified and examined for adherence to SEC regulations. In a similar fashion, ABET, Inc. requires such access to university academic programs in order to accredit systems engineering major programs.

> *External* feedback is the feedback of a system that occurs in the environment outside of the system boundary, acting to alter inputs before they are delivered to a system.

A system "sees" external feedback in the form of its normal input. Without some means of external observation, a system is unaware of the external systems processes and interactions that are using or responding to some portion of its own output, integrating this output into their systems functions, and releasing modified outputs into the environment that becomes part of the system inputs. This underscores a very important point as to why systems engineers add value to customer programs even when the customer is extremely talented and knowledgeable at what they do: It always helps to have a fresh set of eyes on a problem if for no other reason than to provide objective clarity on systems operations.

Feedback complicates systems modeling and analysis. Modern systems, which are highly connected to other systems both to survive competition and to leverage cooperation, have designed processes enabling them to adapt to their surroundings as a means of attaining competitive advantage and improving performance. Systems that modify their output in a manner that actively responds to changes in their environment due to injection of other systems output are called *adaptive* systems. Adaptive systems tend to pose more of a modeling and analysis challenge because the external (and possibly internal) feedback pathways need to be identified in the course of understanding how the system is operating.

It is also possible that while some open systems produce outputs that are simply transformed input as shown in Figure 2.3, this is not always the case. It is possible that the outputs of the system are simply there to allow the system to secure, through a cycle of events, more of the useful inputs it needs to survive.

In the conceptual system of Figure 2.2, the apparent boundary of the engineering management discipline would be the perimeter of the central, dark circle in the figure which sets it apart from the other overlapping systems.

2.5 VISIBILITY

In most broad terms, system structure can be described first in terms of its relationship with the environment and secondly in terms of what it represents and how it is organized to interact. We have seen one form of structure already: open or closed systems that capture how the system interacts with its environment. Here we want to examine two other dimensions of system structure: visibility and mathematical.

From the systems perspective of input, transform, and output, we can describe system structure in terms of the degree of visibility that we have on the internal workings of the system. Are the elements and their interactions readily identifiable, or are some or all of these hidden from view? The answer to this question enables us to specify the system as one of the three basic structural models: a black box, a gray box, or a white box [10]. These three gradations of visibility into the inner workings of a system are illustrated in Figure 2.4.

A *black* box is a structure that behaves in a certain way without providing any visibility into exactly what internal elements are, how they are specifically linked, and how they functionally transform inputs to the system to produce observable system output. Uncovering such information for the purposes of gaining deeper understanding of a black box system consists primarily of repeatedly manipulating

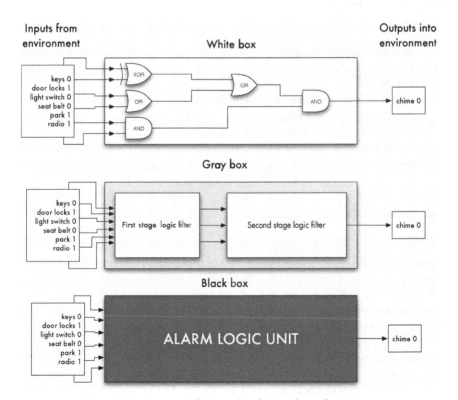

Figure 2.4 Degrees of internal understanding of a system.

the input to get to a point of producing reasonably predictable output(s). Extracting the functional linkage between input and output then becomes the task of design of experiments, regression, response surface methodology, and other data-dependent modeling techniques (see Chapter 4).

A good example of a black box system is human body temperature regulation. If a person's body temperature is too high, placing that person in a bath of cold water lowers it, as does administering loading doses of aspirin. If the person's body temperature is too low, slow heating in a hot tub or with human skin contact will gradually raise the temperature. How exactly the body is reacting to these external stimuli at a subsystem interaction level to make this happen is known, but generally not accessible to discovery without extremely sophisticated instrumentation.

A *gray* box offers partial knowledge of selected internal component interactions and processes (activities). This middle ground perspective is one that is often encountered in practice because even very complex systems have partially accessible and understandable internal processes. For example, in the U.S. economic system which is unperceivable, we could create groups of elements at various levels that would allow a partial view into the inner workings of this system, such as country elements engaged in foreign trade, all legal nonprofit organizations engaged in charity work in the state of New York, and so on. In the gray box illustration of Figure 2.4, while a select number of system elements are visible, their interactions are not. Complete information concerning the system this represents cannot be acquired.

A *white* box systems perspective recognizes complete transparency on a systems internal elements and processes. This ability is rarely achievable when working with existing systems that are complex, but it is very common in newly designed systems such as telecom networks, mechanical devices, business partnerships, athletic teams, and farming operations. It is possible to reasonably fill in gaps in understanding of internal elements and processes of white box systems by making assumptions and then validating these assumptions by measuring output variation in comparison to input changes, as would be done with a black or gray box system.

2.6   IDEF0 MODELS

Black box representations serve a useful purpose early on in the process of designing a system, when system concepts are being explored and requirements are being identified based on system needs and major functions. At this stage, conceptual tools leveraging system thinking are helpful. As will be seen in Chapter 10, once the desired critical functions for achieving the overall systems purpose have been identified, a black box representation for these functions can be created using a method developed by and made public in 1981 by the U.S. Air Force Program for Integrated Computer-Aided Manufacturing (ICAM). More recently, the U.S. Department of Commerce issued Federal Information Processing Standards (FIPS) Publication 183 that defines the Integration Definition for Function Modeling (IDEF0) language, a formal method of describing systems, processes, and their activities.

TABLE 2.1   Family of IDEF Modeling Methods [14]

Method	Purpose	Method	Purpose
IDEF0	Function modeling	IDEF8	User interface modeling
IDEF1	Information modeling	IDEF9	Scenario-driven IS design
IDEF1X	Data modeling	IDEF10	Implementation architecture modeling
IDEF2	Simulation model design	IDEF11	Information artifact modeling
IDEF3	Process description capture	IDEF12	Organization modeling
IDEF4	Object-oriented design	IDEF13	Three schema mapping design
IDEF5	Ontology description capture	IDEF14	Network design
IDEF6	Design rationale capture		

IDEF0 is one member of a family of IDEF methods [15] that can be used to support the SDP during the early phases of a system design effort. IDEF0 models describe the functions that are performed by a system and what is needed to perform those functions. Table 2.1 shows the variety of methods and the purpose for which they were designed [14].

An IDEF0 model is capable of representing the functions, decisions, processes, and activities of an organization or system. It is particularly useful for representing complex systems comprised of a host of processes. The top-level diagram, called Level 0, is the highest level of system abstraction. Using a single box to represent the top level system function, it illustrates in equally high level terms the things that cross the system boundary as inputs and outputs, the physical mechanisms that enable the top level system function, and the controls that determine how this function will operate. Levels 1, 2, and so on, proceed to decompose this Level 0 representation, successively exposing more and more detail concerning the interconnections and interactions of the various subfunctions supporting the overall system function. Proceeding from Level 0 to Level 1, for example, is a bit like lifting the lid off of the black box representation for a system; we begin to see the sequence of processes, activities, or functions linked together to successfully perform the top level function. Even though IDEF0 models are generated using a decomposition strategy, they tend to maintain a "big picture" orientation to the desired system that other approaches such as functional flowcharts can lose as modeling detail increases. It is particularly useful when establishing the scope of a functional analysis as it forces the user to decide on the system boundary in order to display even the highest level of representation for a system.

A process is a systematic series of activities directed toward achieving a goal. IDEF0 uses a single box to represent each activity, combines activities to define a process, and then links system processes to describe a complete system. Creating IDEF0 models of a system generally proceeds by decomposing system functions into layers, the top layer being a system-level model resembling a black box based on the information obtained during stakeholder interviews. In this sense then, it is helpful to decide whether an IDEF0 model is going to be used prior to interviewing

Figure 2.5   Example SIPOC diagram for a community policing process.

stakeholders. Doing so ensures that key information required for the IDEF0 model can be obtained without having to unnecessarily re-interview stakeholders.

IDEF0 representions are similar in structure to the popular SIPOC diagrams commonly used as an intial step taken when mapping processes for Lean Six Sigma applications [16]. SIPOC is an acronym for Suppliers, Input, Process, Output, and Customers. Figure 2.5 illustrates a SIPOC diagram for a hypothetical service process provided by a community-based law enforcement organization.

What differs is the particular level of detail used in the representation. SIPOC diagrams graphically illustrate a simple flow of service or products between suppliers and customers as a first step toward a more detailed value stream mapping that contains process controls and mechanisms along with a host of more details necessary to identify non-value-added components of the process that are then targeted for elimination. Turtle diagrams [17], a far less popular modification of SIPOC diagrams, include information "legs" to the SIPOC diagram that contain details concerning measures, what, how, and who about the process.

Creating an IDEF0 model loosely follows a stepwise procedure. At the start, the overall purpose for the model and the particular stakeholder(s) viewpoint that is going to be reflected in the IDEF0 must be identified on the Level 0 representation. This explicit statement of viewpoint orients a user of the IDEF0 model as to the perspective that should be assumed when interpreting the information in the model. Next, using both stakeholder interviews and independent research, identify the system, its boundary, major raw material inputs and outputs, the top level system function that turns inputs into outputs, and the laws, regulations, operating guidelines, and stakeholder desires that control this function, along with the physical aspects that cause the system to operate. This information will become the top level IDEF0 model.

Following this, identify the major processes that are needed to turn the system inputs into outputs. For each major process, identify the objectives or purpose associated with it. These processes typically are sequentially linked at the highest level. Each of these processes will further break down into sequences of activities organized to achieve the objective(s) of the process. All three of these system structures, the system, the processes, and the activities, are represented as individual IDEF0 models.

Lastly, it is important to decide on the decomposition strategy that will be used. Four common decomposition strategies are [18] as follows:

- *Functional decomposition* breaks things down according to *what* is done, rather than *how* it is done. This tends to be the most common decomposition strategy because it naturally aligns with the strategy used for constructing a functional hierarchy.

- *Role decomposition* breaks things down according to *who* does what.
- *Subsystems decomposition* starts by breaking up the overall system into its major subsystems. If the major subsystems are relatively independent, this is a helpful strategy to initiate the IDEF0, subsequently employing functional decomposition to further decompose the subsystems into processes.
- *Life cycle decomposition* is sometimes used when the stages of the system life cycle are relatively independent and subsystems and their processes generally align with the age of the system.

At the end of these steps, an initial IDEF0 system representation can be formally constructed. From a practical standpoint, an IDEF0 model is a conceptualization that supports functional analysis and ultimately develops a value hierarchy within the SDP. IDEF0 revisions are commonplace during the SDP as new and more accurate information arises with regard to the needs, wants, and desires of stakeholders.

A basic IDEF0 model at any level consists of five possible components [18]:

1. *Activity, Process, or System.* A box labeled by "verb–noun" describing the activity/function that the box represents (e.g., collect intelligence; weld joint; coffee making).
2. *Inputs.* Arrows entering the left side of the box represent the "raw material" that gets transformed or consumed by the activity/function in order to produce outputs (e.g., information; welding rod, electric current, Tungsten Inert Gas (TIG); coffee grounds, water).
3. *Controls.* Arrows entering the top of the box are controls. These specify the conditions required for the function/activity to produce outputs, such as guidelines, plans, or standards that influence or direct how the activity works (e.g., U.S. federal regulations; union safety standards; recipe, coffee machine directions).
4. *Mechanisms.* Arrows connected to the bottom side of the box represent the physical aspects of the activity/function that cause it to operate (e.g., agents; union welder, TIG welding torch, electricity; drip coffee machine, coffee person, electricity). These can point inward or outward of the box. Inward pointing arrows identify some of the means that support the execution of the activity/function. Arrows pointing outward are *call* arrows. These enable the sharing of detail between models or between portions of the same model.
5. *Outputs.* Arrows leaving the box on the right are outputs, which are the result(s) of the activity/function transmitted to other activities/functions within the IDEF0 model, to other models within the system, or across the system boundary to the environment (e.g., intelligence reports, fused metallic bond, pot of coffee).

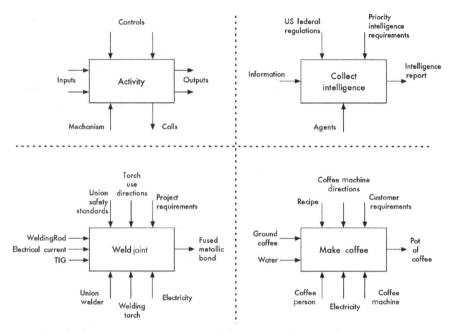

Figure 2.6 Generic IDEF0 model with three functional examples.

The upper left illustration in Figure 2.6 shows the generic structure of an IDEF0 model. The three other models are examples of IDEF0 models for each of the example activities noted in the description above. Constructing even the Level 0 representation of a system can be challenging. Referring to the upper left illustration in Figure 2.6, it is helpful when building each block in a diagram to conceptualize the block as a function that "transforms inputs 1 and 2 into outputs 1 and 2, as determined by controls 1 and 2 using mechanism 1." The verbs in this expression add clarity to the definitions of inputs, outputs, controls, and mechanisms.

Creating an IDEF0 model of a system proceeds in a decomposition fashion similar to that used to create a hierarchy. The highest level model of the system, Level 0, communicates the system boundary in relation to its environment. This top-level representation on which the subject of the model is represented by a single box with its bounding arrows that is labeled as A-0 (pronounced "A minus 0"). The A-0 diagram at Level 0 sets the model scope, boundary, and orientation through the visualization of the box, the purpose and viewpoint expressed in text below the box, and the title of the A-0 diagram at Level 0. The title affords the opportunity to identify the name of the system. The purpose communicates to a reader why they are looking at the IDEF0 model, and whose viewpoint it represents. Remember that a fundamental assumption of system thinking is that perspective matters. If the perspective of a system changes, that is, we look at the system through someone else's eyes, the system might look different. Again, this is precisely why multiple stakeholders must be included in any systems study.

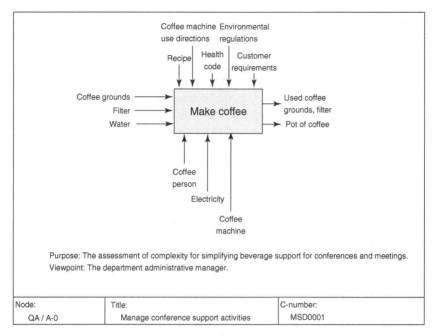

Coffee machine Environmental
use directions   regulations

Recipe | Health | Customer
            code | requirements

Coffee grounds ———→

Filter ———→   Make coffee

Water ———→

Used coffee
grounds, filter

Pot of coffee

Coffee
person

Electricity

Coffee
machine

Purpose: The assessment of complexity for simplifying beverage support for conferences and meetings.
Viewpoint: The department administrative manager.

Node:	Title:	C-number:
QA / A-0	Manage conference support activities	MSD0001

Figure 2.7   An A-0 diagram for the make coffee function.

Figure 2.7 shows a slightly more complicated A-0 diagram for the "Make Coffee" function introduced in Figure 2.6. The boundary of the system, purpose, and viewpoint are shown in the main diagram area. Along the bottom of the A-0 diagram are tracking information specified by the systems engineer who would create the diagram. The "ProjectID" (QA) and "NodeID" (A-0) are unique identifiers. The "Title" block is used to describe the overall activity in the context of the process, subsystem, or system it supports. The "C-number" is a combination of the author's initials (MSD) and document ID number (0001).

We chose this example for three reasons. First, notice that this Level 0 model assumes that a particular mechanism (electric coffee machine) is to be used in any systems solution that might result, which appears reasonable given the viewpoint expressed. However, for most applications of the SDP, we would caution to avoid including solution elements (how to do something) early in the Problem Definition phase, which is when the IDEF0 representation is most likely to be used. Coffee can be made without either electricity or a coffee machine. If the department administrative manager placed a high preference (value) on some functionality provided by an electronic coffee maker, it would be better to include this as stakeholder input for the next level of decomposition (Level 1) rather than as a hard mechanism at this modeling level (Level 0). In this way, solution alternatives are free to creatively satisfy this desired functionality using mechanisms other than an electric coffee maker, if feasible.

Second, even a simple process of making coffee for a meeting can take on complex interactions when viewed from a systems thinking perspective, and the

SDP has useful cues for where to look to recognize these interactions. For example, health code and environmental regulations (legal factor) exert a nontrivial amount of control over this function when the output is to be served in a public forum. Thus, the Make Coffee function interacts with the legal system.

Finally, comparing the output of this IDEF0 model with that of Figure 2.6, notice that we have included the waste products (the unintended consequences) along with the designed product (the intended consequences) as output of the system. In this way, the Make Coffee function as a system interacts with the environmental system existing outside of its system boundary. This interaction is easy to overlook, as some industrial companies operating in the United States have learned the hard way. (See http://www.epa.gov/hudson/)

Figure 2.8 illustrates an A-0 diagram for the SDP introduced in this book. The model shows a top-level view of the SDP, which accepts the current system status as input and produces the desired end-state system solution as output. Stakeholder input and the applicable systems engineering professional standards act as controls for the process execution. The systems engineering project team and organization resources comprise the physical aspects of the SDP.

An IDEF0 model of the SDP then proceeds to decompose the process into the next level, Level 1 (labeled as A-1) of system processes, namely the four phases of the SDP. Figure 2.9 shows a representation at Level 2 of the Problem Definition phase of the SDP. This Level 2 model displays the linkage of the three major SDP

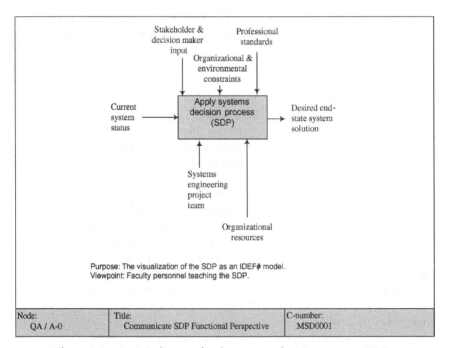

Figure 2.8 An A-0 diagram for the systems decision process (SDP).

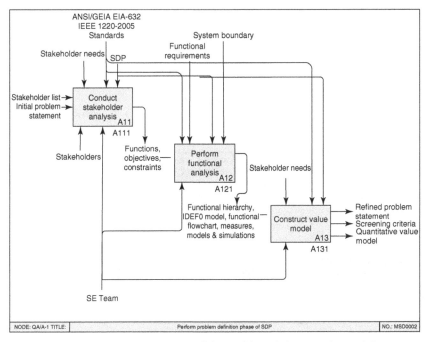

Figure 2.9 Level 2 representation of the problem definition phase of the SDP.

functions performed during this phase. Starting with a stakeholder list and initial problem statement as input, the three system functions (activities, processes) flow from the upper left to the lower right of the Level 2 model following the order in which inputs are handled to create the three necessary outputs required by the decision gate that allows the SDP to progress to the Solution Design phase that follows.

Although sequenced IDEF0 models like this one are commonly linked through inputs and outputs, they can also be linked by controls and mechanisms as well. In Figure 2.9, the professional systems engineering standards (ANSI/GEIA EIA-632 and IEEE 1220-2005) and the SDP act as control on each of the functions shown. Likewise, the SE team links the three functions by acting as a common physical mechanism enabling each of the functions to successfully occur. Each of the subsequent SDP phases can be represented by an IDEF0 diagram as well, which is left as an exercise for the reader.

Figures 2.10, 2.11, and 2.12 illustrate an example of what three levels of IDEF0 models could look like for a hypothetical fast food restaurant in the United States. The Level 1 model in Figure 2.11 presents the top-level system function as a sequence of three subfunctions: "process customer order," "prepare food items," and "deliver food items." Figure 2.12 then shows how the "process customer order" function is decomposed into a sequence of four functions. In turn at Level 2, similar models would be created for the "prepare food items" and "deliver food items"

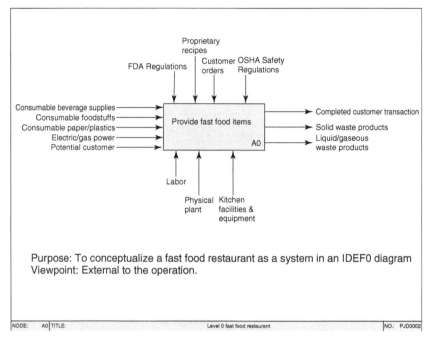

Figure 2.10   Level 0 model of a fast food restaurant.

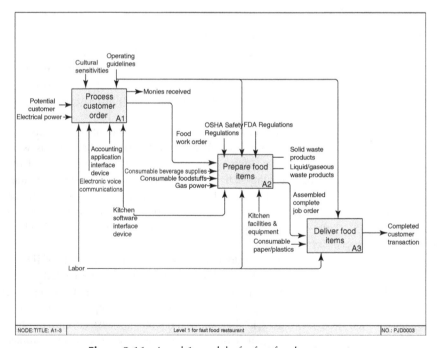

Figure 2.11   Level 1 model of a fast food restaurant.

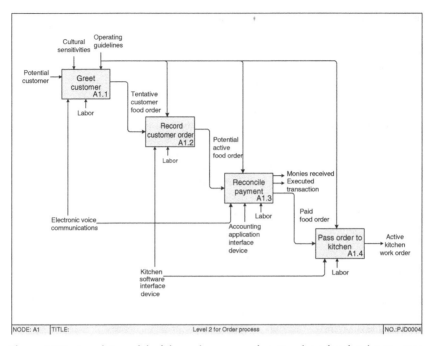

Figure 2.12   Level 2 model of the order process function for a fast food restaurant.

processes shown at Level 1. We leave these as an exercise for the reader. Notice how each level increases the amount of detail being presented concerning the system's operation, which has implications on the amount and quality of information needed from stakeholders in order to properly represent the system at these levels.

IDEF0 diagrams represent one modeling framework that has been implemented in a host of software applications. Since their introduction and with the explosive growth of interconnected systems, other tools strongly leveraging systems thinking have been developed. One such effort resulted in a modeling language based on the object-oriented analysis and design language Unified Markup Language (UML) called SysML.

The SysML (Systems Modeling Language) is a general-purpose modeling language for systems engineering applications that supports the specification, analysis, design, verification, and validation of a broad range of systems and systems-of-systems [19]. Introduced in 2003 by a group of partners interested in improving the precision and consistency of systems diagrams, the development effort split in 2005 into an open-source project (SysML: www.SysML.org) and a commercial software product (OMG SysML: www.sysmlforum.com) offered by the Object Management Group. Where UML is predominantly used for software development, SysML's charter extends into much broader classes of systems, many of which do not include software components.

As can be seen in Table 2.2, SysML contains a host of visualization tools that enable a user to represent a system in a number of ways depending on the need

TABLE 2.2   Open Source SysML Diagrams [19]

SysML Diagrams	Primary Purpose
Activity diagram	Show system behavior as control and data flows. Useful for functional analysis. Compare Extended Functional Flow Block diagrams (EFFBDs), already commonly used among systems engineers.
Block Definition diagram	Show system structure as components along with their properties, operations and relationships. Useful for system analysis and design.
Internal Block diagram	Show the internal structures of components, including their parts and connectors. Useful for system analysis and design.
Package diagram	Show how a model is organized into packages, views, and viewpoints. Useful for model management.
Parametric diagram	Show parametric constraints between structural elements. Useful for performance and quantitative analysis.
Requirement diagram	Show system requirements and their relationships with other elements. Useful for requirements engineering.
Sequence diagram	Show system behavior as interactions between system components. Useful for system analysis and design.
State Machine diagram	Show system behavior as sequences of states that a component or interaction experience in response to events. Useful for system design and simulation/code generation.
Use Case diagram	Show system functional requirements as transactions that are meaningful to system users. Useful for specifying functional requirements. (Note potential overlap with Requirement diagrams.)
Allocation tables	Show various kinds of allocations (e.g., requirement allocation, functional allocation, structural allocation). Useful for facilitating automated verification and validation (V&V) and gap analysis.

being addressed. Many of these tools can augment those introduced in later chapters to support the SDP activities.

2.7   MATHEMATICAL STRUCTURE

The structure of a system can also be described once a symbolic model of the system has been constructed using mathematical notation. Doing so requires us to focus not simply on the elements of a system but also on the relationships existing between elements. These relationships are the binding material of systems, and mathematics provides a means by which these relationships can be captured and analyzed. What is the nature of these relationships? What do they imply about the system itself? How do the relationships or the elements themselves change over time? If allowed to continue in its current operational configuration, what might the state of the system be sometime in the future? Many of the modeling methods that follow in later chapters are chosen based on relationship characteristics.

An assumption at the heart of most mathematical system models is that the observable output of a system is a direct result of the input provided and the controls acting on the system. This relates input to output in a cause-and-effect manner that allows us to think of (a) the system's overall behavior within its boundary and (b) the functions and subfunctions supporting this behavior as mathematical functions.

In each of the IDEF0 model examples introduced in Figure 2.6, the bounding box containing the system function accepts input and transforms this input into output via some known or unknown transformation function that can potentially be expressed as a cause–effect relationship.

Consider the IDEF0 model for a system we label "collect intelligence" in Figure 2.13. Let (c) represent the input (cause) information to the system and let (e) be the output (effect) intelligence report [30]. Suppose that we impose a change in the input information (c), which we denote by Δc. This change in turn causes a change in the output e of the system, which we represent as Δe. The exact translation or transformation of this change, or *perturbation*, is accomplished by some transformation function which we denote as $g(e, c)$. The nature of $g(e, c)$ defines the mathematical structure of the system. Its action moves the system used to produce intelligence reports from one state $S(e, c)$ to another $S(e + \Delta e, c + \Delta c)$. This idea of a *system state* is a general concept that can be used to describe a system's condition, location, inherent health, financial position, and political position, among others.

Is $g(e, c)$ linear or nonlinear? We can determine this by examining the proportionality of the effect response for various changes in input. If Δe is proportional to Δc, then $g(e, c)$ is linear. If not, then $g(e, c)$ is either discontinuous or nonlinear. Determining the best mathematical form of $g(e, c)$ is left up to experimentation and data analysis using methods like linear regression, spline fitting, response surface modeling, and others. Once $g(e, c)$ is identified, is there a best level of output for given ranges of input that the system can be tuned to produce? Optimization techniques such as linear and nonlinear programming, equilibrium models, and related techniques could be used to answer this question and are introduced in Chapter 4.

Does the transformation performed by the system function stay consistent over time? If so, then $g(e, c)$ is likely *time invariant* and time is not explicitly modeled in the mathematical system representation. Otherwise, time should be included in the

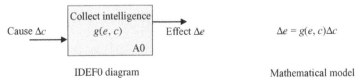

| | IDEF0 diagram | | Mathematical model |

Figure 2.13 Two abstract system models: graphical and mathematical.

system function representation: $g(e, c, t)$, in some manner. If $g(e, c, t)$ is nonlinear, is the system *stable*, or do the effects increase without bound?

Is the relationship between Δc and Δe known with certainty? If not, then $g(e, c)$ should include probability or uncertainty elements. This is typically the case for systems that are best represented by simulations, queuing networks, decision analysis structures, risk models, forecasting methods, reliability models, and Markov processes, among others.

The role played by the function $g(e, c)$ in this abstract system representation is crucial. In mathematical terms, the function $g(e, c)$ is referred to as a *system kernel* [13]. It fundamentally describes the incremental change occurring between input and output to the system, which can occur in discrete steps:

$$g(e, c) \equiv \frac{\Delta e}{\Delta c} \tag{2.1}$$

or continuously:

$$g(e, c) \equiv \frac{de}{dc}. \tag{2.2}$$

The two expressions (2.1) and (2.2) are related by a limit expression as the incremental interval is made infinitesimally small:

$$\lim_{\Delta c \to 0} \frac{\Delta e}{\Delta c} = \frac{de}{dc} = g(e, c). \tag{2.3}$$

Figure 2.14 illustrates a slightly more complicated system structure to represent mathematically. The external feedback loop complicates the input to the system by adding a new input component that represents a manipulated portion of the system output. A common example of a structure of this kind can be envisioned by thinking about the creation of a political speech as an IDEF0 model's function, one element of a much larger political system. In this manner, we can construct a system boundary around the speech writing process from the perspective of a particular speech writer.

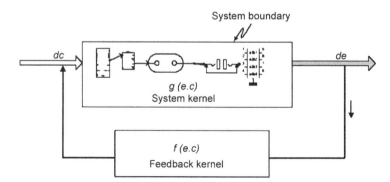

Figure 2.14 A system representation with external feedback.

Constantly changing information flows into this system from the environment, acting as input dc to the speech writing kernel, $g(e,c)$, which processes this input to create the output political message for a speech, de. However, the speech writer also wisely takes into account the public reaction to previous speeches when crafting a new message to be delivered. This consideration of public processing of previous output de creates a new system function called a *feedback kernel* which is part of an external system that we associate with the public. The output of this feedback kernel is *public* opinion relevant to the speech writer.

The public takes the previous output de and uses de as input to the feedback kernel $f(e,c)$. The output of the feedback kernel, $f(e,c)de$, is added to the normal environmental input dc so that the total (new) input to the system becomes $dc + f(e,c)de$ Thus we have

$$de = g(e,c)[dc + f(e,c)de] \qquad (2.4)$$

Finally, by gathering common terms in Equation (2.4) and assuming that $dc \neq 0$, we see that the system kernel alters its form to a new form $g(e,c)_f$, where the subscript f simply designates that the feedback effects have been included in the system kernel:

$$\frac{de}{dc} = g(e,c)_f = \frac{g(e,c)}{1 - g(e,c)f(e,c)} \qquad (2.5)$$

By thinking of the action $f(e,c)$ has on the previous system output de, and noting that the next output from the system kernel $g(e,c)_f$ is again acted on by the feedback kernel, we can easily get a sense that the feedback input is either being left alone $(f(e,c)de = 0)$, amplified $(|f(e,c)de| > 1)$, or dampened $(|f(e,c)de| < 1)$ with each loop negotiation. This example begins to hint at some of the underlying system structure that can introduce system *complexity*, which arises from nonlinear dynamic changes, similar to those exhibited by this example when $g(e,c)$ is nonlinear. Complexity also arises when the system kernel $g(e,c)$ alters its structure in response to changes in input c. This ability of the system to adapt is again a characteristic of complex systems.

In an IDEF0 model representing sequenced functions, activities, or processes, feedback can also be represented in one of the two forms: control feedback and/or input feedback. Control feedback takes the output of a lower bounding box in the sequence order and connects it via an arrow to the control of an earlier bounding box in the sequence. A practical example of this can be seen in the organizational use of after-action reviews and similar activities that use system output to guide the system function. Similarly, input feedback takes the output of a lower bounding box in the sequence order and connects it via an arrow to the input of an earlier bounding box in the sequence. This situation is commonly encountered in rework situations—for example, in manufacturing or report writing.

Large systems are not necessarily complex systems. The presence of complicated and strong element interactions, nonlinear dynamic changes induced by system function kernels, and possibly self-organization behavior can impose complexity on

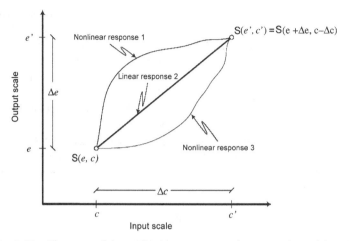

Figure 2.15 Three possible qualitative structures of a system kernel function.

a system; being large in size complicates an already challenging situation. Systems decision problems possessing some or all of these characteristics are recently being referred to as *wicked* problems [21]. Most often, the tool of choice to model and analyze a complex system is simulation.

Early in a system life cycle (see Chapter 3) when input and output data for a system are likely unavailable, it is still possible to gain a qualitative sense of the mathematical structure of a system using this same mathematical structure approach without knowing the exact notational form of either the system kernel $g(e, c)$ or existing feedback functions. Optionally, it may be possible to use a graphical approach in collaboration with the system owner or user(s) to extract a more general, but still useful, approximation of $g(e, c)$ by simply focusing on input and output to the system. Figure 2.15 illustrates three possible resulting function forms for $g(e, c)$ for the case when the system kernel is assumed to be continuous.

The topmost nonlinear curve 1 in Figure 2.15 describes a system kernel $g(e, c)$ that quickly drives the system state $S(e, c)$ to change for small changes in system input c. This shape is called *concave*. As the size of the imposed change on input Δc increases, the output response decreases. The bottommost nonlinear curve, 3, has just the opposite characterization. The system kernel $g(e, c)$ translates large perturbations on the system input into large output responses. This upward opening curve is called *convex*. The middle curve, 2, illustrates a proportional response that remains consistent throughout the range of input changes Δc. The particular shape of the system kernel estimated in this manner provides important clues as to what the mathematical model of this system function should be.

2.8 SPATIAL ARRANGEMENT

The natural state of affairs for systems is that they exist within and among other systems so that for any particular systems decision problem, a systems engineer is inevitably faced with a composite system environment. Determining where and how the specific system under focus is located relative to the other systems it interacts with is a necessary task for setting the scope of the program. Some systems will have extensive interaction with the system under focus, which means that changes occurring in one of these interrelated systems have an impact on the behavior of the system under study. Systems of this kind have a high priority for being included within the scope of the program. As the degree of interaction between systems diminishes, systems falling into this category are more likely to be simply noted for the record and set aside.

One important note to consider is that systems often interact on an abstract level in addition to or in lieu of having linked system elements. Market effects, a consideration related to the financial environmental factor, are a good example of this. Competition for market share between—for example, Linux and Microsoft® WindowsXP operating systems—creates a strong interaction that must be considered when version updates and functionality revisions are being designed and developed. Likewise, two separate countries can be strongly linked politically because of shared vested interests even though no elements of their infrastructure are interconnected. Being sensitive to subtle interactions such as these and others helps identify and characterize lateral systems described in what follows. This is one of the uses of the environmental factors surrounding the four steps of the SDP shown in Figure 1.7 in Chapter 1.

A good visualization of the typical arrangement of systems interrelationships is useful for detecting the connections between various systems. Generally, systems either start out as, or evolve into being, part of other systems in one of three structural relationships: multilateral, lateral, and multilevel hierarchies. A multilevel hierarchical arrangement is one that is very familiar to large organizations because this is typically the manner in which they are structured to accomplish defined objectives. These arrangements are also referred to as *nested* systems. A more commonly encountered term is a "system of systems."

For a systems decision problem, recognizing and understanding the relationships between interacting systems, especially in a system-of-systems situation, is significant because these systems are likely to be at different stages in their individual life cycles. Mature systems tend to have a strong degree of presence in the composite system environment, an extensive network of connections, and significant resource requirements, and they are able to sustain a competitive situation for a long duration of time. In comparison, younger systems tend to have newer technologies, less presence in the composite system environment, less extensive interdependencies with existing systems, may have significant resource requirements, and their ability to sustain competition is less robust. Not recognizing this additional characteristic can result in unsatisfactory systems solutions being constructed. Moreover, notice that these observations, important as they are, say little about system efficiency, effectiveness, ability to leverage resources or arrange cooperative strategic alliances, and a host of other concerns that address the long-term survivability of systems. Each systems decision problem should be approached as being unique, and it is up to the systems team to discern the relevant issues of concern from stakeholder and decision maker input to the SDP.

In a hierarchy, the system under study resides within a broader contextual arrangement called the *metasystem*. It in turn can have groupings of connected elements interacting to accomplish defined objectives. These are called *subsystems*. Within a hierarchical arrangement, like entities exist on the same level. Subsystems particular to these entities exist below them, and systems that they are a part of exist above them. Systems operating on the same level that share common elements are called *multilateral* systems. Systems on the same hierarchical level that are linked only by abstract or indirect effects (e.g., political influence, competition, targeted recruiting at the same population) are referred to as *lateral* systems.

An example of a multilateral arrangement from social network analysis [2, 22] is shown in Figure 2.16. Cindy is a person who exists as a friend to two separate groups of people that do not interact. The bridge relationship connecting the two social groups (systems) through the single individual (shared component) is multilateral systems structure. Changes that occur in one group have the potential for inducing changes in behavior in the other group through this shared component. The two multilateral friendship networks shown in Figure 2.16 would be considered to be in a lateral system arrangement if they did not share the element "Cindy" in common.

The positioning of a system within a hierarchy is relative to the system being examined, being determined in large part by the system purpose or how it functions in its relationship to other systems. It is this idea that motivates the construction of a functional hierarchy for systems decision problems.

Figure 2.17 presents two perspectives of the exact same group of hierarchical objects concerning the suspension system for an automobile. If the system we are concerned with is the vehicle's complete suspension system, we would conceptualize the system hierarchy consistent with Perspective 1: The system would reside

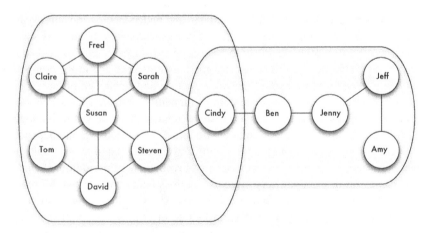

Figure 2.16 Multilateral friendship systems in a social network [1].

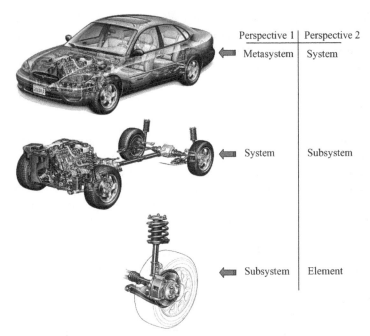

	Perspective 1	Perspective 2
	Metasystem	System
	System	Subsystem
	Subsystem	Element

Figure 2.17 Three hierarchy levels of system spatial placement. (Courtesy of Kevin Hulsey Illustration Inc.).

within the metasystem of the entire automobile and have as one of its subsystems the left rear brake drum and shock absorbing subsystem. If we were instead focusing on the entire automobile as our system of concern, then Perspective 2 would be appropriate. The complete suspension system would be a subsystem and the left rear brake drum and shock absorbing object would be an element of this subsystem. One possible metasystem for Perspective 2 could be the rental car system that uses this automobile as part of its available fleet.

In hierarchical systems, the observable system behavior can easily change, depending on the system being examined. At higher levels in a hierarchy a more abstract, encompassing view of the whole system emerges without attention to the details of the elements or parts. At lower levels, where subsystems and individual elements are evident, a multitude of interacting parts can typically be observed but without understanding how they are organized to form a whole [23].

In a pure multilateral arrangement in which the only connections between systems are shared elements, the relationship between these systems is essentially nonhierarchical. For situations such as this, it is then valuable to conceptualize the arrangement of systems as a *composite multilateral* arrangement [24]. Subsystems and elements can then have the same meaning as in a typical hierarchical structure. The environment shaped by the interactions emanating from the composite multilateral systems takes the place of a metasystem. This is a particularly useful construct for systems decision problems involving strategy and policy, where the dynamics of influence, power, or counteraction are key to shaping the decisions being made. Recent examples of such an arrangement arose when applying systems thinking to U.S. border security policy [25], metropolitan disaster planning [24], and counter-insurgency strategy [26].

2.9 EVOLUTION

While we may observe systems within a limited period of time, and subsequently gather data that we use to design, improve, or change systems, we must always bear in mind the fact that the system does not sit still and wait for us to finish what we are doing. A system is somewhat like an object in a pond. An extremely hard object like a glass ball would change little over time in this environment, yet the environment may radically change around it. Conversely, while the environment in a pond surrounding a saltine cracker may not change much, the internal structure of the cracker would indeed change over time. And so it is with systems.

Systems engineers concern themselves with how system elements interact both within and external to system boundaries at the time they are observing the system. They must also be aware of how these interactions and conditions affecting them might evolve in the time that passes between different team actions. For example, performance data for individual securities on the New York Stock Exchange would have to be continually updated in a systems decision problem concerning portfolio structuring. If this is not done, a potential investment strategy that would work at the time of data collection might not a week, month, or 6 months later. The sensitivity of these solutions to time is very large. The system dynamics are rapid, they have the potential for wide variation in performance, and they are filled with uncertainty. Any system representation and subsequent program planning should take these characteristics into consideration.

Chapter 4 introduces several techniques for modeling and analyzing systems and their effects when evolution is a consideration, whether the changes of interest are considered continuous or discrete. A continuous system is one whose inputs can be continuously varied by arbitrarily small changes, and the system responds to these continuous inputs with an output that is continuously variable. A discrete systems is one whose inputs change in discrete amounts (e.g., steps in time, fixed increases in volume, number of people) that generally have a minimum size below which it does not make sense to reduce. Thus, the output tends to respond in a stepwise fashion as well [13].

2.10 SUMMARY

Systems thinking distinguishes systems engineering from other engineering fields. The holistic viewpoint that embodies this perspective makes the systems engineer a valued and unique contributor to interdisciplinary team efforts. Adopting this world view enables systems to be identified, classified, and represented in such a way as to lead to a deeper understanding of the system's inner workings and interconnected relationships that contribute to system-level behavior. The end result of applying this thinking using the SDP directly contributes to a more comprehensive, more robust, and richer array of possible solutions generated for a systems decision maker to consider.

Systems thinking also affords an ability to establish a proper system boundary, scope, and orientation for a systems decision problem regardless of whether the system is open or closed to its environment. Knowing this information keeps an interdisciplinary systems team focused on the problem at hand and provides an understanding as to where their most significant contributions can be made.

Structure defines the interconnected relationships between system elements and other entities. This structure can be represented in a variety of ways, some mathematical as in input–output functions and systems kernels, and others graphical, as in IDEF0 models. A good start toward understanding this structure can be gained using qualitative methods during stakeholder analysis. The underlying structure of a system can be complicated when internal or external feedback is involved because the natural inputs seen by a system are being preconditioned, possibly without the system's internal elements being aware of it. Moreover, "downstream" output could be intentionally or unintentionally exerting nontrivial controls over early functions in a sequenced process.

Systems naturally contain an internal hierarchical arrangement of elements and processes. Systems also exhibit hierarchical arrangements and ordering with respect to their relationships with other systems. Identifying the structure and form of these hierarchical arrangements during the Problem Definition phase of the SDP leads directly to the critical and necessary understanding that supports the decision gate leading to effective Solution Design, the next phase of the SDP.

2.11 EXERCISES

2.1. For each of the following systems, characterize them as a white box, gray box, or black box system from a particular perspective that you identify. Briefly explain your reasoning and what could be done to increase the level of internal understanding of the system, if appropriate.

(a) Political system in effect governing the activities of the office of Mayor in Providence, Rhode Island.

(b) A magic act you see performed at Lincoln Center in New York City.

(c) A slot machine in operation in Reno, Nevada.

(d) The Dave Matthews Band.

(e) A crime syndicate operating in Newark, New Jersey.

(f) A vacation travel agency.

(g) A Blackberry® device.

(h) Gasoline consumption in the United States and global warming.

(i) A George Foreman® "Champ" grill.

(j) The admission system to college.

2.2. Identify possible sources of internal and external feedback that could or does exist for the systems in the previous question. Identify any possible ways of quantifying this feedback where it exists.

2.3. Using Figure 2.9 and Figure 2.8 as guides, construct IDEF0 Level 1 models for the Solution Design, Decision Making, and Solution Implementation phases of the SDP. Since Figure 2.9 was created using the first edition SDP, change the Level 2 IDEF diagram to accommodate the new SDP elements of the Problem Definition phase shown in Figure 1.7.

2.4. Referencing Figure 2.11, create IDEF0 Level 2 models for the fast food subfunctions "prepare food items" and "deliver food items."

2.5. Create IDEF0 Level 0 and Level 1 models for the following systems:

(**a**) Harley-Davidson® Sportster motorcycle.

(**b**) The U.S. presidential election process.

(**c**) Scuba gear.

2.6. Create a Level 0 model that represents the political speech writing process introduced as an example in Section 2.7. Be sure to include any intended and unintended consequences of the process as output of the model. What controls are imposed on this process?

2.7. What classification would you assign to the following systems:

(**a**) Maslow's hierarchy of needs (Figure 2.18)

(**b**) Algebra

(**c**) The Ten Commandments, Torah, and/or Quran

(**d**) Marriage

(**e**) Motor vehicle operating laws

(**f**) Your study group for a class you are taking

(**g**) A human heart

(**h**) A cell phone

(**i**) An Ipod

Figure 2.18 Maslow's hierarchy of needs.

2.8. Create an IDEF0 Level 0 model for a car purchasing decision as a system from your point of view.

2.9. Why is the Linnaean taxonomy a hierarchy? (www.palaeos.com)

2.10. What type of structure does the Department of Defense's Global Information Grid (GIG) have? Briefly explain. Cite credible resources on the Internet that you use to answer this question.

2.11. In recent times, as organizations have become more complex, corporate leaders have examined the idea of a "flat organization" to explore whether it would operate more effectively and efficiently. In particular, recognizing that the environment around organizations is constantly evolving, one wonders whether a flat organization would outperform a traditional hierarchy in terms of being able to successfully adapt to changes in its environment. Use credible web sources to define a flat organization. Select a major organization that you are familiar with and develop a supported position as to whether it would be better off in some capacity if it transitioned to a flat organization.

2.12. Rensis Likert has conducted extensive research on a nonbureaucratic organization design referred to as System 4 (participative versus democratic). Is System 4 a Multilateral or multilevel arrangement? Explain.

2.13. What type of structure does the U.S. Navy's Trident weapon system have? If you isolate the warhead as a system of concern and were to initiate a systems engineering study on it, what other systems would interact with it?

2.14. The U.S. Department of Homeland Security announced a call for proposals in 2005 for a multi-billion-dollar project called Secure Border Initiative (SBI*net*) to test the ability of technology to control U.S. borders after a succession of failures. What type of structure are they envisioning for the SBI*net*? Suppose that you were going to conduct a systems engineering study that focused on the U.S. border as a system.

   (a) Apply systems thinking to this problem by using the environment factors of the SDP to construct a systems hierarchy that shows the major lateral and multilateral systems, subsystems, and metasystems that define the U.S. border.

   (b) Would a policy of limiting immigration to the United States be a system-level or symptom-level system solution? Explain.

   (c) Which of the systems that you specified in part (a) would you consider to evolve over time?

   (d) For each of the evolving systems you identified in part (c), list one symptom-level and one system-level phenomenon that you think provide evidence that the system is evolving.

2.15. Using a satellite imagery mapping service such as Google Maps or Google Earth, locate the U.S. Customs Port of Entry at the north side of Heroica Nogales. Now, locate the U.S. Customs facility where U.S. Highway 91 becomes Canadian Highway 55 on the northern border of Vermont.

(a) Thinking of these two border crossing locations as systems, how do the systems represented by the environmental factors of the SDP differ between the two locations? For example, does their importance change, depending on which location you are working with?

(b) At which location would a pure technological solution like remote cameras work better?

(c) Compare the stakeholders at both locations. Why might they be different? Which are in common?

(d) For the stakeholders that you identified as being in common at both locations, would you expect their vested interest (stated position) on border security to be the same for both? What does this imply for a single immigration policy for the United States?

(e) For each of the stakeholders you identified in part (c), use the internet to find credible source documentation that enables you to identify their vested interests with regards to the border security issue. Were their any surprises?

2.16. From the viewpoint of a winery owner, the process of making red wine can be conceptualized as a system comprised of three major functions: preparation, fermentation, and aging.

(a) Construct an IDEF0 Level 0 model for a California winery currently in operation that you identify from the Internet. Assume the viewpoint of the winery owner.

(b) Using the three system functions listed, construct an IDEF0 Level 1 model for the winery that properly illustrates the sequence of these functions along with their inputs, outputs, mechanisms, and controls.

(c) Construct an IDEF0 Level 2 model for the "fermentation" function, basing this model's detail on actual information for the winery you choose that is provided by credible Internet references.

(d) Estimate the mathematical representation that could be used for the fermentation function kernel.

(e) Would the Level 0 model change if the viewpoint adopted was that of a distributor or retail wine store operator? Briefly explain.

2.17. Figure 2.19 was created in support of a systems engineering study focusing on information flow of communications on the modern battlefield [27].

(a) There appears to be a natural hierarchy present. How would you represent this in a diagram?

(b) What systems thinking ideas can you apply to this illustration? What structure is attempting to illustrate? Can you identify any elements of complexity? What would a systems component be in this illustration? Does the system behavior change at different levels being observed?

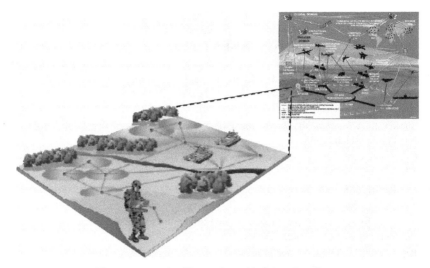

Figure 2.19 An illustration of information flow.

REFERENCES

1. Pidd, M. *Tools for Thinking*. Chichester, West Sussex, England: John Wiley & Sons, 2003.

2. Wasserman, S, Faust, K, Iacobucci, D, Granovetter, M. *Social Network Analysis: Methods and Applications*. Cambridge, England: Cambridge University Press; 1994.

3. Haines, SG. *The Manager's Pocket Guide to Systems Thinking and Learning*. Amherst, MA: Centre for Strategic Management, HRD Press, 1998.

4. Okes, D. *Root Cause Analysis*. Milwaukee, WI: ASQ Quality Press, 2009.

5. Jamshidi, M. *System of Systems Engineering*. Hoboken, NJ: John Wiley & Sons, 2009.

6. Hybertson, DW. *Model-Oriented Systems Engineering Science*. Boca Raton, FL: CRC Press, 2009.

7. Aslaksen, EW. *Designing Complex Systems*. Boca Raton, FL: CRC Press, 2009.

8. Markowitz, HM. Portfolio selection. *Journal of Finance*, 1952;7(1):77–91.

9. McDermott, I. *The Art of Systems Thinking*. Hammersmith, London, England: Thorsons Publishing, 1997.

10. Skyttner, L. *General Systems Theory: Ideas and Applications*. Singapore: World Scientific Publishing Co., 2001.

11. Bullock, A. *The Secret Sales Pitch*. San Jose, CA: Norwich Publishers, 2004.

12. Abbott, EA. *Flatland: A Romance in Many Dimensions*. New York: Dover Publications, 1952.

13. Sandquist, GM. *Introduction to System Science*. Englewood Cliffs, NJ: Prentice-Hall, 1995.

14. Mayer, RJ, Painter, MK, deWitte, PS. *IDEF Family of Methods for Concurrent Engineering and Business Re-engineering Applications*. College Station, TX: Knowledge-Based Systems, 1992.

15. Colquhoun, GJ, Baines, RW, Crossley, R. A state of the art review of IDEF0. *Journal of Computer Integrated Manufacturing*, 1993;6(4):252–264.

16. George, ML, Rowlands, D, Price, M, Maxey, J. *Lean Six Sigma Pocket Toolbook*. New York: McGraw-Hill, 2005.

17. Fox, N. Case Study: Process definitions–Process measuring; how does a small foundry get their ROI off ISO/TS 16949:2002? *AFS Transactions 2005*. Schaumburg, IL: American Foundry Society, 2005.

18. Straker, D. *A Toolbook for Quality Improvement and Problem Solving*. Englewood Cliffs, NJ: Prentice-Hall, 1995.

19. Open source SysML project. Available at http://www.SysML.org. Accessed 2010 April 13.

20. Sandquist, GM. *Introduction to Systems Science*. Englewood Cliffs, NJ: Prentice-Hall; 1995.

21. Yeh, RT. System development as a wicked problem. *International Journal of Software Engineering and Knowledge*, 1991;1(2):117–130.

22. Krackhardt, D, Krebs, V. *Social network analysis*. Semantic Studios Publication; 2002. Available at http://semanticstudios. com/publications/semantics/000006.php]. Accessed 2006 January 6.

23. Heylighen, F. 1998. Basic concepts of the systems approach. Principia Cybernetica Web, (Principia Cybernetica, Brussels). Available at http://pespmc1.vub.ac.be/SYSAPPR. html]. Accessed January 7, 2006 .

24. Driscoll, PJ, Goerger, N. Stochastic system modeling of infrastructure resiliency. ORCEN Research Report DSE-R-0628. West Point, NY: Department of Systems Engineering, U.S. Military Academy, 2006.

25. Driscoll, PJ. Modeling system interaction via linear influence dynamics. ORCEN Research Report DSE-R-0629. West Point, New York: Department of Systems Engineering, U.S. Military Academy, 2006.

26. Driscoll, PJ, Goerger, N. Shaping counter-insurgency strategy via dynamic modeling. ORCEN Research Report DSE-R-0720. West Point, NY: Department of Systems Engineering, U.S. Military Academy, 2006.

27. *Implementation of Network Centric Warfare*. Office of Force Transformation, Secretary of Defense, Washington, DC, 2003.

28. Jackson, MC. *Systems Thinking: Creative Holism for Managers*. West Sussex, England: John Wiley & Sons, 2003.

29. Pidd, M. *Systems Modelling: Theory and Practice*. Chichester, England: John Wiley & Sons, 2004.

30. Driscoll, PJ, Tortorella, M, Pohl, E. Information quality in network centric operations. ORCEN Research Report. West Point, New York: Department of Systems Engineering. 2005.

31. *Rethinking Risk Management in Financial Services*. World Economic Forum Report 110310, New York, April 2010.

Chapter **3**

System Life Cycle

PATRICK J. DRISCOLL, Ph.D.
PAUL KUCIK, Ph.D.

There seems to be a kind of order in the universe, in the movement of the stars and the turning of the earth and the changing of the seasons, and even in the cycle of human life. But human life itself is almost pure chaos.
—Katherine Anne Porter (1890–1980)

A mental note is worth all the paper it is written on.
—Marie Samples, OCME, New York

3.1 INTRODUCTION

All systems have a useful lifetime during which they serve the purpose for which they were created. Just like a human lifetime, the degree to which a system achieves this purpose typically varies with age. New systems start out by hopefully meeting their performance targets. After entry in service, system elements and processes may begin to degrade. Degradation that occurs during a system's useful years motivates a host of specialized maintenance activities, some planned and some unplanned, intended to restore the system to as close to its original state as possible. Creating written lists, calendars, and other memory enhancement techniques are examples of maintenance items we use to restore memory functionality as close as possible to earlier periods of peak performance.

Decision Making in Systems Engineering and Management, Second Edition
Edited by Gregory S. Parnell, Patrick J. Driscoll, Dale L. Henderson
Copyright © 2011 John Wiley & Sons, Inc.

Eventually, most systems degrade to a point where they are no longer effectively meeting consumer needs and are retired. At the retirement decision, the cost of further maintaining a system could be exceeding the cost of replacing the system. Or, perhaps the system is operating as intended but it can no longer provide value to its stakeholders due to changes in the environment within which it exists. The 8Track technology for audio recording and playback is an example of a system that was retired, because it lost its ability to compete in a consumer environment where "smaller is better" drove demand. A similar competition is ongoing between high-definition movie format and Blu-ray™ technology.

In a similar fashion to human physiology, it is useful to think of a system as progressing through a succession of stages known as a *life cycle*. For living systems, this cycle consists of four stages simply described: birth, growth, deterioration, and death [1]. From a systems engineering perspective, there are at least three major reasons why a life cycle structure is an effective metaphor. We can:

1. Organize system development activities in a logical fashion that recognizes some activities must be accomplished prior to others.
2. Identify the specific activities needed to be accomplished in each stage to successfully move to the next stage.
3. Effectively consider the impact that early decisions have on later stages of the system life cycle, especially with regard to cost and risk (the likelihood and consequences of system problems).

As illustrated in Figure 3.1, the system life cycle activities referred to in the first reason have a logical sequencing. They align with the transition of a system from its conceptual birth to eventual retirement. Notice that these activities are not the same as those described in the systems decision process (SDP). The SDP is a cycle of phases to support major systems decisions repeated at critical decision points (gates) typically encountered once in each stage of the life cycle. The four phases of the SDP are described in Part III of this book.

The second reason reinforces that as members of multidisciplinary teams (see Chapter 7), systems engineers maintain an appropriate focus on what needs to be done and when it needs to be done. The specifics of what, when, how, and why associated with these needs are dictated by the life cycle stage of the system. For example, the list of alternative solutions generated in a systems decision process concerning a recently deployed system would consist of process and/or product modifications and enhancements designed to aid the system to better achieve the purpose for which it was created. Later in the stage, the list of alternative solutions might focus on new systems to replace the current system.

Of the three reasons stated as to why a system life cycle metaphor is effective, the third is the most important. Some decisions made during early stages of system development are irreversible; once there is commitment to a particular system concept (e.g., airplane) and a detailed design, consumer needs and resource limitations (including time) usually prevent the design team from switching to alternative

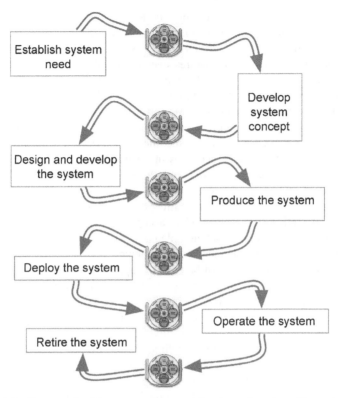

Figure 3.1 Systems decision processes occurring in each system life cycle stage.

design solutions. Other minor design decisions (e.g., paint color) can be altered readily without significant impact on project planning and execution.

However, all decisions have immediate and delayed costs associated with them. These costs can consist of a blend of financial, risk, environmental, technological, legal, and moral factors of direct concern to stakeholders, many of which may not be realized until later stages in the life cycle.

Not considering life cycle costs when making decisions early in the system life cycle could prove to be disastrous to the long term system viability and survivability. The principle underlying this idea goes back to an idea originating during the Scottish enlightenment known as the Law of Unintended Consequences, more recently stated [1] succinctly as "[w]hether or not what you do has the effect you want, it will have three at least that you never expected, and one of those usually unpleasant." The least expensive and most effective hot water pipe insulating material to use in the northeast United States might be asbestos, but deciding to use asbestos without seriously considering the cost factors associated with the operational and retirement stages would not be wise.

This principle reminds us to be careful not to commit to a "whatever is best for right now" solution without examining the degree to which such a solution remains optimal for later life cycle stages as well. Hidden system costs often occur because someone failed to employ systems thinking over a complete system life cycle.

In practice, life cycles are driven by the system under development. They serve to describe a system as it matures over time with regard to its functioning. A life cycle also guides professionals involved with sustaining the value delivered by the system to consumers during this maturation. The life cycle stages need to contain sufficient detail to enable systems engineers, systems managers, operations researchers, production engineers, and so on, to identify where their skill set tasks fit into the overall effort. At the same time, the stages need to be described broadly enough to accommodate related activities in a natural way.

System life cycles are structured and documented for success. Among the myriad of life cycle models in existence, two fundamental classifications arise: predictive and adaptive [2]. Predictive life cycle models favor optimization over adaptability. Adaptive life cycle models accept and embrace change during the development process and resist detailed planning. Once a life cycle for a particular systems decision problem is defined and documented, it is then possible to structure the management activities that will be used to support each stage of the life cycle. This provides the data that are necessary to support major decision gates to move to the next stage. An effective management system prevents system development from occurring in a piecemeal or disjoint basis that has a tendency to increase risk.

The life cycle model we use in this text has the advantage of being able to simply represent stages in a system's lifetime along with the activities within each stage. The structured process used to define and support systems engineering activities within these stages, the systems decision process (SDP), is naturally cyclic, thereby providing a constant feedback mechanism that encourages revision consistent with changes in the system environment, all the while taking full advantage of opportunities to capture and deliver value to the stakeholders. The SDP is typically used at least once during each life cycle stage to determine if the system should advance to the next stage.

As a consequence of separating the system life cycle from the SDP, the SDP provides the essential information for systems decision makers independent of the life cycle stage the system is in. Admittedly, each application of the SDP is tailored to the system and the life cycle stage. Some elements of the SDP may be truncated, while others may be amplified for some systems in some stages. This adaptability feature is perhaps one of the SDPs greatest attributes. How to tailor the SDP is described in Chapter 9.

3.2   SYSTEM LIFE CYCLE MODEL

The life cycle stages listed in Table 3.1 are broadly defined so as to apply to as many systems as possible. As can be seen in the sections that follow, various other life cycle models exist that have specific types of system development models or systems engineering applications in mind. For example, the spiral design model illustrated in Figure 3.3 is frequently used in software system development with an eye toward highlighting risk management throughout the various life cycle stages.

TABLE 3.1 System Life Cycle Model for Systems Engineering and Management

System Life Cycle Stage	Typical Stage Activities
Establish system need	Define the problem
	Identify stakeholder needs
	Identify preliminary requirements
	Identify risk factors and initial risk management plan
Develop system concept	Refine system requirements
	Explore system concepts
	Propose feasible system concepts
	Refine risk factors
	Assess initial performance, schedule, and technology risks
Design and develop system	Develop preliminary design
	Develop final design
	Assess initial development cost, market, and business risks
	Perform development tests to reduce risk
	Refine performance, schedule, and technology risk assessments; include mitigation steps in risk management plan
	Build development system(s) for test and evaluation
	Verify and validate design
	Test for integration, robustness, effectiveness
	Includes production scheduling, economic analysis, reliability assessments, maintainability, and spiral design implementation considerations, among others
Produce system	Produce system according to design specifications and production schedule
	Apply Lean Six Sigma as appropriate
	Refine development cost, market, and business risks; include mitigation steps in risk management plan
	Monitor, measure and mitigate performance, schedule, and technology risk
	Assess initial operational risk to the system
Deploy system	Refine operational risk management plan
	Develop a deployment plan
	Complete training of users and consumers
Operate system	Operate system to satisfy consumer and user needs
	Monitor, measure and mitigate operational risks
	Identify opportunities for enhanced system performance
	Provide sustained system capability through maintenance, updates, or planned spiral developed enhancements
Retire system	Develop retirement plan
	Store, archive, or dispose of the system

3.2.1 Establish System Need

Establishing a clear system need is a critical first step in system management. Successfully addressing the purposes associated with this stage of the life cycle increases the likelihood of a match between the system that is truly needed and the one that is developed. The client facing the problem that generates a system need is typically dealing with the problem's symptoms and resulting effects on a day-to-day basis. It is not unusual for a customer in this situation to communicate an initial system need that is focused on treating these symptoms.

In the short term, treating the symptoms might improve the working conditions associated with the problem but it does not help to resolve the underlying cause(s) of the symptoms, which is much more difficult to uncover. The true cause(s) of observed symptoms typically emerges from a process of intensive interaction with stakeholders using techniques introduced in Chapter 10.

The first stage of the life cycle is about exploration, discovery, and refining key ingredients necessary for a project to get off to a good start. This consumes a large amount of time and effort. Once the actual problem, stakeholder needs, and preliminary requirements are successfully defined, the beginning steps are taken toward effective risk management and the system transitions into the next life cycle stage.

3.2.2 Develop System Concept

The life cycle stage of developing a system concept is centered on applying techniques designed to inspire creative thought contributing to effective, efficient systems for delivering maximum value to the stakeholders. These techniques both generate novel system possibilities that meet stakeholder needs and eliminate those system concepts that are infeasible. Conceptual system models involving graphical illustrations, tables, and charts comparing and contrasting system characteristics, along with a multitude of linked hierarchical diagrams, are used to identify possible system concepts that can meet stakeholder needs.

In light of this initial set of feasible system concepts, various dimensions of system risk are identified and refined, such as performance (will the technology work?), schedule (can it be provided when needed?), and cost (is the system affordable?). The life cycle transitions to the next stage only when a sufficient number of feasible system concepts are identified that possess acceptable levels of risk.

3.2.3 Design and Develop System

The design and development stage involves designing, developing, testing, and documenting the performance of the chosen system concept. Quite often the models produced during this stage take the form of simulations, prototype code modules, mathematical programs, and reduced and full-scale physical prototypes, among others. One must be careful to adhere to professional best practices when testing and analyzing the performance of competitive designs using these models and simulations (see Chapter 4), especially when it comes to verifying and validating model results.

A feasible concept along with a system model enables the system team to develop estimates of program costs along with market and business risks. Of course, the accuracy of these risk estimates depends strongly upon the assumptions made concerning the system deployment environment that will occur in the future. It is wise under these conditions to carefully develop a set of use case scenarios for testing the performance of the system design using models and simulations developed for this purpose. These use cases should reasonably reflect the full span of "What if?" possibilities. This is the only way of identifying system design problems and limitations short of building, deploying and operating a full-scale system. Engineering managers are fully engaged in the system at this point in the life cycle as well, developing plans to address all the dimensions of implementation noted. When satisfactorily completed, the decision gate naturally supports the system going forward into a production stage.

3.2.4 Produce System

Realizing success in the system production life cycle stage is as far from a foregone conclusion as one might imagine. This is a period in the system life cycle that stresses the management team charged with producing a system that meets all of its intended purposes, meets or exceeds design requirements reflecting an acceptable risk across all risk dimensions, and does all this in an effective and efficient manner to achieve competitive advantage for the consumer of system products and services.

Ultimately, the system must deliver value to the stakeholders or it will fail. However, systems can fail through no fault of their own simply because of changing environmental conditions. Remember that throughout the life cycle and all the efforts that have gone into making a concept a reality, time continues to evolve, raising the very real possibility that substantial threats to system success that were not present earlier in the life cycle could exist upon deployment. Thus, during this stage the systems team tries to identify and assess the types and levels of operational risks the deployed system will face.

Depending on the type of system to be delivered—product-focused or service-focused—Lean Six Sigma [3] and other process improvement methods are applied so that the production processes used to make the designed system a reality are as effective and efficient as possible.

A portion of the ongoing risk analysis engages in monitoring, measuring, and mitigating risks identified in the previous life cycle stages. Moreover, a good deal of effort goes into maintaining vigilance for any new risks that might emerge as a result of changes in the environment outside of the control of the team. This sensitivity naturally leads to considering those external factors that could present risk to the system once it is placed into operation.

Operational risk [4] is emerging to be one of the least quantified, less understood, yet potentially largest impact areas of risk for systems engineering. While no single definition is predominant currently, operational risk is generally understood to mean the loss resulting from inadequate or failed internal processes, people, and support systems or from environmental events. During this stage of the life cycle, brainstorming and other ideation techniques are again used to identify an initial list of operational risks that might threaten program success once the system is fielded and in the hands of users and consumers.

3.2.5 Deploy System

Successfully deploying a new or reengineered system is the direct result of executing a well-thought-out deployment plan that provides detailed planning for the activities necessary to place the system into an operational environment. The plan includes, as a minimum, a description of the assumptions supporting the current capabilities and intended use of the system, the dependencies that can affect the deployment of the system, and any factors that limit the ability to deploy the system.

In close coordination with the system users, consumers, and owners, many other deployment details are specified in the deployment plan as well. These include information concerning deployment locations, site preparation, database conversions or creation, and phased rollout sequencing (if appropriate). Training programs and system documentation are created during this life cycle stage. Specific training plans for system users and maintainers play a critical role in achieving a successful system deployment.

The systems team itself transitions into a support role during this life cycle stage as they begin to disengage from primary contact with the system and transfer system functionality to the client. Additional resource requirements for the design team are identified, support procedures designed, and a host of transition activities along with management roles and responsibilities become part of the deployment plan. Operations and maintenance plans are created and specified as well in order to provide explicit guidance to system consumers and users as to how they might capture the greatest value return consistent with the intended design goals.

Finally, contingency plans are developed during this stage as well, some of which are included in the deployment plan while others are maintained internally by the systems engineering and deployment teams in case some of the risks identified and planned for in previous stages become reality. The operational risks identified previously are refined and updated as forecasted information concerning the system environment used to develop an initial list of potential threats to program success becomes reality.

3.2.6 Operate System

The most visible life cycle stage is that of operating the system in the mode it was intended. System users operate systems to provide products and services to system consumers. When we recognize the existence of systems in our environment, we are observing them in this stage of their life cycle. Some everyday systems that fall into this characterization include transportation networks, communications networks, electricity grid supply networks, emergency services, tourism, law enforcement, national security, politics, organized crime, and so on.

Condition monitoring of a system while it is in operation is an activity that has traditionally consisted of use-based measures such as the number of hours of operation, number of spot welds performed, number of patients treated, and so on. Measures such as these have dominated reliability analysis for systems whose wear and tear during periods of nonuse is so small as to render its impact insignificant.

A good example of this can be seen in aircraft systems scheduled maintenance planning, which is based on flight hours of the aircraft in operation and not on the amount of time passed since the aircraft was placed into operation. Many systems have switched to condition-based maintenance for example, in fleet aircraft transportation operations, recognizing that not all pilots fly aircraft in the same manner [5]. For military aircraft, condition-based maintenance assumes that an hour flying routine missions imposes a significantly different level of system stress than does an hour flying in combat missions—at night, in bad weather, amidst hostile fire. Thus, system maintenance planning, which used to consist of executing routine tasks on a preset schedule, is evolving to the point where real-time monitoring of system condition indicators is becoming more commonplace.

The operational risks due to external influences on system elements, services, and performance to meet goals identified in previous life cycle stages are closely monitored during system operation. However, management focus during this life cycle stage is not exclusively centered on potential bad things that might occur. They also maintain a heightened awareness for possible opportunities to enhance system performance that could add value to the stakeholders or increase the competitive advantage of consumers and users.

In fact, when systems engineers are called upon to engage a system during one of its operational life cycle stages, the underlying motivation of the user organization is centered on this very principle of exacting increased performance value out of the existing system. This could mean reengineering the system or its processes, applying optimization techniques to increase the efficiency of some dimensions of the system operation, using reliability methods to better understand and reduce the overall maintenance costs, or perhaps generating new ideas for system replacement that leverage recent developments in technology, knowledge, or the competitive landscape. Some of these advancements or changes in the operating environment may have been predicted and planned for during earlier life cycle stages, in which case system enhancements would be applied using principles of spiral development as well [6] .

3.2.7 Retire System

Finally, when users determine that it is no longer in their best interest to continue operating the system, it is retired from service. While the activities during this stage might be as simple as donating the existing system to a nonprofit organization or placing the system in storage, this life cycle stage can actually be quite complicated. One needs only consider the intricate requirements associated with taking a nuclear power plant offline in order to gain an appreciation for how complex this system retirement stage could be [7].

3.3 OTHER MAJOR SYSTEM LIFE CYCLE MODELS

As mentioned previously, there are several life cycle models in common use today. All of these life cycle models, including the one we use, are based on sets of professionally agreed standards, the primary ones being listed in Table 3.2 [8]. Despite The International Council on Systems Engineering (INCOSE) has taken a lead on setting a common standard for professional practice and education against which programs and businesses can compare their quality levels.

The last standard listed, ISO/IEC 15288, represents a modern interpretation of a system life cycle relative to the complex and heavily stakeholder dependent nature of the typical systems addressed in professional practice today. It represents the evolving standard of systems engineering practice in the United Kingdom. In this vein, the ISO/IEC 15288 system life cycle model is serving a purpose closely aligned with the life cycle model we use, except that, as can be seen in Table 3.3, they combine a life cycle model with various steps in a systems decision process. In this text, we separate the two processes into the system life cycle model discussed earlier and the SDP that supports decision making during all the stages of the system life cycle model.

Given the sequential representation shown in Table 3.3 [9], the ISO/IEC 15288 system life cycle implies that one stage is completed before transitioning into the next. As a result, this life cycle model has been criticized for its lack of robustness

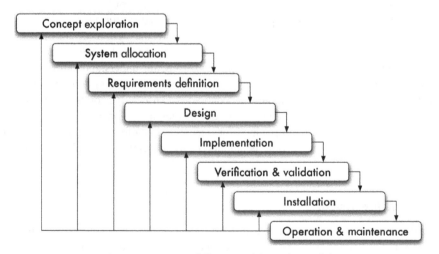

Figure 3.2   Waterfall system life cycle model.

TABLE 3.2   A Comparison of Standards-Driven Life Cycle Models

Standard	Description	System Life Cycle Stages
MIL/-STD/-499B	Focuses on the development of defense systems.	• Preconcept • Concept exploration and definition • Demonstration and validation • Engineering and manufacturing development • Production and deployment • Operations and support
EIA.IS 632	A demilitarized version of MILSTD499B	• Market requirements • Concept definition and feasibility • Concept validation • Engineering and manufacturing development • Production and deployment • Operations and support
IEEE 1220	Introduces the interdisciplinary nature of the tasks involved in transforming client needs, requirements, and constraints into a system solution.	• System definition • Subsystem definition • Preliminary design • Detailed design • Fabrication, assembly, integration, and test • Production • Customer support
EIA 632	Focus is on defining processes that can be applied in any enterprise-based life cycle phase to engineer or reengineer a system.	• Assessment of opportunities • Solicitation and contract award • System concept development • Subsystem design and predeployment • Deployment, installation, operations, and support
ISO/IEC 15288	Includes both systems engineering and management processes at a high level of abstraction.	• Concept process • Development process • Production process • Utilization process • Support process • Retirement or disposal process

TABLE 3.3   The ISO/IEC 15288 Systems Engineering Life Cycle Model

Life Cycle Stage	Purpose	Decision Gates
Concept	• Identify stakeholder needs • Explore concepts • Propose feasible solutions	Execute next stage
Development	• Refine system requirements • Create solution description • Build system • Verify and validate	Continue current stage
Production	• Mass produce system • Inspect and test	Go to previous stage
Utilization	• Operate system to satisfy user needs	Hold project activity
Support	• Provide sustained system capability	Terminate project
Retirement	• Store, archive, or dispose of system	

in dealing with a wide variety of system problems. The waterfall model (Figure 3.2) is more robust because system problems arising in any stage can lead the systems team to recycle back through earlier stages in order to resolve them.

In contrast to the waterfall, the spiral life cycle model shown in Figure 3.3 formalizes the notion of repeated cycling through a development process. Each spiral produces increasingly more complex prototypes leading to a full-scale system deployment. In essence, the spiral model executes a series of waterfall models for each prototype development.

One attractive feature of the spiral model is the explicit recognition of the important role that risk plays in system development. This same consideration is intentionally incorporated in the system life cycle model we use. In both life cycle models, various types of risks are identified during each prototype development cycle—for example, investment risk, performance risk, schedule risk, and so on. If these risks are successfully mitigated, the systems team evaluates the results of the current cycle, presents the results and conclusions in support of the decision gate, and, if approved, proceeds to enhance the prototype in the spiral model or moves to another stage in our life cycle model. Failing to resolve important risks can cause the program to terminate during any stage.

Several other specialized models are used for system development, although none as prevalent as the two already mentioned. Rapid applications development,

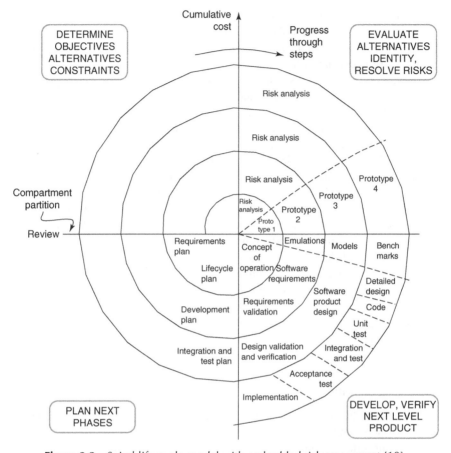

Figure 3.3 Spiral life cycle model with embedded risk assessment [10].

a methodology created to respond to the need to develop software systems very fast, strives to deploy an 80% system solution in 20% of the time that would be required to produce a total solution [11], and agile life cycle models [12] are among these specialized models.

3.4 RISK MANAGEMENT IN THE SYSTEM LIFE CYCLE

As the complexity of systems and their environment increases, the number, type, likelihood, and impact of events that can and might occur to threaten the well-being of systems becomes increasingly more difficult to identify.

A *risk* is a probabilistic event that, if it occurs, will cause unwanted change in the cost, schedule, or value return (e.g., technical performance) of an engineering system [13]. The goal of risk management is to identify and assess risks in order to enact policy and take action to reduce the risk-induced variance of system technical performance, cost, and schedule estimates over the entire system life cycle. In other words, risk management describes a collection of conscience and deliberate actions to protect the system from the adverse effects of specific events that have a nonzero probability of occurring in the future.

System complexity works against accomplishing this goal in an easy manner because by increasing the number and type of interconnections, vested interests, and uncertainty, it becomes more and more difficult to effectively apply risk management. This is especially true if the risk management activities lack formal, repeatable organization. The situation today is that it is simply impossible to "fly by intuition" in this regard. Moreover, while risks associated with specific components, so-called *nonsystemic* risks, might be identifiable by ad hoc procedures based on experience alone, the more subtle and elusive *systemic* risks, those inherent in the entire system (shared across components), will routinely avoid detection without some organized, repeatable process. This latter group can, if left unattended to, take down an entire financial, communications, transportation, or other system when they occur [14].

Risk management, which is comprised of three main activities—risk identification, risk assessment, and risk mitigation—is an ongoing process applied throughout the life cycle of a systems engineering project. In this section, we take a broad view of risk management [15], focusing on core principles and concepts that set the stage for a more in-depth exploration in later chapters.

3.4.1 Risk Identification

The process of identifying risks consists of determining any sources of risk and the scenarios under which they may occur. Risk identification seeks to discover and categorize uncertain events or conditions whose occurrence will have a negative impact on system cost, schedule, value, technical performance, or safety. The focus of this effort often changes during a systems decision problem. A team could be initially concerned about the risk associated with having the project proposal approved, shifting then to possible risk impediments to the SDP effort and, finally, shifting to addressing threats to the successful implementation and sustained health of the selected system solution. Ideally then, techniques used to identify risks need to be flexible or general enough to apply throughout the life of a systems decision problem and its resulting solution. In what follows, we refer to the systems decision problem effort as "the project."

Two convenient techniques for identifying possible risks to systems are prompt lists and brainstorming. Both techniques involve extensive interaction with the system stakeholders. Their unique insights arising from their extensive knowledge of the operating environment of the needed system are critical information elements needed to develop comprehensive risk categories.

A *prompt list* is simply a listing of possible categories of risks that are particular to the current systems decision problem. They function as word recall prompts during stakeholder interviews, helping participants to think of as many risks to its success as possible. As a technique, a prompt list can be used on an individual basis or in a group setting with a facilitator from the systems team in control.

For example, when identifying risk elements during a life cycle stage that focuses on establishing the need for a system, the team could use a prompt list consisting of the SDP environmental factors—technological, health & safety, social, moral/ethical, security, cultural, ecological, historical, organizational, political, economic, and legal—in order to develop a preliminary list of major risks associated with developing a system to meet the needs of stakeholders. The risk elements emerge in subsequent discussions as the details required to document risks in a risk register are identified. Executive board objections to the overall decision support, potential financial problems with funding the effort to completion, knowledge deficiencies due to venturing into new competitive territory, political backlash from government administrators or the general public, and so on, are examples of the types of risk that arise.

Brainstorming (see Chapter 11) is a technique that works much in the same manner as a prompt list except that a neutral human facilitator from the systems team serves a similar purpose as the prompt list: to elicit without judgment from the stakeholders any possible risks to successful project completion they might identify from their experience. Brainstorming is also performed almost exclusively in a group setting. The facilitator might employ project schedules, graphical illustrations, data tables, or even a prompt list to help the participants identify risks.

A successful brainstorming session depends heavily on the participation of key stakeholders (see Chapter 10). As with many senior leaders of organizations who have constant demands on their time, these key stakeholders may not be able to assemble as a single group for any significant length of time. When stakeholder access is limited, prompt lists are a better technique to use for risk identification because they allow decentralized participation in the risk identification process while maintaining a common frame-of-reference provided by the logical structure of the list. In either instance, a good practice is to plan on at least two complete iterations of stakeholder interviews so that the results of the first set of interviews might be leveraged as prompts for all stakeholders during the second session.

There are six common questions [16] that are commonly used to capture various dimensions of risk to the system during brainstorming sessions with key stakeholders. The answers to these questions provide the data needed to begin to analyze risks and plan for their mitigation during a systems decision problem. The six questions are:

1. What can go wrong?
2. What is the likelihood of something going wrong?
3. What are the consequences?
4. What can be done and what options are available?
5. What are the tradeoffs in terms of risk, costs, and benefits?
6. What are the impacts of current decisions on future options?

TABLE 3.4   Techniques for the Identification of Potential Sources of Risk in the NSTS Program

Hazard analysis	Design and engineering studies
Development and acceptance testing	Safety studies and analysis
FMEAs, CILs, and EIFA	Certification test and analysis
Sneak circuit analyses	Milestone reviews
Failure investigations	Waivers and deviations
Walk-down inspections	Mission planning activities
Software reviews	Astronaut debriefings and concerns
OMRSD/OMI	Flight anomalies
Flight rules development	Aerospace safety advisory panel
Lessons-learned	Alerts
Critical functions assessment	Individual concerns
Hot line	Panel meetings
Software hazard analysis	Faulty tree analysis
Inspections	Change evaluation
Review of manufacturing process	Human factors analysis
Simulations	Payload hazard reports
Real-time operations	Payload interfaces

To make an important point clear: Identifying project risks is a demanding task that consumes a good deal of time, effort, and brainpower to do it right. Using a structured, repeatable method that is easy to understand is a key ingredient to success. As an example of how important this process is to successful systems decision problems and how systems engineers attempt to address this concern as comprehensively as possible, consider the listing of techniques used by NASA scientists and risk specialists to identify risks to the National Space Transportation System (NSTS) [17] shown in Table 3.4. These tasks represent thousands of work hours by a host of people across a broad range of system stakeholders.

As each risk is identified, it is categorized, to ensure that risks are not double-counted and that the identification of risks is comprehensive. The intent is to group risks into mutually exclusive and collectively exhaustive categories. INCOSE recognizes four categories of risk that must be considered during a system decision problem: technical risk, cost risk, schedule risk, and programmatic risk [18, 19].

Technical risk is concerned with the possibility that a requirement of the system will not be achieved, such as a functional requirement or a specific technical performance objective, because of a problem associated with the technology incorporated into the system, used by the system, or interfacing with system input and output. One component of the SDP described in Chapter 10 is a functional analysis, which is used to develop a functional hierarchy that illustrates what any feasible system solution must do to be considered successful. For a host of modern systems, technology is the main driver of these functions. By considering the risk to accomplishing each function, a comprehensive treatment of technical risk ensues.

Cost risk is the possibility of exceeding the planned design, development, production, or operating budgets in whole or in part. For any system, estimates of future life cycle costs are subject to varying degrees of uncertainty due to uncontrollable environmental factors, time, and the source of information used to develop these estimates. The further forward in time these costs are anticipated to occur, the more uncertainty is associated with their estimates. While objective cost data with similar systems decision problems is desirable, subjective expert opinion is often used to create cost estimates for items less familiar to the project team and stakeholders. This injects additional uncertainty that must be taken into account, as we show in Chapter 5. Cost risk planning is more complicated than simply accounting for program spending and balancing any remaining budget. Cost risk extends over the entire system life cycle. Decisions made throughout the system life cycle are assessed for their downstream impact on the total system life cycle costs. It becomes necessary to identify major cost drivers whose variability can cause the projet to "break the budget" rapidly, thus causing a termination of the effort in the worst case. Properly eliciting the information needed to model and analyze cost uncertainty requires careful thought and consideration [20].

Schedule risk is the possibility that a project will fail to achieve key milestones agreed upon with the client. Scheduling individual tasks, duration, and their interrelationships is critical to sound project planning. Doing so directly identifies those system activities that lie on a critical path to project success (see Chapter 13). Systems engineers and program managers should focus a large amount of their effort on these critical path tasks because when these tasks fail to achieve on-time start and completion times, the overall project schedule and delivery dates are directly effected. While the more common method of identifying critical path activities is deterministic, recent developments have demonstrated significantly improved benefits for analyzing cost, schedule, and risk *simultaneously* via Monte Carlo simulation [21].

Programmatic risk arises from the recognition that any systems decision problem takes place within a larger environmental context. Thus, it is an assessment of how and to what degree external effects and decisions imposed on the projet threaten successful system development and deployment. This last form of risk is closely related to the concept of operational risk emerging from the banking industry [22]. Increased levels of critical suppliers, outsourcing specific engineering tasks, budget reductions, personnel reassignments, and so on, are all examples of programmatic risk.

The INCOSE risk categories provide a useful framework for facilitating risk identification and ensuring a comprehensive treatment of risks. It should be noted that the above risk categories interact with each other throughout a system life cycle. While standard in a systems engineering environment, these are not the only grouping categories that are used. Commercial banks, for example, divide their risk categories into financial, operational, and, more recently, systematic risks in order to track the most common undesirable future events they face.

The systemic risk category is worth emphasizing because of its recent realization in global securities markets. The Counterpolicy Risk Management Group [23] suggests an effective definition for our use. A *systemic risk* is the potential loss or damage to an entire system as contrasted with the loss to a single unit of that system. Systemic risks are exacerbated by interdependencies among the units often because of weak links in the system. These risks can be triggered by sudden events or built up over time with the impact often being large and possibly catastrophic.

Systemic risk is an interesting phenomenon gaining growing attention across all risk concerns with systems. Recently, the impact of unmitigated systemic risk events occurring within the financial markets was felt across the globe. The U.S. Congressional Research Service (CRS) describes systemic risk in the following manner:

> All financial market participants face risk—without it, financial intermediation would not occur. Some risks, such as the failure of a specific firm or change in a specific interest rate, can be protected against through diversification, insurance, or financial instruments such as derivatives. One definition of systemic risk is risk that can potentially cause instability for large parts of the financial system. Often, systemic risk will be caused by risks that individual firms cannot protect themselves against; some economists distinguish these types of risks as a subset of systemic risks called systematic risks. Systemic risk can come from within or outside of the financial system. An example of systemic risk that came from outside of the financial system were fears (that largely proved unfounded in hindsight) that the September 11, 2001 terrorist attacks on the nation's financial center would lead to widespread disruption to financial flows because of the destruction of physical infrastructure and death of highly specialized industry professionals. Systemic risk within the financial system is often characterized as contagion, meaning that problems with certain firms or parts of the system spill over to other firms and parts of the system [24].

The CRS report emphasizes several characteristics of systemic risk that all systems experience: shared risk due to system interconnectivity of people, organizations, equipment, policy, and so on. Systems engineering teams should be aware that systemic risks loom large on complicated projects. As the system solution structure grows, so does the likelihood that the activities supporting its development within the SDP will be subdivided among groups of the team with specialized knowledge and experience. While both effective and efficient, the project manager (PM) must maintain an integrated, holistic perspective of the overall project. Without this perspective and sensitivity to systemic risk, the project could be doomed to failure. A recently release report of the World Economic Forum strongly emphasized this point by bringing together a wide range of systems thinking experts to assist the financial services industry to develop just such a perspective. During the financial crisis of 2007 and 2008, no one regulatory authority or organization in the financial services industry had system-wide oversight that might have identified the rising systemic risk of over-leveraging that occurred [14].

A common and effective means of documenting and tracking risks once they are identified is through the use of a *risk register*. A risk register holds a list of key risks that need to be monitored and managed. When used properly, it is reviewed and updated regularly and should be a permanent item on any project meeting agenda. Figure 3.4 shows an example risk register for the rocket problem using several of the risk categories noted earlier. The values shown in the impact, likelihood, and risk level columns are developed using the techniques described in what follows.

Risk	Category	Impact	Likelihood	Risk Level	Current	Mitigation	Risk Owner
Government failure to set aside contingency funds	Financial	Medium	Low	Amber	None	Monthly monitoring of contingency funds by design team	Client
Breach of legislation	Legal	Medium	Medium	Amber	Compliance audit	Peer review by legal advisors	Team internal legal
Substandard composite material used in multiple component housings	Systemic	High	Low	Green	Periodic material sampling	Material engineering review during IPRs	Project lead engineer

Figure 3.4 Example risk register used during the SDP.

3.4.2 Risk Assessment

Once risks have been identified and categorized, the next challenge is to determine those risks that pose the greatest threat to the system. This *risk assessment* process involves assessing each hazard in terms of the potential, magnitude, and consequences of any loss from or to a system. When there exists historical data on these losses or the rate of occurrence for the risk event, the risk analysis is directly measured from the statistics of the loss. Otherwise, the risk event is modeled and predicted using probabilistic risk analysis (PRA) techniques [25]. This latter option has become the norm in modern risk analysis because for complex systems, especially those involving new or innovative technologies, such historical loss data rarely exists. Because some of the hazards to the system may involve rare events that have never occurred, estimates of the probability of occurrence can be difficult to assess and often must be based on a subjective estimate derived from expert opinion. When this occurs, techniques such as partitioned multiobjective risk method (PMRM) that use conditional risk functions to properly model and analyze these extreme events are employed [26].

The consequence imposed on system success when a risk event does transpire can involve increased cost, degradation of system technical performance, schedule delays, loss of life, and a number of other undesirable effects. With complex systems, the full consequence of a risk may not be immediately apparent as it might take time for the effects to propagate across the multitude of interconnections. These "downstream" effects, often referred to second- third-, and higher-order effects, are very difficult to identify and assess, and can easily be of higher consequence than the immediate ones. The risk of the Tacoma-Narrows bridge on Highway 16 in Seattle collapsing can be assessed from structural engineering information and historical data existing from its previous collapse in 1940. However, suppose that when this bridge collapses, the express delivery van carrying human transplant organs does not make it to the regional hospital in time to save the patient. The patient happens to be a U.S. senator who is the current champion of a new bill to Congress authorizing direct loans to Washington State residents suffering under the collapse of the mortgage industry. The bill fails to pass and thousands of people lose their homes, and so on. The middle English poet John Gower captured this domino effect in his poem *For Want of a Nail*, the modern nursery rhyme version of which goes:

For want of a nail, the shoe was lost;

For want of the shoe, the horse was lost;

For want of the horse, the rider was lost;

For want of a rider, the battle was lost;

For want of the battle, the kingdom was lost;

And all for the want of a horseshoe nail.

Probability–impact (P–I) tables [27], also known as probability–consequences tables, are a straightforward tool that can be used both to differentiate between and help prioritize upon the various risks identified and to provide clarifying summary information concerning specific risks. In concept, P–I tables are similar to the matrix procedure described in Military Standard (MIL-STD) 882 [28], elsewhere adapted to become a bicriteria filtering and ranking method [18].

P–I tables are attractive for use early in the system life cycle because as a qualitative technique they can be applied using only stakeholder input. Later, as risk mitigation costs become available, a third dimension representing the mitigation cost range can be imposed on the P–I table, thereby completing the trade space involved with risk management. Stakeholders are asked to select their assessed level of likelihood and impact of risks using a constructed qualitative scale such as very low, low, medium, high, and very high. If the actual probability intervals are difficult to assess at an early stage, a similar constructed scale can be used to solicit stakeholder input as to the likelihood of risks: unlikely, seldom, occasional, likely, and frequent [18]. The point is to start the risk management process early in the system life cycle and not to delay risk consideration until sufficient data are available to quantify risk assessments.

Typically, each of these qualitative labels is defined with a range specific of outcomes for the risks that helps the stakeholder to distinguish between levels. Using ranges, such as those illustrated in Figure 3.5 for five qualitative labels, helps normalize estimates among stakeholders. Ideally, what one stakeholder considers very high impact should correspond to what all stakeholders consider very high impact. When this is not possible to achieve, other methods such as swing weighting (see Chapter 10) become useful.

It is very important to understand what can go awry with subjective approaches such as that used in the P–I table approach, and nearly all of these considerations

	Scale	Prob	Impact on project		
			Schedule delay	Cost increase	Performance
Value ranges	Very high	40–50	>6 Days	>20%	Multiple major failures
	High	30–40	3–5 Days	15–20%	Limited major failures
	Medium	20–30	2–3 Days	10–15%	Single major failure
	Low	10–20	1–2 Days	5–10%	Multiple minor failures
	Very low	0–10	<1 Day	<5%	Limited minor failures

Figure 3.5 Example of constructed value range scales.

are based on the fact that stakeholders are involved [29]. Among these, three are important to highlight: Stakeholders can have very different perceptions of risk and uncertainty [30]; qualitative descriptions of likelihood are understood and used very differently by different stakeholders; and numerical scoring schemes can introduce their own source of errors. Straightforward techniques such as calibration tests [29] can help move stakeholders to a common scale while helping the systems team translate stakeholder input for use in risk modeling and assessment. The swing weighting technique introduced in later chapters can easily be modified and used for eliciting reasonably accurate stakeholder numerical scores. Its basis in decision analysis mitigates many of the scoring error concerns noted in the literature.

Since each risk element is characterized in terms of its likelihood of occurrence and subsequent impact should the event occur, a two-dimensional P–I table as shown in Figure 3.6 can be used to categorically match each risk with its pairwise correlation to the two characteristics. The resulting table enables the systems team to prioritize its risk management efforts appropriate to the threat level posed by specific risk elements.

For example, risk 5 has been estimated by stakeholders to have a very low likelihood of occurring and, if it does occur, will have very low impact on the system. Although it would continue to be monitored and measured throughout the system life cycle stages in which it was present, it more than likely would receive very little mitigation effort on the part of the systems team. Risk 4 on the other hand, has been estimated by stakeholders to have a high likelihood of occurring and, if it does occur, will have a high (and serious) impact on the success of the project, would command a good degree of attention throughout the life cycle of the system.

Figure 3.7 shows that the stakeholders consider risk element 3 to have three different impacts on the system: schedule (S), technical performance (P), and cost ($), each with varying estimations on their likelihood of occurring and their potential impact should they occur. In this example, the likelihood of violating cost limits for the program is estimated to be very low, but if it does occur it has the possibility of potentially terminating the program because of its very high impact. This is an example of an extreme event described earlier. Its low probability of occurrence does very little to allay the fears associated with this risk, should it occur.

P–I tables provide an important perspective on the anticipated risks that a system or project will face. To form a comprehensive understanding, P–I table

Probability impact table for project x risk elelments					
Very high	3				
High				4	
Medium					
Low		1,2	6		
Very low	5				
	Very low	Low	Medium	High	Very high
	Probability of occurrence				

(Impact on vertical axis)

Figure 3.6 Example P–I table for six risk elements.

Probability impact table for risk element 3					
	Very high	$		5		
Impact	High					
	Medium					
	Low		P			
	Very Low					
		Very Low	Low	Medium	High	Very High
		Probability of occurrence				

Figure 3.7   Specific P–I table for risk element 3.

results should be combined with other methods as appropriate. These include: capturing the frequency of occurrence, estimating correlation to other risks, estimating "time to impact" if a risk were to come to fruition, using decision analysis, and incorporating simulation experiments to assess the dynamic effects associated with risk. Generally, the analysis proceeds through increasing levels of risk quantification, beginning with a qualitative identification of the risk, followed by an understanding of the plausible range of each parameter, a "best estimate" of each parameter, and finally an estimate of the probability distribution of each parameter and the effect on the overall program. The size of the system, the severity of the risks, and the time available will determine the appropriate degree of quantification.

Assessment of technical risk, which involves the possibility that a requirement of the system will not be achieved, is enabled by functional analysis, introduced in Chapter 2, and further discussed in Chapter 10. Through functional analysis, the systems engineer defines the functions that the system must perform to be successful. Technical risk is assessed by considering each function and the likelihood and consequence of hazards to that function.

It is important to consider any required interactions between functions and risks to these interactions. Being sensitive to the connections between all elements of a systems solution forces a systems team to pay attention to a common source of failure in complex systems: *the seams of a system*. These seams afford interface compatibility and sharing protocols between systems. They are system boundaries rather than system components. Because of this, they are easily overlooked during risk identification sessions with stakeholders who have not internalized a systems thinking perspective of their operational environment. Accounting for the importance of each function and the degree of performance required enables a prioritization and comprehensive treatment of technical risk.

Assessment of cost risk, which involves the possibility of exceeding the design, development, production, or operating budgets in whole or in part, consists of examining (a) the various costs associated with a system solution or project, (b) their uncertainties, and (c) any possible risks and opportunities that may affect these costs. The risks and opportunities of interest are those that could potentially increase or decrease the estimated costs of the project, this includes decisions made

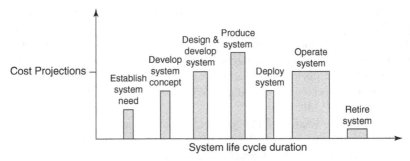

Figure 3.8   Example system life cycle cost profile.

throughout the system life cycle, which may have downstream effects on the total system life cycle costs. These risks are projected for all stages of the life cycle of a system project. Assessment of cost risk is enabled by an understanding of the uncertainty involved with the major cost drivers. The resulting analysis produces a projected system life cycle cost profile as shown in Figure 3.8. This profile varies by the type of system. A software program, for example, has high design and development costs, but generally lower production and deployment costs. These estimates are less certain and more likely to vary the more into the future they occur. Chapter 5 discusses life cycle costing in detail and describes the use of Monte Carlo simulation analysis to assess cost risk.

Assessment of schedule risk, which involved the possibility that the system or project will fail to achieve a scheduled key milestone, examines the time allotted to complete key tasks associated with the project, the interrelationships between these tasks, and the associated risks and opportunities that may affect the timing of task accomplishment. Schedule risk analysis relies on analytical methods such as Pert charts to unveil the sometimes complex logical connections between the tasks. Chapter 13, which describes the Solution Implementation phase of the SDP, addresses scheduling of program tasks, duration, and their interrelationships, as well as identifying system activities that lie on a critical path to project success. These critical path tasks should be a primary focus of schedule risk assessment, because a delay in the completion of any of these tasks will result in a delay in the overall program schedule.

Programmatic risk assessment considers all threats to successful system development and deployment resulting from external effects and decisions imposed on the system or project. This assessment is informed by an understanding of the system and its relation to lateral systems and the metasystem within which it is spatially located. A thorough stakeholder analysis, discussed in Chapter 10, will also enable an assessment of the programmatic risks.

The nature and methods of risk assessment vary somewhat across the risk categories described. In addition to assessing the risks in each category, a systems engineer must consider the seams here as well: possible interactions between risk categories. These interactions can impose correlations that should be included in Monte Carlo simulation models [31]. For example, schedule delays could result in monetary fines for not meeting agreed-upon contractual deadlines. Also, there may exist correlation between risks, with the occurrence of one risk increasing (or decreasing) the likelihood of other events happening. These dependencies can again be modeled using simulation to analyze the effect of *simultaneous* variation in cost, schedule, and value (technical performance) outcomes. Dependencies between cost elements can be accounted for using correlation coefficients [32] or they can be explicitly modeled [33].

By assessing the relative likelihoods and consequences of each risk across and among each category, risks can be prioritized, policy can be set, and actions can be taken to effectively and efficiently mitigate risks.

3.4.3   Risk Mitigation

With the knowledge gained through risk identification and risk assessment, project managers and systems engineers are equipped to reduce risk through a program of *risk mitigation* designed to monitor, measure, and mitigate risk throughout the system life cycle. Risks should be continuously monitored once identified, even if their assessed threat to the success of the program is minor. Time and situational factors beyond the control of the systems team and stakeholders can dramatically increase (or decrease) the potential threat posed by risk factors. Maintaining a watchful eye on the system environment throughout a systems decision problem helps to identify these risks early, thereby reducing the likelihood of unwelcome surprises.

The goal of risk mitigation is to take action to decrease the risk-based variance on performance, cost, value, and schedule parameters over the entire system life cycle. Figure 3.9 shows a graphical illustration of the variance of a project's total cost

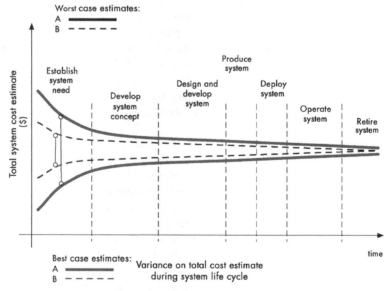

Figure 3.9   Estimate of system cost variance over life cycle.

estimate before effective risk management (A) and after (B). The spread between worst-case and best-case estimates is reduced earlier in the life cycle, yielding more accurate estimates of total system costs and dramatically reducing the threat of cost overruns to project success. Effective risk management has a likewise effect on value, technical performance, and schedule.

Once risks are identified, are actively being monitored, and are being measured, systems teams should be proactive in taking action to mitigate the potential threats to the system or project. Simply being aware of potential system or project risks is insufficient to properly manage or control the degree of their presence or impact. The primary means of deciding how to do this is through a risk management plan that clearly prioritizes the risks in terms of their relative likelihoods and consequences. To be successful, the risk management plan must be supported by organizational and project leadership. By properly aligning incentives, technical expertise, and authority, these leaders can help facilitate the greatest likelihood of overall success.

Once a risk has been identified, assessed, and determined to require mitigation, there are several options available to mitigate system risk. It may be possible to *avoid* the risk, if the organization can take action to reduce the probability of occurrence to zero or completely eliminate the consequences of the risk. It may be appropriate to *transfer* the risk to another organization through a contract; an insurance policy is one example of this approach. An organization may *reduce* risk by taking action to reduce the likelihood of the hazard occurring or reduce the severity of consequences if the hazard does occur. Finally, an organization may choose to *accept* risk if it has little or no control over the risk event and the overall system threat is considered to be very low. Each risk should be considered individually and within the context of the larger system as management decides on the appropriate approach (avoid, transfer, reduce, or accept) and the subsequent actions to take as a result.

All system activities involve risk; therefore, risk management must be a continuous process applied throughout the system life cycle of any systems engineering project. To illustrate this concept as it applies to a large-scale project, Figure 3.10 graphically displays the degree of involvement for major program teams throughout the shuttle life cycle stages identified by the Johnson Space Center. Of particular note is the one category that engages in risk management right from the start: the design and development teams. This just happens to be the principal location of systems engineers for the program.

3.5 SUMMARY

It is impossible to successfully complete systems decision problems without integrating comprehensive planning across an entire system life cycle. Current system life cycle models are based on standards set by professional organizations.

The system life cycle provides an effective metaphor for structuring the critical activities performed by a systems team during a project. The system life cycle introduced in this chapter consists of seven stages: establish system need, develop

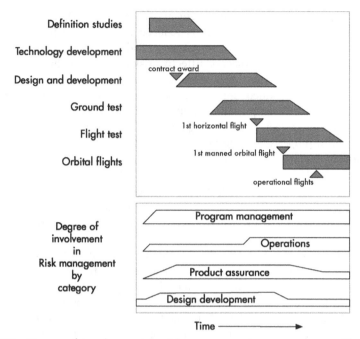

Figure 3.10 Degree of involvement in risk management across program life cycle (NASA JSC).

system concept, design and develop system, produce system, deploy system, operate system, and retire system. As described in Chapter 2, there are six principal stakeholder groups associated with every systems decision problem. All six of these groups are completely imbedded in life cycle stage activities to maximize the potential contributions their expertise has on a program.

The life cycle is separate from the SDP that systems teams execute during each stage of a system life cycle. The SDP is described in Part III.

The system life cycle model structures risk management necessary for continued successful operation. Effective risk management is crucial to success. Identifying risks is a challenging task, further complicated by the need to focus attention on the seams of systems, boundaries at which required systems interactions often fail. Awareness of these seams and the ability to identify and anticipate second-, third-, and higher-order consequences of risk events is enhanced through systems thinking. Failing to adequately manage risks will cause problems for a system. Risk management is an ongoing activity throughout the entire system life cycle.

3.6   EXERCISES

3.1. For each of the following systems, use the system life cycle model introduced in this chapter to identify the life cycle stage you believe the system to be in based on your current level of knowledge and understanding of the system.

(a) The Vonage internet communications system.

(b) Trade unions in the United States.

(c) Satellite radio in the United States.

(d) Al-Qaeda terrorist network.

(e) The relationship existing between you and your current group of friend.

(f) Cancer research in the United States.

(g) Apple iPhone.

(h) Your parent's automobile(s).

(i) Mono Lake ecosystem in California.

(j) The aquifer system supporting Phoenix, Arizona.

(k) Amtrak rail system in the United States.

(l) MySpace.com.

(m) Legal immigration process.

(n) Eurail pass system.

(o) Professional cricket in Pakistan.

(p) The musical group Gorillaz.

(q) Professional soccer in the United States.

(r) The U.S. interstate highway system.

3.2. Given the stages you identified for each of the systems in the previous question, we now want to consider risk relative to these stages. For each of the systems:

(a) List the risks you believe exist that threaten its successful continued operation.

(b) Construct a probability–impact (P–I) table using the scales shown in Figure 3.6 to place each of the risks.

(c) From the risks that you identified, pick the two highest and two lowest assessed risks and construct individual P-I tables for each of these following the example in Figure 3.7.

(d) For the four risks that you constructed P–I tables, identify two actions that could be taken by system owners or users to manage these risks. Additionally, explain whether these management actions would eliminate, mitigate, or have little effect on the level of risk that you assessed?

(e) Lastly, if the systems associated with these risks advance to their next life cycle stage, will the risks you assesed still be relevant? Explain.

REFERENCES

1. Merton, RK. The unanticipated consequences of purposive social action. *American Sociological Review*, 1936;1(6):894–904.

2. Archibald, RD. Management state of the art. Max's Management Wisdom. Available at http://www.maxwideman.com. Accessed January 20, 2005.

3. George, ML. *Lean Six Sigma*. New York: McGraw-Hill, 2002.

4. Hoffman, DG. *Managing Operational Risk: 20 Firmwide Best Practice Strategies*. New York: John Wiley & Sons, 2002.

5. Amari, SV, McLaughlin, L. Optimal design of a condition-based maintenance model. Proceedings of the Reliability and Maintainability, 2004 Annual Symposium—RAMS, pp. 528–533.

6. A Survey of System Development Process Models, CTG.MFA-003. Center for Technology in Government. New York: University of Albany, 1998.

7. Nuclear Regulatory Legislation, NUREG-0980, 1(6), 107th Congress, 1st Session, Office of the General Council, U.S. Nuclear Regulatory Commission, Washington, DC, June 2002.

8. Sheard, SA, Lake, JG. Systems engineering standards and models compared. Software Productivity Consortium, NFP, 1998. Available at http://www.software.org/pub/externalpapers/9804-2.html.

9. Price, S, John, P. The status of models in defense systems engineering. In: Pidd, M, editor. *Systems Modeling: Theory and Practice*. West Sussex, England: John Wiley & Sons, 2004.

10. Boehm, B. *Spiral Development: Experience, Principles, and Refinements*. CMU/SEI-2000-SR-008. Pittsburgh, PA: Carnegie Mellon Software Engineering Institute, 2000.

11. Maner, W. 1997: Rapid applications development. Available at http://csweb.cs.bgsu.edu/maner/domains/RAD.htm. Accessed June 7, 2006.

12. The Agile Manifesto. Available at http://www.agilemanifesto.org. Accessed July 8, 2006.

13. Garvey, PR. *Analytical Methods for Risk Management*. Boca Raton, FL: Chapman & Hall/CRC Press, 2009.

14. Rethinking Risk Management in Financial Services. Report of the World Economic Forum, New York. Available at http://www.weforum.org/pdf/FinancialInstitutions/RethinkingRiskManagement.pdf. Accessed April 20, 2010.

15. Vose, D. *Risk Analysis: A Quantitative Guide*. West Sussex, England: John Wiley & Sons, 2000.

16. Haimes, YY. Total risk management. *Risk Analysis*, 1991;11(2):169–171.

17. Post-Challenger evaluation of space shuttle risk assessment and management. The Committee on Shuttle Criticality Review and Hazard Analysis Audit, Aeronautics and Space Engineering Board. Washington, DC: National Academy Press, 1988.

18. Haimes, YY. *Risk Modeling, Assessment and Management*. New York: John Wiley & Sons, 1998.

19. INCOSE-TP-2003-016-02. *Systems Engineering Handbook*. Seattle, Washington.

20. Galway, LA. Subjective probability distribution elicitation in cost risk analysis: A review. RAND Techical Report: Project Air Force. Santa Monica, CA: RAND Corporation, 2007.

21. Primavera Risk Analysis, ORACLE Data Sheet. Available at http://www.oracle.com. Accessed April 20. 2010.

22. Operational Risk. Report of the Basel Committee on Banking Supervision Consultative Document, Bank for International Settlements, January 2001.

23. Containing systemic risk: The road to reform. Report of the Counterparty Risk Management Policy Group III. New York, NY: available at http://www.crmpolicygroup.org, August 6, 2008.

24. Labonte, M. Systemic risk and the Federal Reserve. CRS Report for Congress R40877, Washington, DC: October 2009.

25. Modarres, M. *Risk Analysis in Engineering: Techniques, Tools and Trends*, Boca Raton, FL: CRC Press, 2006.

26. Haimes, YY. *Risk Modeling, Assessment, and Management*. Hoboken, NJ: John Wiley & Sons, 2004.

27. Simon, P. *Risk Analysis and Management*. London, England: AIM Group, 1997.

28. Roland, HE, Moriarty, B. *System Safety Engineering and Management*, 2nd ed. New York: John Wiley & Sons, 1990.

29. Hubbard, DW. *The Failure of Risk Management*. Hoboken, NJ: John Wiley & Sons, 2009.

30. Kahneman, D, and Tversky, A. Subjective probability: A judgment of representativeness. *Cognitive Psychology*, 1972;3:430–454.

31. New horizons in predictive modeling and risk analysis. ORACLE White Paper, ORACLE Corporation, Redwood Shores, California, 2008.

32. Book, SA. Estimating probable system cost. *Crosslink*, 2001;2(1):12–21.

33. Garvey, PR. *Probabilistic Methods for Cost Uncertainty Analysis: A Systems Engineering Perspective*. New York: Marcel Decker, 2000.

Chapter **4**

Systems Modeling and Analysis

PAUL D. WEST, Ph.D.
JOHN E. KOBZA, Ph.D.
SIMON R. GOERGER, Ph.D.

All models are wrong, some are useful.

—George Box

All but war is simulation.

—U.S. Army

4.1 INTRODUCTION

Thinking about a system's essential components, attributes, and relationships in abstract ways occurs in various forms throughout the system life cycle. This was described in Section 2.8 during the discussion on the spatial placement of systems and their components. It may also take the form of a mathematical representation, such as $E = mc^2$, that examines the relationship between available energy, an object's mass, and the speed of light. At times it is useful to create a physical mockup of an object and put it in the hands of a user to obtain feedback on how the system performs or reacts to change. All of these ways of thinking about systems are using *models* to better understand the system. This chapter explores models—what they are and how they're used—and a unique type of model, a *simulation*, which exercises a system's components, attributes, and relationships over time.

Decision Making in Systems Engineering and Management, Second Edition
Edited by Gregory S. Parnell, Patrick J. Driscoll, Dale L. Henderson
Copyright © 2011 John Wiley & Sons, Inc.

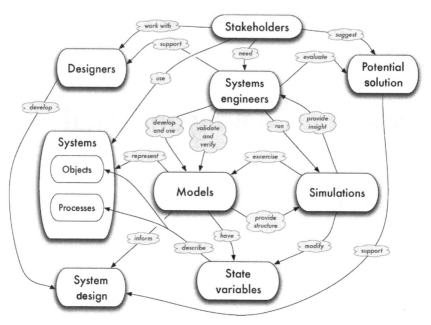

Figure 4.1   Concept map for Chapter 4.

This chapter introduces four of the tools a systems engineer needs to master to fully understand the system life cycle, introduced in Chapter 3, and to develop and evaluate candidate solutions to present to a decision maker. These tools are: *measures of system performance and effectiveness*, *models*, *simulations*, and *determining sample size*. Each tool builds on the previous one. Figure 4.1 captures the fundamental relationships between systems engineers and the modeling and simulation (M&S) process.

4.2   DEVELOPING SYSTEM MEASURES

System measures enable systems engineering teams to evaluate alternatives. They are vital for deciding which models or simulations to use and are therefore key considerations throughout the M&S process. It is essential that measures *based on stakeholder values* be identified before any modeling and simulation effort begins. A fatal flaw in system analysis occurs when modelers force a system into favorite analysis tools, observe dominant outcomes, and attempt to retrofit measures to outcomes. Almost always these measures do not reflect the values of the stakeholders or even the modeler and thus fail to provide meaningful support to the decision maker.

Measures are tied either directly or indirectly to every system objective identified in the problem definition process. Although most objectives can be evaluated directly, others, such as "To develop a sound energy policy," can be assessed only in terms of percentage of fulfillment of other objectives [1]. Measures can take several forms, but the most prevalent fall into the categories of *measures of effectiveness* (MOE) and *measures of performance* (MOP).

A *measure of performance* is a quantitative expression of how well the operation of a system meets its design specification.

A *measure of effectiveness* is a quantitative expression of how well the operation of a system contributes to the success of the greater system.

For example, the measure *miles per gallon* reflects the reliability of how well an engine performs. It is not directly related to the success of the greater system's mission, which may be to transport people across town. An effectiveness measure for such a vehicle may better be stated in terms of *passengers delivered per hour*.

A military weapon system may meet a specification of firing 1000 *rounds per minute* (an MOP), but if the *probability of hitting* a target with a single round (an MOE) is 0.9 and the *probability of destroying the target, given a hit* (an MOE), is 0.9, then the measure of performance, rounds per minute, is not very helpful in evaluating the effectiveness of the greater system in its mission of destroying targets.

Performance measures are developed by systems engineers during the Problem Definition phase of the SDP and are described in the system specification. Effectiveness measures are also based directly on stakeholder values, but these measures view the system in a larger context and therefore selecting meaningful MOE can be more difficult. In the 1970s, the U.S. Army produced a seven-step format for defining MOE that is helpful both in the Problem Definition phase of the SDP and for describing the relevance of a measure to stakeholders. In its discussion of MOE, the Army noted several characteristics of a good MOE [2]. These are:

- A good MOE reflects and measures functional objectives of the system.
- A good MOE is simple and quantifiable.
- A good MOE measures effectiveness at echelons above the system (how it contributes).
- A good MOE involves aggregation of data.
- A good MOE can be used to determine synergistic effects of a system.

Defining a good MOE has seven steps, two of which are optional. These are:

1. Define the measure. Include both a narrative description and an equation.
2. Indicate the dimension of the measure. Is it a ratio, an index, a time interval?
3. Define the limits on the range of the measure. Specify upper and lower bounds.
4. Explain the rationale for the measure. Why is this measure useful?
5. Describe the decisional relevance of the measure. How will this measure help the decision maker?

Time To Estimate Range

1. *Definition of the Measure.* Time to estimate range is the elapsed time from detection of a target to estimation of range. Input data are the moment of detection and the moment estimation of range is complete. Relation of output to input is:

   time to estimate range = time of estimation − time of detection.

2. *Dimension of the Measure.* Interval-elapsed time in terms of seconds. If the measure is taken at different times or under varying circumstances, it can be used in the form of mean time to estimate range or median time.

3. *Limits on the Range of the Measure.* The output can be zero or any positive value. The resolution of the measure is limited by the precision of taking start time and end time. The data cannot be disassociated from the definition of computed estimation used, whether it is the first estimate stated regardless of accuracy or is the final in a series of estimates which is used for firing.

4. *Rationale for the Measure.* This measure addresses a component of target acquisition time. Problems in estimation are assumed to contribute to the length of estimation time.

5. *Decisional Relevance of the Measure.* This measure can be used to compare estimation times of means of range estimation (techniques, aids, range finders, trained personnel) with each other or a standard. It would not ordinarily be used alone, but would be combined with accuracy of estimation or accuracy of firing in most cases.

6. *Associated Measures.*

   • Accuracy of range estimation
   • Firing accuracy
   • Time to detect
   • Exposure time
   • Time to identify

Figure 4.2   A fully defined measure of effectiveness (MOE).

6. List associated measures, if any, to put this measure in context (optional).
7. List references (optional). Give examples of where it has been used.

An example MOE using this format is shown in Figure 4.2.

Thinking about how a system will be evaluated occurs continuously as the understanding of the system evolves. This brief introduction on measures will be expanded in depth in Chapter 8.

4.3   MODELING THE SYSTEM DESIGN

This section introduces the concept of modeling. *Modeling* is more of an art than a science. Just as with playing a musical instrument or painting a picture, some people have more talent for modeling than others. However, everyone can improve their skill with knowledge of their discipline and practice building models. This section provides insight into what makes good models and presents guidelines for building them.

4.3.1 What Models Are

A *model* is an abstract representation of a system. Consider a model airplane. It could be a 1/72 scale plastic model that comes in a kit. The assembled model has some of the features of the real aircraft, such as the shape of the fuselage and the angle of the wings. However, it does not have a working engine or controllable flight surfaces.

Another airplane model is the ready-built radio controlled (R/C) airplane. It flies with the user controlling the elevator, ailerons, and speed. Electric ducted fans even replace jet engines! However, this model is also not the real aircraft. The weight and balance are different, the rudder is fixed, and the flight dynamics are different than the real aircraft.

Both of these are useful representations of the real aircraft system, but each has a different purpose. The model kit illustrates the shape of the fuselage in relation to wing angle. The R/C model demonstrates how the aircraft would look in the sky. Neither model is useful for predicting the maximum altitude or flying range of the actual aircraft. Their usefulness, as in all models, depends on what aspect of the actual system is being studied.

A model captures essential attributes and relationships between system components, but it does not capture them all.

A model is an abstract representation of a system.

The *modeler* must choose what to put in and what to leave out of a model, as well as how to represent what is put in. This is the essence of modeling. It is also what makes modeling an art. It can be very difficult to choose the key aspects of the system and incorporate them into a workable model. A set of differential equations may accurately represent the relationships among system variables, but it is not workable if it cannot be solved. Modelers build models using their knowledge of the system (or experience with similar systems) and their understanding of available modeling techniques.

4.3.2 Why We Use Models

One of the greatest ideas the Wright brothers had was to model the wing instead of building and flying different wing shapes. In the following excerpt, Wilbur Wright describes how they developed their understanding of the wing. It illustrates many of the reasons to use a model.

> It took us about a month of experimenting with the wind tunnel [Figure 4.3] we had built to learn how to use it effectively. Eventually we learned how to operate it so

Figure 4.3   Replica of Wright brothers' wind tunnel, 5-10-39 [3].

that it gave us results that varied less than one-tenth of a degree. Occasionally I had to yell at my brother to keep him from moving even just a little in the room because it would disturb the air flow and destroy the accuracy of the test.

Over a two-month period we tested more than two hundred models of different types of wings. All of the models were three to nine inches long. Altogether we measured monoplane wing designs (airplanes with one wing), biplanes, triplanes and even an aircraft design with one wing behind the other like Professor Langley proposed. Professor Langley was the director of the Smithsonian Museum at the time and also trying to invent the first airplane. On each little aircraft wing design we tested we located the center of pressure and made measurements for lift and drift. We also measured the lift produced by wings of different aspect ratios. An aspect ratio is the ratio or comparison of how long a wing is left to right (the wing span) compared to the length from the front to the back of the wing (the wing chord). Sometimes we got results that were just hard to believe, especially when compared to the earlier aerodynamic lift numbers supplied by the German Lillienthal. His numbers were being used by most of the early aviation inventors and they proved to be full of errors. Lillienthal didn't use a wind tunnel like Orville and I did to obtain and test our data.

We finally stopped our wind tunnel experiments just before Christmas, 1901. We really concluded them rather reluctantly because we had a bicycle business to run and a lot of work to do for that as well. It is difficult to underestimate the value of that very laborious work we did over that homemade wind tunnel. It was, in fact, the first wind tunnel in which small models of wings were tested and their lifting properties accurately noted. From all the data that Orville and I accumulated into tables, an accurate and reliable wing could finally be built. Even modern wind tunnel data with the most sophisticated equipment varies comparatively little from what we first discovered. In fact, the accurate wind tunnel data we developed was so important, it is doubtful if anyone would have ever developed a flyable wing without first developing these data. Sometimes the nonglamorous lab work is absolutely crucial to the success of a project [3].

A couple of bicycle mechanics from Ohio used models to become the first to achieve controlled powered flight. Their use of models saved them time, money, and, probably, their lives. Otto Lillienthal's data was gathered from over 2000 glider tests over more than 20 years [4]. He was killed in a glider crash in 1896. After three months of studying physical models in a wind tunnel, the Wright brothers understood how wings behave. This understanding allowed them to create a flyable wing design for their plane.

This example illustrates the three major reasons we use models.

Models are Flexible. By changing parameter values, a single model can represent the system across a broad set of conditions or represent a variety of related systems.

Models Save Time. It is usually much faster to build or change a model of a system than the system itself. Testing the design using a model could reveal flaws that can be fixed more quickly and less expensively before the system is built.

Models Save Money. Building prototypes of a car or a production line is expensive. It is much cheaper to make changes to a mathematical or computer model than to build different versions of the actual system.

4.3.3   Role of Models in Solution Design

The primary purpose of a model is to understand how the actual system design will or does perform. This understanding can then be used to design the system in a way that improves or optimizes its performance. The Wright brothers used their wind tunnel to gain an understanding of how a wing behaves. They then used this knowledge to design a wing for their aircraft.

There are three different types of system designs: *satisficing, adaptivising,* and *optimizing* (Figure 4.4). *Satisficing* is a term that comes from a concatenation of *satisfying* with *sacrificing.* Satisficing, as a goal for an existing system, means the current performance is satisfactory. For a new system, satisficing means that any good feasible solution with acceptable intrinsic cost tradeoff will be satisfactory. If the goal is to move up a hillside, a satisficing solution may be to crawl uphill through some bushes.

The primary criterion for *adaptivising* solutions is cost effectiveness. Improved system performance is the goal, but only when it can be obtained at a reasonable cost. Satisficing accepts any solution that works. Adaptivising looks for "good" solutions. For the example of moving up the hill, an adaptivising solution may be to walk a path that leads up the hill. Ascent may not be as rapid as crawling up a steep slope, but it is a lot easier and will probably get us there faster.

Optimizing solutions are at least as good and often better than all other solutions. They are the best according to performance measures. However, they may require

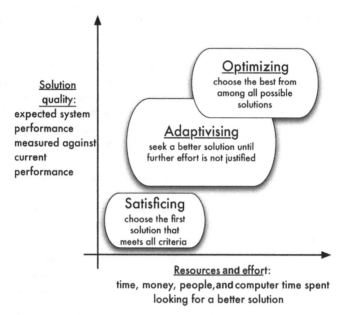

Figure 4.4   The relationship between effort and solution quality.

much effort or expense. If the goal is to get to the top of the hill in the quickest manner, the optimizing solution may be to land on top of the hill in a helicopter.

4.3.4   Qualities of Useful Models

> Art is a selective recreation of reality. Its purpose is to concretize an abstraction to bring an idea or emotion within the grasp of the observer. It is a selective recreation, with the selection process depending on the value judgments of the creator [5].

The above definition of art sounds perilously close to our definition of a model. *Aesthetics* is the branch of philosophy that relates to art and includes the study of methods of evaluating art. Just as art can be evaluated, so can models. Just as the artist must make value judgments when capturing an idea or emotion in a concrete form, so the engineer must make value judgments when representing the key elements of a system in a usable form. Just as art can be beautiful, so can models. This section presents the characteristics used to judge models.

Parsimony. One of the primary characteristics of a model is its level of complexity. The principle of parsimony (also known as Occam's Razor) states that when choosing among scientific theories that make exactly the same predictions, the best one is the simplest [6]. The same is true of models—simpler is usually better. Given several models of the system with comparable output, the best model is the simplest one. It is typically easier to understand. It also may require fewer and more reasonable assumptions. For these reasons, it is usually easier to explain to stakeholders.

Simplicity. Another characteristic is expressed by the relationship between the complexity of the model and that of the system it represents. The model's level of complexity typically increases in response to the system's level of complexity. This is not always the case, but a complex model for a simple system usually represents a lack of effort on the part of the modeler. It takes time, effort, and creativity to simplify a model. Sometimes this can result in a simple model of a complex system, which is truly a thing of rarest beauty.

Accuracy. Accuracy is another important characteristic of models. Accuracy is "the degree to which a parameter or variable, or a set of parameters or variables, within a model or simulation conforms exactly to reality or to some chosen standard or referent" [7]. An inaccurate model is not much good to anyone. Accuracy must often be considered hand-in-hand with model complexity. A "quick-and-dirty" solution results from a simple model with low accuracy. Sometimes this level of accuracy is enough to gain sufficient insight into the system behavior. At other times, more accurate, and perhaps more complex, models are needed.

Robustness. Robustness characterizes the ability of a model to represent the system over a wide range of input values. A model that is not robust only represents the system for a narrow range of inputs. Beyond this range, other models must be used or the structural parameters of the existing model must substantially change. This is not necessarily a problem. For example, if the narrow range of input values covers all of the alternatives from the Solution Design phase of the SDP, the model is good. However, if the narrow range only covers a portion of the design space, other models will be needed; or a more robust model must be developed.

Scale. The scale of the model is an important characteristic that refers to the level of detail used in the problem. If a model contains both insect life cycles and continental drift, it may suffer from inconsistent scale. Both ends of the spectrum may exist in the actual system, but there are orders of magnitude difference in how they affect the system. If you are modeling the life cycle of a flower, the effects of continental drift are negligible. However, if you are modeling the spread of a flower species over geologic periods, continental drift is much more important than insect life cycles.

Fidelity. Fidelity is an overall characterization of how well the model represents the actual system. Just as the high-fidelity recordings of the late vinyl age were touted as being truer to the actual sound produced by an orchestra, a high-fidelity model is closer to representing the actual system. Fidelity is an aggregate characteristic that brings together complexity (parsimony and simplicity), accuracy, robustness, and scale. A 100% replication of an actual system represents a duplicate or copy of the actual system.

Balance. Just as an artist must choose which aspects of an idea to convey and how to represent it in an art form, a systems engineer must choose which aspects of a system are essential and how to represent them in the concrete form of a model. The modeler must balance and blend complexity (parsimony and simplicity), accuracy, scale, and robustness. (One definition of an expert is knowing what to ignore.) Among the choices the modeler must make is the type of model (or models) to use. Section 4.5 introduces some of the different types of models.

4.4 THE MODELING PROCESS: HOW WE BUILD MODELS

Just as there is no "correct" way to create art, there is no 'correct' way to create a model. However, there are steps or processes that have been used in the past to develop models. This section presents only one possible modeling process. Think of it is a list of things to consider while developing a model, rather than a rigid step-by-step procedure. In fact, some of the steps listed in Figure 4.5 may need to be repeated several times in different orders as the model is developed.

Typically, a model begins in the mind and is then rendered in a form that is tested or examined for usefulness. The experience gained may then be used to revise and improve the model. This sequence is repeated until an acceptable model is created, verified, validated, and accredited.

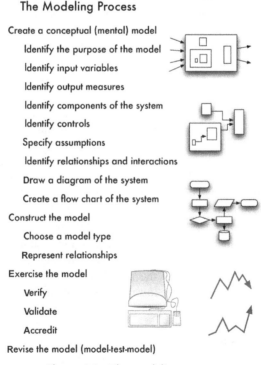

The Modeling Process

Create a conceptual (mental) model
   Identify the purpose of the model
   Identify input variables
   Identify output measures
   Identify components of the system
   Identify controls
   Specify assumptions
   Identify relationships and interactions
   Draw a diagram of the system
   Create a flow chart of the system
Construct the model
   Choose a model type
   Represent relationships
Exercise the model
   Verify
   Validate
   Accredit
Revise the model (model-test-model)

Figure 4.5 The modeling process.

4.4.1 Create a Conceptual Model

Not surprisingly, the *conceptual model* begins in the modeler's mind as the realization of a key relationship or an idea about representing a key feature of the system. For example, the modeler may notice that inventory goes up as demand decreases or production increases. How can this relationship be represented in a model?

> A *conceptual model* is a theoretical representation of the system based on anecdotal or scientific observations. "It includes logic and algorithms and explicitly recognizes assumptions and limitations" [8].

Many of the items or steps in Figure 4.5 related to the conceptual model have already been considered as part of the systems decision process (see Chapter 2). These include identifying system elements or components, specifying assumptions, analyzing data to identify relationships and interactions, and drawing diagrams and flowcharts for the system. Other items need to be considered in the context of the desired model.

Identify the Purpose of the Model. The most important consideration is the purpose of the model. One purpose of a model might be to predict how a system will perform under a given set of conditions. Consider the following examples: How many satellites can be placed into space if two space shuttles or payload rockets are purchased? What is the worst-case response time to a distress call from manned lunar station? What penetration will a specific rocket thrust or certain projectile shape achieve through a concrete block? Another possible purpose could be to determine the set of design choices that optimizes performance as in the following examples. How many machines should be purchased to maximize profit? What balance of forces should be used to maximize the probability of success? What is the best fin configuration for a rocket system? The purpose of the model will influence the choice of modeling tools used.

Identify Input Variables. The purpose of the model affects the choice of model *input variables*. These could represent aspects of the design that the systems engineer can choose—sometimes called controllable or independent variables. These could include such factors as the number of space shuttles, payload delivery rockets, or emergency response teams. They can also represent parameters that the user of the model may be unsure of or wishes to vary across a range of values as they test the model exploring "what if?" scenarios. These may include the rate of inflation, the cost of rocket fuel, rocket fin configuration, the amount of thrust produced by a rocket engine, or the thickness of the asbestos shielding on the space shuttle.

Identify Output Variables. Output variables represent what the modeler wants to learn from the model. These are sometimes called uncontrollable or dependent variables, and may include the amount of payload that can be taken into space, the response time for an emergency response team or the number of rockets to purchase.

Identify Controls. Identifying inputs, outputs, controls, and components (mechanisms) of a system provide the elements necessary to build in IDEF0 model. Of those, controls are often the most difficult to understand. The IEEE Standard for IDEF0 (IEEE Std 1320.1-1998) defines a control as "a condition or set of conditions required for a function to produce correct output." This may include a wide variety of conditions, including regulations, instruction manuals, or "trigger" events. Section 2.6 lists the family of IDEF models and points out that IDEF0 focuses on functions. IDEF0 models must have at least one control and at least one output, although typically a function will also include at least one input, which is transformed into an output by the function.

Interestingly, the same variable could be an input variable for one model and an output variable for another. For example, suppose a model is created to predict the probability of detecting an explosive device in a bag that is screened by Machine A and Machine B. An input value could be the detection rate for Machine B at an airport terminal. However, if the purpose of the model is to determine the detection rate for Machine B that maximizes the probability of detecting the device, then the detection rate for Machine B is an output variable.

After the modeler understands the system and has created a conceptual model, the model can be built.

4.4.2   Construct the Model

One of the key decisions in constructing a model is choosing a type of model to use. Different types of models and some of their implications will be discussed in Section 4.5. For now, we will focus on some of the different factors that lead to the choice of a model type.

A *constructed model* is an implementation of a conceptual model—mathematical, physical, or codified.

Choose a Model Type. Sometimes assumptions about the system lead naturally to a choice of model. If it is assumed that there is no randomness in the system, then deterministic models are appropriate (Section 4.5). The availability of data can also drive the choice of a model type. For example, lack of data for a certain input variable may lead to another model type that does not need that input. The type of data is also a factor. For example, quarterly data may lead to a discrete-time model that moves in 3-month steps. In addition, the goal of the model can be the primary factor in choosing a model type. For example, optimization models are likely to be used if the purpose is to find the best solution.

Represent Relationships. After the model type is chosen, the system relationships relevant to the goal must be represented in the model. Each type of model has specific ways of representing relationships. For example, a linear model $y = mx + b$ is defined by the choice of m and b. A continuous-time Markov chain is described by the definition of the state variables and the transition rates.

Issues may arise during the M&S process which require rethinking the conceptual model. For example, perhaps there is insufficient data to determine the value of m in a linear model or perhaps the range of data is too narrow relative to the desired range of system operation.

4.4.3 Exercise the Model

Once a model is constructed, the modeler must determine how good the model is and how and when it is appropriate to use. This is referred to as exercising or testing the model. There are several concepts related to this testing: verification, validation, and accreditation.

Verification. Verification determines whether the constructed model matches the conceptual model [9]. That is, was the conceptual model correctly implemented? First, simple tests are done—sometimes referred to as sanity checks. Are there special cases where the behavior or condition of the conceptual model is known? For example, in the conceptual model if there is a limited payload and no rocket launch, there should be no satisfied demand for resupplies needed at a manned lunar base. Does this happen in the constructed model?

If the model passes the simple tests, more complicated ones are tried until the model developer and accreditation agent are confident that the constructed model performs as the conceptual model states it should. Input variables and parameters may be varied over a range of values to determine the model's accuracy and robustness. If a conceptual model predicts that inventory will rise as demand decreases, does this occur in the constructed model? Actual data from the system or a similar system may be used as input to the model, and the output can then be compared to the actual behavior of the system.

During this process, failure can be more enlightening than success. Parameters may need to be adjusted or extra terms or components added to the model that were initially overlooked because of tacit assumptions that were not intended. Sometimes the model does not represent the system well, and a new type of model must be developed.

Validation. Validation determines whether the conceptual model (and constructed model if it is verified) appropriately represents the actual system [9]. For example, a conceptual model may assume a linear relationship between thrust and speed; that is, speed increases proportionately as thrust increases. If the actual system includes both drag and thrust, then there may be times when speed declines because drag from the payload and atmospheric conditions are greater than the force generated by thrust. In this case, the conceptual model is not valid.

Validation is concerned with building the right model. Verification is concerned with building the model right. Both concepts are important and both need to be checked when testing a model. Testing is usually not separated for verification or validation. Tests are conducted and failures are investigated to determine if the problem is with the conceptual or constructed model. Previous steps in the process may need to be repeated to correct the problems.

Validation is building the right model. Verification is building the model right.

Accreditation. Accreditation is an "official" seal of approval granted by a designated authority [9]. This authority is frequently the decision maker, but may be an accreditation agency, such as ABET, Inc., which accredits academic engineering programs. This signifies the model has been verified and validated for an intended purpose, application, or scenario. Accreditation represents the process of users accepting the model for use in a specific study. This can be very informal, for example, if the same people who develop the model will be using it. On the other hand, if the model is developed by one organization and used by another, accreditation can be a lengthy and involved process. Normally, models are accredited by a representative of the using organization. All models and simulations used by the Department of Defense must be verified, validated, and accredited [9].

It may be tempting to believe that a model is only judged by its usefulness. However, the process used to create the model and the testing involved in its development are both very important when judging a model. What characteristics were considered for inclusion in the model? What assumptions were made while developing the model? What value ranges or conditions were tested? The answers to such questions build confidence in the model's capabilities.

4.4.4 Revise the Model

As has already been discussed, test results may cause changes in the model. This is sometimes called the *model–test–model* concept. The idea is to use knowledge gained about the system or the model to continually revise and improve the conceptual or constructed model. Revising the model is always an option and typically happens quite frequently during model development.

Recall the story of the Wright brothers' wind tunnel. Their experience flying gliders and data from others helped them to know what they wanted to measure and what they wanted to vary. Lift and drift were their output variables. The different types of wing shape and aspect ratios were their input variables. The wind tunnel gave them the means to do it. They spent over a month experimenting with the wind tunnel. This corresponds to verification and validation. They wanted to be confident that their model would yield useful results and they understood the conditions under which it would. At some point they had a discussion and agreed they were ready to collect data. This corresponds to accreditation. Once they were confident of how to use their model, they systematically explored the system by trying different wing shapes and aspect ratios.

4.5   THE MODEL TOOLBOX: TYPES OF MODELS, THEIR CHARACTERISTICS, AND THEIR USES

There are several types of models including physical, graphical, and mathematical. Even though these are distinct types of models, they are often used together as will be discussed below.

Physical. Physical models involve a physical representation of the system, such as the wing designs the Wright brothers tested in their wind tunnel. These models can be time-consuming and expensive to produce. However, they can provide valuable data that can be studied to learn about the system. Sometimes physical models are the only alternative because mathematical models may not be practical. This was the case for many aerospace models until high-speed computers allowed real-time solution of mathematical formulas for fluid flow dynamics to replace, or reduce the number of, physical wind tunnel tests.

Physical models need not always be miniature versions of the real system. Sometimes a physical representation is used to take advantage of a physical property. For example, soap film takes a shape that minimizes surface area. Before the age of computers, some optimization problems were solved using soap films and wire frames or glass plates with wooden pegs. Surprisingly, this technique is currently a topic of discussion concerning a fundamental research question related to computational complexity theory [10].

Sometimes physical models are based on analogies; something in the physical model represents a quantity of interest in the system. An interesting example of this is the analog computer. An analog computer uses electrical circuits to represent differential equations. Combinations of operational amplifiers, resistors, and capacitors are used to construct circuits that perform such mathematical operations as integration, differentiation, summation, and inversion. The Norden bombsight used during World War II was a mechanical analog computer that calculated the trajectory of the bomb and determined when to release it.

Physical models have always had limited areas of application, and the use of digital computers has further limited their use. However, they can still effectively demonstrate system performance. Seeing a truss design made out of toothpicks support a large weight can be pretty dramatic!

Graphical. Graphical or schematic models use diagrams to represent relationships among system components. Examples are causal loop diagrams that represent feedback relationships or Petri networks that represent a sequence of conditions over time. Figure 4.6 shows a network model of a discrete event simulation (see Section 4.6). These can be used to show alternatives.

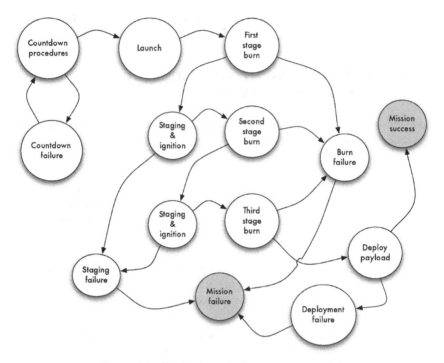

Figure 4.6   Rocket launch discrete event model.

Concept maps, such as those used at the start of each chapter in this book, can represent how different system characteristics or components affect one another. Sometimes these diagrams are the end product of the systems decision process and can provide an understanding of some aspect of the system being studied. Sometimes they are graphical front-ends for computer packages that generate and execute code. The code may represent mathematical equations based on the diagrams, which are numerically solved to provide quantitative results. System dynamics software is an example of such a package. The relationships in the diagram are represented as differential equations that are numerically solved by the computer. In other packages, such as the discrete-event simulation languages Arena® and ProModel®, the graphical model represents the logical structure of simulation model (see Section 4.6). The computer uses this to generate code that is run to produce both numerical and graphical output. For example, the computer may show people or jobs moving through the system.

Mathematical. Mathematical models use quantitative relationships to represent systems. It can be helpful to view mathematics as a language, such as Latin or Spanish. Learning a new language involves vocabulary and syntax. Vocabulary refers to the meaning of words. Syntax is learning how to put words together to convey ideas. Mathematics also has a vocabulary and syntax. Just as words in a novel convey complex thoughts, emotions, and relationships, mathematics can convey the complex relationships and structure of a system. Part of the challenge of constructing a mathematical model is translating from English to math.

Mathematical models can provide exact or approximate solutions. Approximations are models that provide a solution close to the actual solution. Although not as accurate as exact solutions, approximations are usually much easier to solve. As an example, suppose we wish to determine the area under the function $f(x) = x^2$ between points a and b. The exact solution is

$$\int_a^b x^2 \, dx = \frac{1}{3}x^3 \bigg|_a^b = \frac{b^3 - a^3}{3}. \qquad (4.1)$$

An alternative is to use an approximate model to represent the area with a trapezoid. The approximate solution would be

$$\left(\frac{a^2 + b^2}{2}\right)(b - a) = \frac{b^3 - ab^2 + a^2b + a^3}{2}. \qquad (4.2)$$

If $a = 1$ and $b = 2$, the exact solution is 7/3 and the approximate solution is 5/2. The approximation error is 1/6. The advantage of the approximate solution is that it did not require integration.

Mathematical models can be solved analytically or numerically. An analytical solution is a mathematical expression that yields a value after appropriate parameter values are substituted. For an analytical solution, the hard work is usually done before the parameter values are given. They are represented as variables or constants in the final expression. The values are substituted into the expression and some arithmetic is performed to obtain the solution. A numerical solution requires a (sometimes lengthy) series of calculations after the parameter values are given.

Again consider the area under the function x^2 between points a and b. Both of the solutions, Equations (4.1) and (4.2) in the previous paragraph are analytical solutions. Instead of using constants such as a and b, a numerical (or computational) solution needs numbers from the beginning. If $a = 1$ and $b = 2$, a numerical solution approach may divide the interval between 1 and 2 into some number of steps—in this case, 10. For each step, the area under the curve could be approximated as a rectangle (see Figure 4.7). Adding the area of each rectangle gives the solution.

Numerical solutions often use approximations, which can introduce approximation error. They may also have errors due to computation, such as rounding or truncation. For example, consider the expression 3(4/3). If it is evaluated analytically, the result is 4. If it is evaluated numerically on a computer, the result is 3.9999999, because the computer cannot represent the exact value of 1/3.

Computers can be used to obtain either analytical or numerical solutions. Mathematical analysis software such as Mathematica®, Matlab®, and

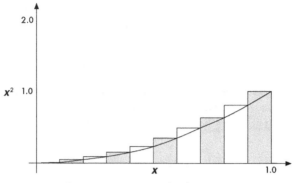

Figure 4.7   Area under the curve.

Maple™ can solve many problems analytically. They also provide a variety of numerical tools to solve problems.

4.5.1   Characteristics of Models

Models have a number of *characteristics* that can be used to determine when they are appropriate to use. Some of these characteristics are shown in Figure 4.8 and are discussed below. These characteristics will be presented as either/or possibilities—a model is either static or dynamic—however, models often have elements of both.

> *Descriptive versus Prescriptive. Descriptive* models describe or predict how a system will behave. Their outputs give information about system behavior. *Prescriptive* models prescribe a course of action—they dictate what the best thing to do. A descriptive model of a payload rocket launch could represent

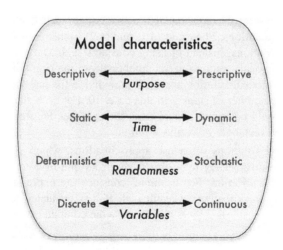

Figure 4.8   Model characteristics.

the pre-launch, launch, and post-launch activities at the launch pad to compare the different possible locations for launch observers, security forces, emergency response teams, and launch monitoring cameras. A prescriptive model would be used to determine the proper mixture of rocket fuel and payload. Consider a scheduling problem in which a number of different jobs must be processed on a machine. There are different types of jobs, and a changeover on the machine is needed when switching from one type of job to another. A descriptive model would be used to determine the total production time for a given sequence of jobs. A prescriptive model would be used to determine the sequence with the shortest total production time.

How many boxes are needed to cushion the fall of a motorcycle and rider jumping over an elephant? Several models are needed to solve this problem. A descriptive model is needed to describe the speed and momentum of the motorcycle and the rider over time. The values of these quantities at the moment of impact (hopefully with the boxes not the elephant) would be an input to a prescriptive model that would determine the number of boxes.

Static versus Dynamic. Static models do not change. *Dynamic* models represent change. This change usually occurs over time, but could also be over distance (e.g., gravitational or magnetic attraction) or some other measure. As an example of a problem that is static, suppose that a company has a development budget and a large number of potential products, each of which has a development cost and a potential profit. Which products should be developed to maximize the potential profit? The parameters of this problem do not change. They are static. Change in the model usually refers to the primary output of the model and whether or not it changes. In this problem, there is only one list of products; it does not change. If the potential profit for each product changed over time to reflect changes in consumer needs or tastes, the primary output of the model, the list of projects to invest in today, is still static. However, if the company had annual budgets for each of the next five years and wanted to know which projects to invest in each year, the output becomes dynamic because the list changes each year.

As another example, suppose there is a rocket manufacturing plant. A dynamic simulation model could represent arriving raw materials (metal alloys, fiber-optic cable, wiring harnesses, computer processors, instrument gages, etc.), storing these materials, modifying them into rocket components, assembling the components into a rocket, and delivering the rockets to the customer. The simulation could monitor inventory levels over time, the time in system for the different materials, or the occurrence of shortages.

Dynamic models can represent transient or equilibrium behavior. *Transient-focused* models concern behavior that depends on the initial conditions of the system. They are usually appropriate over short time intervals. *Equilibrium-focused* models represent system behavior after it is no longer affected by initial conditions. They are usually appropriate over longer time intervals. The amount of fuel left in the launch rocket after the initial launch phase depends a lot on the amount of fuel and payload in the rocket prior to ignition. This is transient behavior. The amount of fuel in the space shuttle prior to reentry does not depend on how heavy the initial payload was (except in extreme cases) since another rocket subsystem launched the shuttle into space.

Deterministic versus Probabilistic. *Deterministic* models use quantities that are known with certainty. *Probabilistic* (or *stochastic*) models have at least one quantity with random values. Suppose the chairs of a chair lift arrive at the top of a ski slope every 30 seconds. If every other chair has one skier and the rest have two, how many skiers are at the top of the slope after 10 minutes? All the quantity values are known with certainty, so a deterministic model is appropriate here. If the arriving chair has one skier with probability 0.5 or two skiers with probability 0.5, how many skiers are at the top of the slope after 10 minutes? The number of skiers on a particular chair is not known and the number of skiers after 10 minutes cannot be known with certainty. A probabilistic model is appropriate here. A deterministic model of the rocket production plant's inventory would use fixed quantities for the times between material deliveries, production service times, and pickup times. The model would still be dynamic and inventories would change over time, but there would be no variation and, eventually, a repeating pattern will appear. A probabilistic model would introduce random variation into at least one of the quantities, which would result in continually changing behavior over time.

Discrete versus Continuous. The variables in a problem can be discrete, continuous, or mixed. One way to tell is to put all the possible values for the variable in a set. If the number of values is countable (i.e., a one-to-one correspondence exists with the positive integers), the variable is discrete. *Discrete* means separate, distinct, or unconnected. In this case, the values are separate so we can distinguish and count them. Obvious examples are the number of rocket fins in the rocket assembly plant inventory at noon each day or the number of cars passing through an intersection on a given day. Another example of a discrete problem we have already discussed is the set of projects to fund. Different sets of projects are distinct from each other and we can count how many different possibilities there are. The sequence of jobs to process is another example.

Continuous variables represent quantities that take on values from an interval, such as $0 \leq x \leq 10$ or $-501.6 \leq y$. How many values are between 0 and 1? How do you separate them? Is 0.5 different than 0.50000001? An example of a continuous quantity is the total production time for a sequence of jobs or the average time it takes a rocket to enter near space once the launching rockets have ignited. Mixed variables take on values from both countable and uncountable sets.

4.5.2 The Model Toolbox

Several different tools can be used to fix one board onto another. A nail gun uses compressed air to drive the nail and is ideal for applications requiring many nails.

However, it may be too expensive or take too much time to set up to drive a single nail. A claw hammer is relatively inexpensive, but may require some expertise to use without bending the nail or damaging the wood when missing the nail. If neither of these quality defects is a concern, a heavy wrench or even a rock could be used to drive the nail. All of these tools are options for accomplishing the objective. Each has its advantages and disadvantages, which must be considered when choosing which one to use. The same is true of modeling tools.

This subsection surveys the modeler's toolbox and presents some of the options available when deciding which type of model to use. To aid in understanding how these tools relate to each other, the toolbox is divided into 16 compartments based on the characteristics previously presented (see Table 4.1). Organizing the toolbox this way should help you determine the appropriate tool to use. Similar to the Myers–Briggs Type Indicator, the "personality" of each compartment will be represented by a four-letter code. The sequence represents Static (S) or Dynamic (D), Deterministic (D) or Probabilistic (P), Descriptive (D) or Prescriptive (P), Discrete (D) or Continuous (C).

Ideally, this section would provide a thorough presentation of the tools in each compartment of the toolbox and how to use them. However, since entire courses are dedicated to accomplishing this task for many of these tools, we will, instead, give examples of the problems associated with each compartment, list some of the appropriate tools and suggest helpful references. References that survey many of the compartments include references 11–13. An online source for all things related to operations research and management science is reference 14. Two good references for using spreadsheets are references 11 and 15.

The majority of the following discussion concerns example problems containing characteristics associated with each of the compartments. To further explain each of the compartments, consider the following problem setting. A news vendor sells newspapers on a street corner. The newspapers are purchased from the publisher at $0.50 and sold for $0.75. This is called the news vendor problem. We will look at different versions of the problem as we explore each of the compartments.

The first two compartments describe problem characteristics that should be familiar to you. Arithmetic, algebra, and geometry are key tools used throughout the course of secondary and college education. Spreadsheets have helpful built-in functions that automate many of the associated mathematical tasks and are frequently used for large models [11, 15].

TABLE 4.1   The Modeler's Toolbox

		Static		Dynamic	
Models		Deterministic	Probabilistic	Deterministic	Probabilistic
Descriptive	Discrete	1-SDDD	5-SPDD	9-DDDD	13-DPDD
	Continuous	2-SDDC	6-SPDC	10-DDDC	14-DPDC
Prescriptive	Discrete	3-SDPD	7-SPPD	11-DDPD	15-DPPD
	Continuous	4-SDPC	8-SPPC	12-DDPC	16-DPPC

Compartment 1. SDDD (Static, Deterministic, Descriptive, Discrete). For the news vendor problem, what will the profit be if the news vendor purchases 50 newspapers and sells 35 of them? This problem is static; the context is a specific point in time. The data are discrete since the purchases and sales are in whole newspapers. The data are also deterministic because all the quantities are known and fixed. Finally, the problem is descriptive. No decision is needed.

Recall the company that has a development budget and a large number of potential products, each of which has a development cost and a potential profit. Given a subset of the products, compute the total development cost and potential profit for the subset. A spreadsheet model could be used.

Compartment 2. SDDC (Static, Deterministic, Descriptive, Continuous). Some simple examples include computing the number of gallons of water needed to fill a pool that is 15 feet long, 10 feet wide, and 4 feet deep or computing the monthly payment for a $1000 loan at 10% interest compounded annually and amortized over 10 years. The formulas $V = IR$ and $F = ma$ are other examples from this compartment. Suppose the time it takes until the next newspaper sells is 1.5 times the time it took the last newspaper to sell. If it takes 0.1 hour to sell the first newspaper, then what is the total time needed to sell 10 papers? The quantity of interest is now continuous rather than discrete.

Compartment 3. SDPD (Static, Deterministic, Prescriptive, Discrete). As an example of a problem associated with this compartment, let us consider the company that has a development budget and a large number of potential products, each of which has an estimated development cost and a potential profit. Which products should be chosen to maximize the potential profit? A prescriptive solution is needed to identify the best combination of products. Prescriptive problems are usually complicated by constraints. Here, the development budget limits or constrains the possible combinations of products.

Suppose the news vendor also has the option of selling the magazine *Systems Engineering Today*, which can be purchased for $0.75 and sold for $1.50. Suppose up to 20 customers will purchase only a newspaper, 15 customers will purchase both a newspaper and magazine, and five customers will purchase only a magazine. If the news vendor has $15.00 to purchase supplies, how many newspapers and magazines should she purchase to maximize her profit? The problem is now prescriptive instead of descriptive.

A similar problem structure can be seen in a crew scheduling problem. An airline wishes to schedule its pilots and flight attendants for the next month. There are many conditions that the schedule must satisfy. There are union requirements for the number of hours a crew member can be on duty per day or away from their home station. Each crew member's next flight must leave from the destination of their last flight (or they need to be flown as a passenger to their next starting point). The crew members submit work requests, and their seniority must be considered when constructing the schedule. Costs increase if crews stay overnight away from their home stations or fly too many hours in a day. Of course, it is more costly to have flights without an available crew! The airline wants a schedule that is feasible (i.e., satisfies all the conditions) and has the lowest cost (or at least lower than most of the other schedules).

This is a complicated problem, why does it belong in this compartment? First, it is discrete. A solution is a schedule assigning crew members to flights. Each possible schedule could be numbered and counted. The number of solutions is countable and, therefore, discrete. The airline wants the schedule with the lowest cost. Finding a better solution or the best solution requires a prescriptive model. There is no randomness in the problem, so it is deterministic. Finally, the problem is static; the solution does not change over time. The schedule is only for one month.

Small problems in this compartment can be solved by enumerating all the possible solutions, using a descriptive model to evaluate each one, then choosing the best one. Large problems of this type require discrete optimization tools. It is often difficult to find the best possible solutions for very large problems, so heuristics are often used. These are algorithms that find solutions that are good, but not necessarily the best. Some discrete optimization topics include combinatorial optimization and integer programming. References for tools in this area include references 12 and 13. Note that many prescriptive tools include the term programming (e.g., integer programming or linear programming). In this context, programming means planning. A concert program (or play list depending on the type of concert) is a plan for the music to be performed.

Compartment 4. SDPC (Static, Deterministic, Prescriptive, Continuous). Interestingly, continuous problems are often easier to optimize than discrete ones. Suppose a quantity of interest is described by the formula $y = (x - 3)2$. What value of x gives the minimum value of y? Calculus can solve this problem analytically. Numerical techniques include line-search algorithms such as the bisection search or the golden mean search.

There are many different types of problems in this compartment. The function to be optimized may be linear or nonlinear. There may or may not be constraints. If there are, they may be linear, nonlinear, or both. As a result, many different tools have been developed to model and solve these types of problems.

Linear programming is a tool that can very efficiently solve some problems in this compartment. The function to be optimized and the constraints to be satisfied must all be linear functions. The simplex method [16–18] is an efficient algorithm to find optimal solutions for linear programs and is now widely available in spreadsheet packages such as Microsoft® Excel.

As an example, consider the following product mix problem. A gravel company has two types of aggregate mixes it sells. Each consists of a different ratio of two component aggregates and has different profit margins. The quantities of component aggregates are limited. If the company can sell all it produces, how much of each mix should be produced to maximize profit? This is the continuous version of the news vendor problem with newspapers and magazines. Instead of whole numbers of newspapers and magazines, the quantities can be in any amount. The problem is still static and deterministic. It is also prescriptive since the best solution is needed—one that will maximize profits.

A primary difficulty faced with most nonlinear problems is distinguishing between *local optimal* solutions and *global optimal* solutions. Figure 4.9 shows a function of x that has several points where the first derivative is zero and the second derivative is positive. These points are local optimal solutions (for a minimization problem); they are better than the nearby points that surround them. A global optimal solution is a local optimal solution with the best value. There are a number of techniques to identify local optimal solutions [19, 20]. *Metaheuristic techniques* [21] such as tabu search, simulated annealing, genetic algorithms, and generalized hill climbing try to move past the local optimal solutions to find the global optimal solutions.

These metaheuristic algorithms are flexible prescriptive tools and are appropriate (with varying degrees of effectiveness) for any compartment with a prescriptive element.

Compartments 5 through 8 are similar to Compartments 1 through 4, respectively, except one or more of the input variables or parameter values are now uncertain quantities. They are random variables with a probability mass function (if they are discrete) or a probability density function (if they are continuous). This randomness could be the result of measurement error or the inability to know with certainty what will happen. One approximation technique is to model and solve the problem as a deterministic one by fixing each random variable at its expected value. This approach can be misleading because the result is typically not the expected value of the solution.

Analytical techniques may be available for these compartments depending on the type of distributions and relationships involved (e.g., sums of normal random variables) [22, 23]. If the discrete random variables have only a few

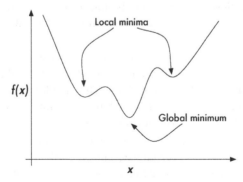

Figure 4.9   Local and global minima of $f(x)$.

possible values, probability tree diagrams can be a useful tool [16]. Statistical tools such as hypothesis testing or regression analysis can be helpful [24, 25]. An easy-to-apply numerical technique is Monte Carlo simulation [11]. Suppose the distribution functions of two random variables X and Y are known and the distribution of $Z = XY$ is needed. The functional relationship is specified in a program or a spreadsheet. Random number generators produce values for X and Y, which are substituted into the function. The resulting value of Z is saved. The process is repeated, perhaps millions of times. The resulting data represent the distribution of Z. Spreadsheet add-ins for this technique include @Risk® and Crystal Ball®.

Compartment 5. SPDD (Static, Probabilistic, Descriptive, Discrete).

Compartment 6. SPDC (Static, Probabilistic, Descriptive, Continuous). The previous versions of the news vendor problem were deterministic and did not consider the fact that the number of newspapers that can be sold on a particular day is not known when the vendor buys her newspapers. Demand is uncertain and dependent on factors such as the weather, local traffic conditions, and world events. The vendor could make more profitable decisions about the number of papers to order if uncertainty were reduced. It may be possible to use a probability distribution to quantify this uncertainty and determine probabilities for the number of newspapers wanted on a particular day. If so, an example of a discrete descriptive problem is to compute the number of newspapers the news vendor must buy so there is at least a 95% probability that all will be sold that day. A continuous descriptive problem is to determine the probability the news vendor has unsold newspapers at the end of the day if she buys 25 newspapers. Another example is to find the average time it will take to sell 10 papers if the time between sales is random with a given distribution. If an airplane has 50 seats, an airline wants to know the probability that more than 50 people show up for the flight if 55 tickets are sold.

Compartment 7. SPPD (Static, Probabilistic, Prescriptive, Discrete).

Compartment 8. SPPC (Static, Probabilistic, Prescriptive, Continuous). Prescriptive problems seek the best answer among a number of alternatives. The vendor wants to know how many papers to purchase to maximize expected profit given a distribution for daily demand. Linear programs with random coefficients are an example of a continuous problem in this compartment. Decision analysis under risk or uncertainty [16] and stochastic programming [26, 27] are probabilistic prescriptive tools for Compartments 7 and 8. Monte Carlo techniques can also be used.

The remaining compartments are dynamic; the model parameters or outputs change, typically over time.

Compartment 9. DDDD (Dynamic, Deterministic, Descriptive, Discrete).

Compartment 10. DDDC (Dynamic, Deterministic, Descriptive, Continuous). Physics examples of continuous systems are falling bodies, spring systems, and voltage across capacitors or current through inductors in electrical circuits. Calculus and differential equations are often-used tools for these systems. Fourier and Laplace transforms can also be helpful. These systems are deterministic. Engineers and scientists planning a space mission can predict where the planets will be years in advance; there is no uncertainty.

In discrete-time systems, time passes in discrete increments (e.g., whole hours or days). Discrete-time systems are common if digitized signals are involved. For example, a digital cell phone samples the speaker's voice 8000 times per second. Each sample results in an 8-bit codeword that is transmitted through the telephone network. The voice information is converted from analog (continuous information) to digital (discrete information). Since the information is periodic (one codeword every 1/8000 seconds), the network is synchronized with high-speed transmission channels interleaving codewords from hundreds or thousands of conversations. The listener's phone receives a codeword 8000 times per second. The phone converts the codeword (discrete information) into an analog (continuous information) signal that is (usually) heard by the listener. Compact disc and DVD players also receive periodic (i.e., one codeword every fixed time interval) digital data from the disc, which is converted into an analog signal for playback. Coding systems often attempt to predict the next codeword to reduce the number of bits needed to carry information. This prediction is an example of a dynamic, deterministic, descriptive discrete system.

Difference equations, discrete-time Fourier transforms, and z-transforms are primary tools for discrete-time systems [28, 29]. Electrical engineers will become intimately familiar with these in courses on digital signal processing, digital communications systems, and digital control systems.

Compartment 11. DDPD (Dynamic, Deterministic, Prescriptive, Discrete).

Compartment 12. DDPC (Dynamic, Deterministic, Prescriptive, Continuous). Continuous and discrete-time control systems are examples of these compartments [29, 30]. Consider a missile locked onto a target. The target may move and winds may deflect the missile. A prescriptive solution that changes over time is needed to guide the missile to its target. Another example is an automobile cruise control. A target speed is set and the throttle is adjusted to meet the target. Hills will affect the speed and the throttle must adjust to keep the target speed. These types of systems can be modeled in continuous time or, using sampling and digital electronics, in discrete time.

Compartment 13. DPDD (Dynamic, Probabilistic, Descriptive, Discrete).

Compartment 14. DPDC (Dynamic, Probabilistic, Descriptive, Continuous). Stochastic or random processes [23], discrete-event simulation [31] and time series analysis [32] are primary tools for systems with randomness that changes over time. These dynamic probabilistic systems are all around us: the number of people in line at the coffee shop throughout the day, the gas mileage over different segments of a trip home, whether or not the next person through the checkpoint is a smuggler. A company designing a new factory may wish to determine where the bottlenecks for material flow will be or predict the pick time for orders if the layout of a warehouse is changed. A continuous-time model represents the system at all points in time. A discrete-time model may only represent the system at fixed points in time (e.g., quarterly).

Compartment 15. DPPD (Dynamic, Probabilistic, Prescriptive, Discrete).

Compartment 16. DPPC (Dynamic, Probabilistic, Prescriptive, Continuous). A call center wants to determine staffing levels so there is a 95% probability an incoming call is answered by a human operator. Call volume changes over time and the company wastes money if too many operators are on duty, so the optimal number of operators changes over time.

The price of an airline ticket changes over time. The airline wants to maximize revenue while selling a perishable commodity; after the flight leaves, an open seat has no value. Suppose an airline has 100 seats on a flight from New York to Phoenix departing at 9:20 a.m. next February 5th. Consumer demand for tickets that will occur 8 months prior to departure will be different than the demand that will occur 1 month prior. Demand fluctuates over time, but can be influenced by price. If the airline sets the price too low, all the seats are sold months in advance. If the price is too high, the flight leaves with many empty seats. Neither solution maximizes revenue. The airline needs a pricing strategy that sets the best ticket price to maximize revenue conditioned on the time remaining until the flight departs and the number of seats that have been sold at different pricing levels. Finding the optimal price for each ticket over time is an example of a dynamic, probabilistic, prescriptive problem. This is a yield management problem and is common in the travel industry for companies such as airlines, hotels, rental car operators, and cruise lines. A model that sets the ticket price every day would be a discrete-time model; one that sets the price at the time a quote is requested would be a continuous-time model.

Tools for these compartments include probabilistic dynamic programming [16] and Markov decision processes [23]. Metaheuristic optimization techniques can also be used with stochastic descriptive models [21].

This section presented the "big picture" of modeling. It began with a discussion of what models are and why they are used. Qualities were introduced as a way to evaluate or compare models. A sample modeling process was provided that included the different aspects of building a model. Finally, model characteristics were used to classify different types of models.

It is important to remember that, just as "a map is not the territory it represents," a model is not the system it represents—"all models are wrong." They have limitations, drawbacks, and faults, and unpleasant consequences can result if they are used haphazardly.

4.6 SIMULATION MODELING

The last section discussed how models can be relatively simple yet powerful tools often used by systems engineers. In the next section we will discuss another set of tools used by systems engineers and which employ models: simulations. If models are analogous to claw hammers used to pound in nails or screw drivers to install screws, then simulations can be thought of as nail guns and drills which allow a carpenter to drive in numerous nails and screws much more rapidly. As we will see, each set of tools, models and simulations, has its appropriate place in the systems engineer's toolbox.

In its simplest form, a *simulation* is a model operated over time. In more mathematical terms, it is the numerical execution of a model to see the potential effect of the variables in question on the model's output [31]. In other words, a simulation is the execution of a model over time to produce varied outputs for use in analysis of the system.

A simulation is a model operated over time.

We use simulations to see the potential effects of time or some other variable(s) on the system a model is representing. Often, vast amounts of data representing the states of a system need to be generated in order to determine the statistical significance of change on the system. This data may be too difficult or costly to collect on the actual system in abundance to calculate the most probable effects of modifications to the system. A simulation of a viable model of a system can be used to generate this data.

4.6.1 Analytical Solutions versus Simulation; When It Is Appropriate to Use Simulation

It is appropriate to use simulation when a system is sufficiently complex that the possibility of a simple analytical solution is unlikely [31]. Many times we model systems which on the surface appear simple. However, once we begin to alter variable values, the mathematical representation or model becomes too complex to solve using a closed form analytical solution.

In general, choosing to use a simulation is based on six factors [33]:

- An operational decision based on a logical or quantitative model is needed.
- The system being analyzed is well-defined and repetitive.
- System/model activities, variables, and events are interdependent and change over time or condition.
- Cost/impact of the decision is greater than the cost of developing the model, executing the simulation, and assessing the data.
- Experimenting with the actual system costs more than developing the model, executing the simulation, and assessing the data.
- System events occur infrequently, not allowing for the system to achieve a steady state of performance where multiple adjustments to the system events can be made and assessed in an attempt to create a more efficient system.

The first three factors address system characteristics and structures. The next two address resource concerns. The last factor deals with the frequency of system usage. The first five factors can be used to help determine if a simulation is appropriate for assessing the alternatives generated in the solution design phase of the SDP. The last factor is a possible mitigating factor used to balance the risk of making or not making changes to a system which is used infrequently.

4.6.2 Simulation Tools

Analytical Solutions versus Simulation Simulation tools are categorized in different ways depending on their intended use. In this section, we will present both industry and military simulations to provide the broader scope of preexisting simulation tools, which currently exist and need not be developed. They do require the development of the scenarios but the basic simulation engine exists alleviating the need to program a theoretical model. Also, their user interfaces allow the rapid development of a simulation scenario. From industry, we will limit our discussion to ProModel®. With regard to military simulations, we will describe various military categories of simulations and provide a general overview of some of the existing simulations used in the analytical community.

Complex Queuing Simulations (ProModel®) ProModel® is a discrete-event simulation tool used by industry to model manufacturing, transportation, logistics, and service related systems. It allows systems to be represented as a series of queues and servers in which an entity, typically a customer product being produced/processed, is either being serviced or awaiting service. The simulation tool allows one to test various alternative layout designs and service processes prior to implementing the layout or process in the system. This tool has the fidelity to model "resource utilization, production capacity, productivity, inventory levels, bottlenecks, throughput times, and other performance measures" [33].

ProModel® is used by many industries to simulate their manufacturing and/or transportation system(s). ProModel® can also be used to simulate military systems. Figure 4.10 is an example of a student-lead systems engineering project to simulate alternative solutions for a proposed missile defense system. Similar simulation tools currently available on the market are Arena® and AutoMod.

Military Simulations *Military simulations* are normally used by the U.S. Department of Defense to simulate a myriad of aspects related to military operations and systems. These include, but are not limited to, issues related to: environmental factors, vehicle performance, communications capabilities, weapons effects, human behavior, and so on. There are fundamentally four categories of military simulation: simulation typologies, real-time versus non-real-time simulations, hierarchy of models and simulations, and military simulations versus games. These categories are illustrated in Figure 4.11.

Simulation Typologies. There are four *simulation typologies*: live, virtual, constructive, and augmented. The right side of Figure 4.11 is a visual depiction

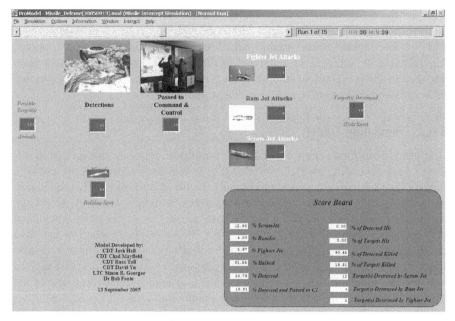

Figure 4.10   ProModel® anti-ballistic missile simulation example [34].

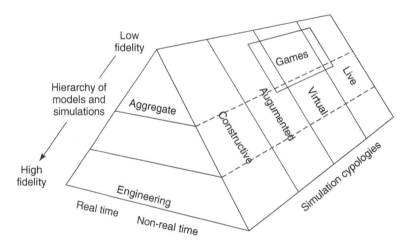

Figure 4.11   Simulation types [35].

of the relationship between the first three simulation typologies and games. *Live simulations* involve real people using real systems in a simulated environment. An example of a live simulation is an emergency response team conducting a full dress rehearsal on the launch pad prior to the launch of a manned rocket. *Virtual simulations* include the additional complexity of real users interacting with the simulated equipment. An astronaut conducting

		Environment	
		Real	Synthetic
People	Real	Live	Virtual
	Synthetic	Augmented	Constructive

Figure 4.12 Simulation–reality relationships.

flight training in a simulator is an example of a virtual simulation. Each typology can be considered in terms of people and environment. Each may be either real or synthetic, as shown in Figure 4.12. A common augmented reality technique is to display computer-generated people in optical systems through which real people look. This allows the use of real weapons against synthetic criminals or combatants in a training environment without the risk of bodily harm or death.

The third typology is *constructive simulation*. In this typology, simulated entities operate in a simulated environment. Constructive simulations are normally used to simulate activities in rapid succession in order to provide insight into emerging system behavior or possible outcomes based on changes in equipment or processes. Agent-based simulations (see Section 4.6.2) of human behavior or multifaceted relationships between entities and which replicate hundreds, if not thousands, of entities interacting in a complex virtual environment are examples of constructive simulations. For example, NASA could use a constructive simulation to assess the effect(s) of asteroid activity on a space station. It could use other constructive simulations to predict the oxygen exchange rate in a space station based on the predicted growth of vegetation in the station, air filtering capacity of the ventilation system, and the number of personnel in the station.

Real-Time versus Non-Real-Time Simulations. Simulations can be categorized based on their ability to provide "real-time" performance or "non-real-time" performance. *Real-time* simulations are usually associated with virtual simulations that provide human participants with timely information, reactions, and effects. *Non-real-time* simulations are usually associated with constructive simulations that run uninterrupted for their duration. During solution design and decision-making phases of the SDP, systems engineers frequently use non-real-time simulations to generate data for use in analysis of alternatives. The use of non-real-time simulations allows users to conduct studies using the highest fidelity algorithms available with little concern for how long it takes to run the simulation because the system activities modeled in this type of simulation are not constrained to actual clock time. Non-real-time simulations can work faster or slower than actual clock time. A real-time simulation could be used to conduct flight training or to test the reaction times of emergency response teams. A non-real-time simulation could be used to identify the disturbance of the air flow (lift and drag) as it streams over the fins of a new rocket design.

Hierarchy of Models and Simulations. Military simulations are also categorized according to the simulation's level of *fidelity* (see left side of Figure 4.11). The greater the fidelity, normally the greater the detail of the item(s) you are attempting to simulate. For example, a low-level fidelity model of a rocket system might represent the rocket as a solid object where a higher fidelity model might include the fins, fuel system, guidance system, and load capacity of the rocket. Fidelity can be seen as a continued scale but is routinely organized into smaller groupings to make it easier to group simulations into consistent hierarchy categories. Some organizations place simulations into a hierarchy consisting of five levels: campaign, theater, mission, engagement, and engineering [36]. The Army Modeling and Simulation Office's publication, "Planning Guidelines for Simulation and Modeling for Acquisition Requirements and Training," describes a four-level hierarchy where the hierarchy combines theater and campaign into one level [37]. Hughes refers to these four levels as campaign or theater, battle or multi-unit engagement, single engagement, and phenomenological [38].

The front triangle of Figure 4.11 is an illustration of a three-tier *hierarchy*: aggregate, entity, and engineering. The *aggregate* level consists of simulations or models that combine entities into groups based on functionality or association. Examples of these are boxes of gages or reels of wire in a rocket manufacturing plant warehouse or the payload capacities of the rockets, respectively. The aggregation of entities helps to reduce the computational requirements for modeling and simulating large-scale scenarios. Aggregate simulations routinely have less fidelity than is normally found at the entity and engineering level simulations. Aggregate simulations would represent a group of vehicles, such as a convoy, as it moves packaged goods from one side of a country to another.

Entity-level simulations reside between aggregate and engineering level simulations. Entity-level simulations represent individual platforms and the effects created by or acting on them. An entity is a platform, system, product, or individual that is creating effects or receives the effects of the system, such as the entity in ProModel®. As such, a "distinguishable person, place, unit, thing, event, or concept" simulation must maintain or track information about each entity in the model [39]. An entity-level simulation of an asteroid field would replicate each individual asteroid, its relevant behaviors, and positional data as a rocket passes through the asteroid field.

Engineering-level simulations normally deal with the components of one or two individual systems or subsystems. These simulations may go into more detail (higher fidelity) to model the physical aspects of a system. Modeling the subcomponents, components, and behaviors of a rocket's guidance system is an example of an engineering-level simulation. Organizations use this form of modeling and simulation prior to building a new piece of equipment or integrating new technologies into existing facilities or systems before testing the changes using live simulations.

Systems engineers choose the appropriate level of simulation fidelity based on what aspect(s) of a system they are assessing.

Simulations versus Games. The military and industry produce numerous simulations to provide training for their people or to conduct analysis of operations/systems. The needs of the research and training communities place different constraints on the validity of the model or simulation. However, in the gaming industry the overriding concern is to manufacture a product people are interested in purchasing and playing. A notable exception is the game America's Army™, which was originally built as a recruiting tool for the U.S. Army.

Games are "activit(ies) engaged in for diversion or amusement" [40]. As such, the focus of the gaming industry has been to produce simulations which deliver entertainment value to its customers; however, developing or adapting games for use in education or training is becoming more popular. The belief that individuals will learn more if they are actively engaged in the learning activity is one of the fundamental reasons for using games. Most computerized games are simulations primarily designed for entertainment but may be used for learning, individual or team training, or analytical studies seeking to gain insight into human behavior.

One can categorize games as aggregate or entity-level simulations. Aggregate games are strategic in nature (e.g., Risk®, Axis & Allies, Railroad Tycoon™, Kohan II: Kings of War, etc.). In contrast, first shooter and role-playing games can be considered entity-level games (e.g., America's Army®, Air-Sea Battle, Alex Kidd in the Enchanted Castle, etc.). No matter which category of simulation a game resides, game developers' primary concern is entertainment not realism or the accurate portrayal of human performance during the design and coding of a game. This lack of realism can place games at the lower end of the fidelity spectrum (see Figure 4.11).

Simulation Behaviors A simulation executes based on the prescribed set of *behaviors* outlined for each entity. These behaviors can be complex or very simple. They can be reactive or proactive in nature. Simulation behaviors are most often limited based on the model architecture used to represent these behaviors. Generally speaking, the five most popular cognitive model representations in use in the late twentieth and early twenty-first centuries are agent-based, Bayesian-network, multiagent system, neural-networks, and rule-based. We will only address rule-based, agent-based, and multiagent representations in this section.

Rule-Based Simulations. A *rule-based* (knowledge-based) simulation replicates human behavior using a catalog of actions with causal if/then association to select and execute an appropriate action [41, 42]. This causal representation often requires an extensive effort to identify and code all relative possible conditions an entity may encounter along with viable entity actions for those conditions. Subject matter experts are routinely used to establish and validate these data prior to its use. Rule-based simulations are best used to model systems, which are physics-based, or for replicating systems with a relatively limited (countable) number of states and actions.

Agent-Based Simulations. Agent-based representations model intelligence through codified objects that perceive characteristics of the environment and act on those perceptions [41]. There are several types of agent-based cognitive architectures. Two of these are *reactive* and *rational* agents. A reactive agent bases its actions solely on the last set of sensory inputs. Often the approach uses a simple condition action rule (e.g., if this is my perceived state of world, then I choose this action). A rational agent uses sensors to perceive its environment and performs actions on the environment using effectors. Rational agents maintain a state of situational awareness based on their past knowledge of the world and current sensory inputs [41].

Multiagent System Simulations. The *multiagent system* (MAS) is a relatively new representation for modeling and simulating behaviors based on the complex adaptive system (CAS) theory. Developed in the late 1970s, MAS is a system with autonomous or semiautonomous software agents that produce adaptive and emergent behaviors. The model uses a bottom-up approach where software agents have independent micro decisions that generate group and system-level macro behaviors. A MAS can use any form of agent-based software technology (reactive, rational, goal-based, utility-based, etc.) that has agents characterized as possessing intentions that influence their actions. Multiagent systems are used in large domains where nonlinearity is present [43]. The MAS, limited only by the physics constraints of the simulation boundaries, uses an indirect approach to search the large domain for viable results. Another feature of MAS is its ability to allow agents to evolve to create new agents which, in general, are better suited to survive or prosper in the simulated environment [44].

Agent-based and MAS simulations are often used to explore a wide spectrum of possible system effects based on an extensive range of variables inputs. This allows systems engineers to assess the likelihood of possible system-level behavior in the context of various system constraints. Agent-based tools often lack the higher fidelity levels found in most physics-based simulations. However, these lower fidelity agent-based models allow for more varied behavior interactions. This makes them highly useful for simulating system alternatives in which complex, nonlinear interactions can occur between entities that are difficult, if not impossible, to specify in the closed-form expressions required of physics-based models and simulations.

4.7 DETERMINING REQUIRED SAMPLE SIZE

A convenience of deterministic models is that given a known set of inputs, the same output is ensured. For example, using $E = mc^2$ with a given mass, if the speed of light was unknown or uncertain and consequently represented by a random variable between 299 and 300 million meters per second(mmps), instead of a constant 299,792,458 mmps, then a single calculation would not necessarily predict the true amount of available energy. Instead, some number of calculations, each with a different value for the random variable, would be required to gain a certain level of confidence that the true value had been captured. These calculations, interchangeably called *replications*, *trials*, or *runs*, may significantly add to the cost of an experiment, so the modeler needs a mechanism to understand the trade-offs between replication cost and certainty. Deterministic models of complex systems may also benefit from testing only a sample of the possible combinations instead of the entire population.

Two methods are usually used to determine a reasonable number of replications, the *required sample size*, on which to base an estimate. The first, *power analysis*, is an approach that determines how much "power" a test has to detect a particular effect. It is often referred to as the *power of the test*. This approach is not explored in this chapter.

The second method is based on the logic of a statistical confidence interval (CI). In this case, the goal is to have a certain degree of confidence that the true population mean has been captured by determining the sample mean, given a certain number of observations and an acceptable probability of error. The formula for determining a confidence interval is shown in Equation (4.3) as

$$CI_n = \bar{x}_n \pm t_{n-1,1-\alpha/2}\sqrt{\frac{s_x^2}{n}}. \tag{4.3}$$

This says that we can believe, with only an α probability of error, that the true mean is the same as the sample mean, $\bar{x}_n$, give or take some margin of error. The margin of error—everything to the right of the $\pm$ symbol—is made up of the variance of the sample, s_x^2, a desired level of confidence, $1 - \alpha$, and the t statistic for $n - 1$ degrees of freedom where n is the sample size.

For example, a health inspector may take 10 samples of drinking water to measure the amount of a contaminant and records the following levels, in parts per million (ppm):

$$29, 34, 20, 26, 20, 35, 23, 30, 27, 34.$$

The mean, $\bar{x}_n$, is found to be 27.8; the variance, s_x^2, is 31.51; n is 10; and α is 0.10 for a 90% confidence interval. Inserting these numbers into Equation (4.3), we can expect the true mean value of contaminant in the drinking water to be 27.8 ± 3.254, or an interval of between 24.546 and 31.054 ppm.

Since everything to the right of the $\pm$ symbol represents margin of error, the number of samples required, n, can be calculated if the modeler is willing to specify that error up-front. This is done routinely, for example, when someone declares some value, "give or take a couple." This is the same approach taken in power analysis when the modeler selects a level of effect to be detected. The modeler is saying, "I want to know how many samples/trials/replications/runs are required to detect a certain margin or error." These calculations can be done easily in a spreadsheet, although the modeler must rely on experience and judgment to avoid certain pitfalls. Specifically:

- Select a meaningful measure to use for sample size estimation.
- Generate a reasonable number of pilot runs to establish a solid mean and variance.
- Beware of the assumption that the pilot variance represents the true variance.
- Always round up the estimated n to the next higher number. Err on the side of caution.
- Generate an estimated n from the pilot runs, then incrementally increase the actual runs, rechecking the estimate for n, until the estimated n equals the actual n and your professional judgment is comfortable with the results and they are defensible to a decision maker.
- The larger the n, the greater the likelihood that the sample mean represents the true mean. Find the balance between replication cost and result accuracy.

Consider a golf club manufacturer interested in the performance of a new club. A swing machine hits 10 balls that fly the following distances, in yards:

$$220, 210, 189, 201, 197, 200, 205, 198, 196, 200$$

Engineers want to use this series of pilot replications to determine how many balls need to be hit to find the mean distance, plus or minus two yards. By saying that, they acknowledge that they are willing to have a margin or error of two yards. Since it has been shown that the margin of error is represented in Equation (4.4), the engineers can set the margin of error equal to two and solve for n, as in Equations (4.5) through (4.9):

$$\text{Margin of error} = t_{n-1,1-\alpha/2}\sqrt{\frac{s_x^2}{n}} \tag{4.4}$$

$$2 = 1.833\sqrt{\frac{72.27}{n}} \tag{4.5}$$

$$\Rightarrow 2 = \frac{(1.833)\,(8.5)}{\sqrt{n}} \tag{4.6}$$

$$\Rightarrow \sqrt{n} = \frac{(1.833)\,(8.5)}{2} \tag{4.7}$$

$$\Rightarrow n = \left(\frac{15.58}{2}\right)^2 \qquad (4.8)$$

$$\Rightarrow n = 60.68 \qquad (4.9)$$

Hitting 61 more golf balls is not likely to be a problem, unless they replace the swing machine with a professional golfer like Tiger Woods, whose per-swing fee may be very costly to the company. In that case, they may have Tiger hit another 10 balls and recalculate the required n. They can continue to incrementally move toward the 61-ball goal until the revised n equals the actual n. With each increment, the recorded variance approaches the true variance for the club and is therefore a better representation of reality.

Although this approach provides only an estimate, it also provides the modeler with an easy and defensible tool for determining how many replications are required to achieve a desired result.

4.8 SUMMARY

Modeling a system's essential components, attributes, and relationships, and understanding how they change over time, provides systems engineers with critical insights into all stages of the system's life cycle. While Chapter 2 appropriately frames the system thinking perspective on how systems can be thought of and represented, this chapter has introduced tools that reveal the inner workings of a system. Chapter 4 highlights the fundamentals of system measures, models, simulations, and required sample size to an understanding of the life cycle. It starts with three key reasons for understanding the system life cycle, all of which are supported directly with the tools and techniques from this chapter. The three motivating reasons are as follows:

1. We can organize system development activities in a logical fashion that recognizes some activities must be accomplished prior to others.
2. We can identify the specific activities needed to be accomplished in each stage to successfully move to the next stage.
3. We can effectively consider the impact that early decisions have on later stages of the systems life cycle, especially with regard to cost and risk.

Knowing which tools to use at given stages of the system life cycle is part of the art of systems engineering illustrated throughout this book. There is rarely a single right answer, though there are always many wrong ones. The system modeler must always keep in mind that as abstract representations of objects or processes, models at best provide only insights into understanding a system and at worst lead decision makers astray. The opening quote of this chapter, "All models are wrong, some are useful," must never be far from the systems engineer's mind.

4.9   EXERCISES

4.1. Use model qualities to compare and contrast the following models used at a fast food restaurant to order burger supplies for the next week:

    (a) Fit a distribution to weekly data from the past year, then order an amount that would satisfy demand for 95% of the weeks.

    (b) Order supplies based on how many burgers were sold last week.

4.2. Use model qualities to compare and contrast the following models used to predict gas mileage for an automobile:

    (a) Randomly choose a number of different destinations within 100 miles of your present location. Gather experimental data by driving from your location to each destination and back. Record the average speed and average mileage. Fit a line to the data.

    (b) Develop a formula to predict gas mileage by using formulas for kinetic energy and chemical–mechanical conversion with assumptions about wheel bearing friction and engine torque.

4.3. Describe an example of a robust and a nonrobust system or model. Justify your labels.

4.4. Describe an example of a high-fidelity and a low-fidelity system or model. Justify your labels.

4.5. Describe an example of a system or model that is dynamic and deterministic.

4.6. Describe an example of a system or model that is deterministic and probabilistic.

4.7. Describe an example of a system or model that is static and predictive.

4.8. Describe an example of a system or model that is descriptive and dynamic.

4.9. A security checkpoint screens passengers entering the passenger terminal of an airport. List the probabilistic elements in the system. Describe how you could make these elements deterministic.

4.10. For each of the following problems, categorize the following system or model using each of the four characteristics (i.e., the modeler's toolbox).

    (a) A rubber duck floating in a bathtub.

    (b) A NASCAR pit stop.

    (c) The latest version of the computer game Halo.

    (d) $E = mc^2$.

    (e) The Illinois Lottery Mega Millions game.

    (f) The Federal Emergency Management Agency (FEMA) wishes to develop a model to predict the resiliency of a metropolitan area; how quickly an area can return to normal following a natural or man-made disaster.

    (g) A state department of transportation wants to determine how long the left-turn lane should be at a busy intersection.

(h) An artillery officer needs to determine the angle of launch for a given type of artillery round.

(i) The U.S. Air Force must determine which planes to fly which routes to which destinations. Each type of plane has its own range, speed, load size, and weight capabilities. The location of supply bases and the locations of receiving bases and their materiel requirements are given.

(j) The FEMA is pre-positioning disaster relief supplies in regional areas. They need to determine where these supplies should be located so they can be quickly deployed in an emergency. The likelihood of different types of disasters (e.g., floods, hurricanes, earthquakes, tornadoes) and their severity must be considered. The budget is limited and FEMA wants to maximize the impact they have following a disaster.

(k) An investment company has $1 million to invest in five different asset classes (e.g., short-term treasury notes, stocks, bonds, etc.). The company must consider the return and risk of each asset class. Return is usually measured as the expected value of the return over some time horizon (e.g., a year). Risk is usually measured as the standard deviation of return over the same time horizon. The company wants to maximize its return while limiting risk to a set level.

4.11. What is the difference between a model and a simulation?

4.12. Why would a systems engineer use a model?

4.13. List and describe the four qualities you feel are the most important for a model to be useful. Why did you pick these four?

4.14. List and describe the three types of models. Give examples of each (different than those found in this chapter).

4.15. What are the four major steps in the modeling process?

4.16. List the three means of exercising a model and explain how they differ.

4.17. What is the difference between a deterministic and a probabilistic model? Give an example of each using a fast-food restaurant as your system.

4.18. When and why is it appropriate for a systems engineer to use a simulation?

4.19. List and describe the three simulation typologies. Give an example of each using a fast-food restaurant as your system.

4.20. Which of the three simulation hierarchies would be used to describe a model of traffic flow of individual vehicles down an interstate highway? Why?

4.21. What is the difference between a game and a simulation? Can a game be a simulation? Why?

4.22. Construct a seven-paragraph measure of effectiveness for an unmanned aerial vehicle (UAV) designed for U.S. border surveillance.

4.23. Pilot runs of a simulated emergency response system results in response times of 25, 42, 18, 36, 28, 40, 20, and 34 min. Calculate the required sample size necessary to be 90% confident that the sample mean represents the true mean, given or take 3 min.

REFERENCES

1. Sage, AP, Armstrong, JE, Jr. *Introduction to Systems Engineering*. Wiley Series in Systems Engineering. New York: Wiley-Interscience, 2000.

2. *The Measures of Effectiveness*, USACDC Pamphlet No. 71–1. U.S. Army Combat Developments Command, Fort Belvoir, VA, 1973.

3. http://www.wrightflyer.org/WindTunnel/testing1.html. Accessed April 4, 2010.

4. Porter, N. 1996. The Wright Stuff. WGBH Boston. Available at http://www.lilienthal-museum.de/olma/ewright.htm. Accessed April 4, 2010.

5. http://www.importanceofphilosophy.com/Esthetics_Art.html. Accessed April 4, 2010.

6. http://pespmc1.vub.ac.be/OCCAMRAZ.html. Accessed April 4, 2010.

7. Simulation Interoperability Standards Organization (SISO), Fidelity Implementation Study Group (ISG). Fidelity ISG Glossary. V 3.0, 16 December 1998.

8. "A Glossary of Modeling and Simulation Terms for Distributed Interactive Simulation (DIS)," Office of the Under Secretary of Defense (Acquisition and Technology), Washington, DC. August, 1995.

9. Department of Defense Instruction. 2003. DoD Instruction 5000.61: DoD Modeling and Simulation (M&S) Verification, Validation and Accreditation (VV&A). Available at https://www.dmso.mil/public/library/policy/policy/i500061p.pdf. Accessed June 7, 2006.

10. Aaronson, S. Guest column: NP-complete problems and physical reality. *ACM SIGACT News Archive*, 2005;36(1):30–52.

11. Ragsdale, C. *Spreadsheet Modeling and Decision Analysis*, 4th ed. Mason, OH: South-Western College Publishing, 2003.

12. Foulds, LR. *Combinatorial Optimization for Undergraduates*. Undergraduate Texts in Mathematics. New York: Springer, 1984.

13. Wolsey, L. *Integer Programming*. New York: Wiley-Interscience, 1998.

14. Institute for Operations Research and the Management Sciences. Available at http://www2.informs.org/Resources/. Accessed April 4, 2010.

15. Powell, SG, Baker, KR. *The Art of Modeling with Spreadsheets: Management Science, Spreadsheet Engineering, and Modeling Craft*. New York: John Wiley & Sons, 2003.

16. Hillier, FS, Lieberman, GJ. *Introduction to Operations Research*, 8th ed. New York: McGraw-Hill, 2005.

17. Taha, HA. *Operations Research: An Introduction*. 6th Ed. Upper Saddle River, NJ: Prentice-Hall, 1996.

18. Winston, WL, *Operations Research: Applications and Algorithms*, 4th ed. Belmont, CA: Duxbury Press, 2003.

19. Gill, P, Murray, W, Wright, MH. *Practical Optimization*. New York: Academic Press, 1981.

20. Nocedal, J, Wright, S. *Numerical Optimization*. New York: Springer-Verlag, 1999.

21. Glover, FW, Kochenberger, GA. *Handbook of Metaheuristics*. International Series in Operations Research & Management Science. New York: Springer, 2003.

22. Ross, S. *A First Course in Probability*, 6th ed. Upper Saddle River, NJ; Prentice-Hall, 2001.

23. Ross, S. *Introduction to Probability Models*, 8th ed. New York: Academic Press, 2003.

24. Devore, J. *Probability and Statistics for Engineering and the Sciences*, 6th ed. Belmont, CA: Duxbury Press, 2004.

25. Hayter, A. *Probability and Statistics for Engineers and Scientists*, 2nd ed. Belmont, CA: Duxbury Press, 2002.

26. Birge, JR, Louveaux, F. *Introduction to Stochastic Programming*. New York: Springer, 2000.

27. Spall, JC. *Introduction to Stochastic Search and Optimization*. New York: John Wiley & Sons, 2003.

28. Proakis, JG, Manolakis DG. *Discrete Signal Processing, Principles, Algorithms, and Applications*, 3rd ed. Upper Saddle River, NJ: Prentice-Hall, 1996.

29. Franklin, GF, Powell, JD, Workman, ML. *Digital Control of Dynamic Systems*, 3rd ed. Upper Saddle River, NJ: Prentice-Hall, 1997.

30. Norman, S, Nise, NS. *Control Systems Engineering*, 4th ed. New York: John Wiley & Sons, 2003.

31. Law, AW, Kelton, WD. *Simulation Modeling and Analysis*, 3rd ed. New York: McGraw-Hill, 2000.

32. Chatfield, C. *The Analysis of Time Series: An Introduction*, 6th ed. Boca Raton, FL: Chapman & Hall/CRC, 2003.

33. Harrell, C, Ghosh, BK, Bowden, RO. *Simulation Using ProModel®*, 2nd ed. New York: McGraw-Hill, 2000.

34. Foote, BL, Goerger, SR. Design considerations for simulating ABM systems. Presentation to Huntsville Simulation Conference, Huntsville, Alabama, 2005.

35. Goerger S. R. Validating Computational Human Behavior Models: Consistency and Accuracy Issues. Doctoral dissertation, Department of Operations Research. Monterey, CA: Naval Postgraduate School, 2004.

36. Combat Modeling Overview. Class notes, Introduction to Joint Combat Modeling (OA/MV4655), Operations Research Department, Naval Postgraduate School, Monterey, California, July 2002.

37. Army Modeling and Simulation Office (AMSO) Planning Guidelines for Simulation and Modeling for Acquisition Requirements and Training (Change 1). September 15, 2002. Available at http://www.amso.army.mil/smart/documents/guidelines/guidelines-revisedsep02.doc. Accessed June 7, 2006.

38. Hughes, WP. *Military Modeling for Decision-Making*, 3rd ed. Alexandria, VA: Military Operations Research Society, 1997.

39. *Military Handbook for Joint Data Base Elements for Modeling and Simulation (M&S)*. August 5, 1993.

40. Merriam-Webster Online [WWW Document]. Available at http://www.m-w.com/cgi-bin/dictionary]. Accessed June 28, 2006.

41. Russell, S, Norvig, P. *Artificial Intelligence: A Modern Approach*. Upper Saddle River, NJ: Prentice-Hall, 1995.

42. Dean, T, Allen, J, Aloimonos, J. *Artificial Intelligence: Theory and Practice*. Redwood City, CA: Benjamin/Cummings Publishing Company, 1995.

43. Holland, JH. *Hidden Order: How Adaptation Builds Complexity*. Cambridge, MA: Perseus Books, 1995.

44. Freedman, A. *The Computer Desktop Encyclopedia (CD Rom—Ver. 12.1)*. Point Pleasant, PA: Computer Language Company, 1999.

Chapter 5

Life Cycle Costing

EDWARD POHL, Ph.D.
HEATHER NACHTMANN, Ph.D.

The last 10 percent of performance generates one-third of the cost and two-thirds of the problems.

—Norman R. Augustine

5.1 INTRODUCTION TO LIFE CYCLE COSTING

In Chapter 1, a system is defined as "an integrated set of elements that accomplishes a defined objective. System elements include products (hardware, software, firmware), processes, people, information, techniques, facilities, services, and other support elements" [1]. As part of a system, these integrated elements have a system life cycle consisting of seven stages: conceptualization, design, development, production, deployment, operation, and retirement of the system. This system life cycle was described in Chapter 3. Throughout each of these seven stages, various levels of life cycle costs occur, including development, production, support, and disposal costs.

Life cycle costing (LCC) should be used for solution design (Chapter 11) and is required for systems decision making (see Chapter 12). LCC is used by a systems engineering team to estimate whether the new system or the proposed system modifications will meet functional requirements at a reasonable total cost over the duration of its anticipated life. When successfully employed and managed, life cycle

Decision Making in Systems Engineering and Management, Second Edition
Edited by Gregory S. Parnell, Patrick J. Driscoll, Dale L. Henderson
Copyright © 2011 John Wiley & Sons, Inc.

costing is also a tool for systems decision making. The Society of Cost Estimating and Analysis [2] defines a life cycle cost estimate in the following way:

> Life cycle cost estimate is an estimate that covers all of the cost projected for a system's life cycle, and which aids in the selection of a cost-effective total system design, by comparing costs of various trade-offs among design and support factors to determine their impact of total system acquisition and ownership costs.

The concept map in Figure 5.1 provides a pictorial overview of life cycle costing. Life cycle costing centers around the development of a system cost estimate, which in conjunction with the schedules is used by a program manager to manage the system's acquisition, operation, or disposal. System design and operational concepts drive the key cost parameters, which in turn identify the data required for developing a system cost estimate. As part of a systems engineering team, cost analysts rely on historical data, subject matter experts (SME), system schedules, and budget quantities to provide data to use life cycle costing techniques.

System risk depends on the life cycle stage of the system, and this risk affects the key cost parameters that drive the system cost estimate. Cost estimating is a critical activity for successful public and private organizations. Cost estimates are required to (a) develop a budget to obtain a new system, (b) prepare a bid on a project proposal, (c) negotiate a price for a system, and (d) provide a baseline from which to track and manage actual costs. Major cost estimating errors can dramatically impact the credibility and long-term viability of the organization.

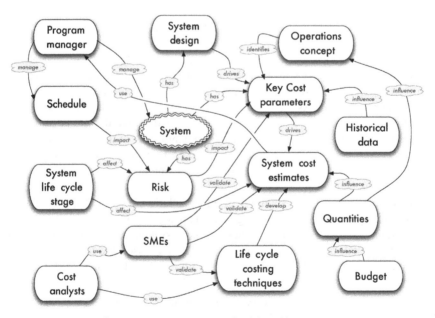

Figure 5.1   Concept map for life cycle costing.

Selection of the most appropriate LCC technique depends on the amount and type of available data and the perceived system risks. As each stage of the life cycle progresses, additional information concerning system requirements, system design and system performance becomes available, and some dimensions of uncertainty are resolved. Therefore, which LCC technique is most useful depends on the stage of the system life cycle. Recommendations of LCC techniques by life cycle stage are provided in Table 5.1 along with suggested references for each technique.

The Association for the Advancement of Cost Engineering (AACE) International prescribes a cost estimation classification system that can be generalized for applying estimate classification principles to system cost estimates [3]. Under this classification system, the level of system definition is the primary characteristic for classifying cost estimates. Other secondary characteristics shown in Table 5.2 (adapted from reference [4]) include the end usage of the estimate, estimating methodology, expected accuracy range, and effort to prepare the estimate.

Estimates are grouped into classes ranging from Class 1 to Class 5. Class 5 estimates are the least precise because they are based on the lowest level of system definition, while Class 1 estimates are the most precise because they are closest to full system definition and maturity. Successive estimates are prepared as the level of system definition increases until a final system cost estimate is developed.

The *level of systems definition* provides ranges of typical completion percentages for systems within each class, which can provide information about the maturity and extent of available input data. The *end usage* describes the typical use of cost estimates generated at that level of system definition, such that Class 5 estimates are generally used only for low-level screening or feasibility analysis. The *methodology* contains the typical estimating methods that are employed to generate each class of estimate. The less knowledge we have of the system, the more appropriate it is to provide a cost estimate range instead of a single number.

The *expected accuracy range* indicates the degree to which the final cost outcome for a given system is expected to vary from the estimated cost. The values in this column do not represent percentages as generally given for expected accuracy but instead represent an index value relative to a best range index value of 1. For example, if a given industry expects a Class 1 accuracy range of $+15$ to -10, then a Class 5 estimate with a relative index value of 10 would have an accuracy range of $+150$ to -100 percent. The final characteristic, *preparation effort*, provides an indication of the cost, time, and other resources required to prepare a given estimate, which is again a relative index value.

5.2 INTRODUCTION TO COST ESTIMATING TECHNIQUES

Once definitions have been determined for all the cost elements forecasts to occur during a system's life cycle, the systems engineering team begins cost estimating—defined by Stewart [5] as "the process of predicting or forecasting the cost of a work activity or work output." This process is divided into 12 major steps discussed in more detail in the *Cost Estimator's Reference Manual* [5].

TABLE 5.1 LCC Techniques by Life Cycle Stage

LCC Techniques	Life Cycle Stages						
	Concept	Design	Development	Production	Deployment	Operation	Retirement
Expert judgment Cost estimating relation-ship [5]	Prepare initial cost estimates	Estimate by analogy Refine cost estimates	Estimate by analogy	Estimate by analogy Create production estimates			
Activity-based costing [6]				Provides indirect product costs		Use for operational trades	
Learning curves [7, 8]			Provides development and test unit costs	Provide direct labor production costs			
Breakeven analysis [9]			Use in design trades	Provides production quantities		Use for operational trades	
Uncertainty and risk analysis [10]	Use with CER estimates	Use with analogy or CER estimates	Affects development cost	Affects direct and indirect product cost	Affects deployment schedules	Affects O&S cost projections	
Replacement analysis [4]							Determines retirement date

TABLE 5.2 AACE Cost Estimate Classification Matrix [4]

	Primary Characteristic	Secondary Characteristic			
		End usage	Methodology	Expected accuracy range	Preparation effort
Estimate class	Level of system definition expressed as percentage of complete definition	Typical purpose of estimate	Typical estimating method	Typical ± range relative to best index of 1[a]	Typical degree of effort relative to least cost index of 1[b]
Class 5	0% to 2%	Screening or feasibility	Stochastic or judgmental	4 to 20	1
Class 4	1% to 15%	Concept study or feasibility	Primarily stochastic	3 to 12	2 to 4
Class 3	10% to 40%	Budget authorization or control	Mixed but primarily stochastic	2 to 6	3 to 10
Class 2	30% to 70%	Control or bid tender	Primarily deterministic	1 to 3	5 to 20
Class 1	50% to 100%	Check estimate or bid tender	Deterministic	1	10 to 100

[a]If the range index value of "1" represents +10 − 5%, then an index value of 10 represents +100 − 50%.
[b]If the cost index value of "1" represents 0.005% of project costs, then an index value of 100 represents 0.5%.

This manual is an excellent source for developing life cycle cost estimates. The book contains extensive discussion and numerous examples of how to develop a detailed life cycle cost estimate. The 12 steps are:

1. Developing the work breakdown structure
2. Scheduling the work elements
3. Retrieving and organizing historical data
4. Developing and using cost estimating relationships
5. Developing and using production learning curves
6. Identifying skill categories, skill levels, and labor rates
7. Developing labor hour and material estimates
8. Developing overhead and administrative costs
9. Applying inflation and escalation (cost growth) factors
10. Computing the total estimated costs
11. Analyzing and adjusting the estimate
12. Publishing and presenting the estimate to management/customer

While all of these steps are important, the earlier steps are critical since they define the scope of the system, the appropriate historical data, and the appropriate cost models to use. In addition, the identification of technology maturity for each cost element is critical since many cost studies cite technology immaturity as the major source of cost estimating errors [11].

The use of these steps is dependent upon the phase of the life cycle, which impacts the level of detail required for the cost estimate. Many of these steps are by-products of a properly executed systems engineering and management process. In the project planning phase (see Chapter 13, Solution Implementation), a work breakdown structure for the systems is established. The low-level activities are then scheduled to develop a preliminary schedule.

Once all the activities have been identified and scheduled, the next task is to estimate their costs. The best approach is to estimate the cost of the activities based on past experience. Specifically, one would like to be able to estimate the costs associated with the activities using historical cost and schedule data. Finding these data and organizing them into a useful format is one of the most difficult and can be the most time-consuming step in the process. Once the data are found and organized, the analysts must ensure that it is complete and accurate. Part of this accuracy check is to make sure that the data are "normalized." This is accomplished to ensure that the analyst is not making an "apples to oranges" comparison.

One form of data normalization is to use the proper inflationary/deflationary indices on estimates associated with future costs (Step 9). Once the data have been normalized, it is then used to develop statistical relationships between physical and performance elements of the system and cost. Steps 4 and 5 are used to establish baseline cost estimating relationships and adjust the costs based on the quantities purchased. Steps 6, 7, and 8 are used when a detailed "engineering" level estimate

(Class 2 or Class 3 estimate) is being performed on a system. This is a very time-consuming task, and these steps are necessary if one wants to build a "bottom-up" estimate by consolidating individual estimates of each of the work activities into a total project cost estimate. Like the earlier techniques in Steps 4 and 5, these steps are even more dependent on collecting detailed historical information on activities and their associated costs.

Finally, Steps 11 and 12 are necessary elements to cost estimating. Step 11 provides the analyst the opportunity to revise and update the estimate as more information becomes available about the system. Specifically, this may be an opportunity for the analysts to revise or adjust the estimate based on the maturity of the technology [11]. Additionally, it provides the analyst the opportunity to assess the risk associated with the estimate. The analyst can account for the data uncertainty quantitatively by performing a Monte Carlo analysis on the estimate and creating a distribution for the systems life cycle cost in a similar fashion as is done for value measures in Chapter 12. Step 12 is one of the most important steps, it does not matter how good an estimate is if an analyst cannot convince the management that they have done a thorough and complete cost analysis.

All assumptions should be clearly articulated in a manner that provides insight on the data sources and key assumptions used. The foundation for a cost estimate is the basic list of ground rules and assumptions associated with that estimate. Specifically, all assumptions, such as data sources, inflation rates, quantities procured, amount of testing, spares provisioning, and so on, should be clearly documented up front in order to avoid confusion and the appearance of an inaccurate or misleading cost estimate.

In this chapter, we will highlight a few of the key tools and techniques that are necessary to develop a reasonable preliminary cost estimate. Specifically, we will focus on developing and using cost estimating relationships and learning or cost progress curves. The details associated with developing a comprehensive detailed estimate are extensive and cannot be given justice within a single textbook chapter. Interested readers are referred to Stewart et al. [5] and Ostwald [7].

Once the estimate is developed and approved, it can be used to develop a budget, create a bid on a project or proposal, establish the price, or to form a baseline from which to track actual costs. A second benefit is that it can be used as a tool for cost analysis for future estimates in the organization.

5.2.1 Types of Costs

First, although most people think of "costs" in terms of dollars, cost can refer to any term that represents resources to an organization—for example, hours, man years, facilities space, and so on. These measures are an important factor in making meaningful tradeoff decisions that affect the organization whose system is being studied. These resource measures can often be converted to dollars and usually are in order to provide senior management a unifying measure they can easily assess.

There are a variety of costs associated with developing new systems or modifying existing ones. These costs vary based on where in the life cycle they occur and the type of system being developed, constructed, or acquired. These cost classifications are useful in identifying the various sources of cost as well as the effect those sources have on the system life cycle cost. Costs are partitioned into four classes [5]: acquisition, fixed and variable, recurring and nonrecurring, direct and indirect.

Acquisition Cost. The total cost to procure, install, and put into operation a system, a product, or a specific piece of infrastructure (e.g., building, bridge, tunnel, transportation system). These are the costs associated with planning, designing, engineering, testing, manufacturing, and deploying/installing a system or process.

Fixed and Variable Costs. Fixed costs are those costs that remain constant independent of the quantity or phase of the life cycle being addressed in the estimate. Typical fixed costs include such items as research, lease rentals, depreciation, taxes, insurance, and security. Variable costs are those costs that change as a function of the number of systems or system output. Variable costs increase or decrease as the amount of product or service output from a system increases or decreases. Typical variable costs include direct labor, direct material, direct power, and the like. In other words, any cost that can be readily allocated to each unit of system output can be considered a variable cost.

Recurring and Nonrecurring Costs. Recurring costs are costs that repeat with every unit of product or every time a system process takes place. Like variable costs, they are a function of the quantity of items output. Nonrecurring costs are those costs that occur only once in the life cycle of a system. For example, the costs associated with design, engineering, testing, and other nonproduction activities would be normally classified as nonrecurring when developing a system cost estimate, because they are not anticipated to repeat once they occur.

Direct and Indirect Costs. Direct costs are those costs that are associated with a specific system, end item, product, process, or service. Direct costs can be further subdivided into direct labor, direct material, and direct expenses. Labor costs are those costs associated with the labor used directly on an item. Direct material costs are those costs associated with the bills of material purchased for the manufacture of the item and direct expense may be subcontracted work for part of the system or product. Indirect costs are those costs that cannot be assigned to a specific product or process; are usually pooled into an overhead account, which is applied to direct costs as a burden. Examples of indirect costs may include items like security, accounting and finance labor, janitorial services, executive management, training, and other activities and costs that are not directly related to the specific product or process but are an integral part of the organization that is responsible for the product, process, or service. Activity-based costing is a life cycle costing technique that can provide more accurate indirect cost analysis based on the premise that indirect costs should be allocated according to important functional activities that are performed during the system life cycle [6]. Indirect costs associated with these activities are identified and grouped into multiple cost pools based on similar cost drivers. Resulting system life cycle costs are thus based on a more detailed analysis than traditional indirect costing.

5.3 COST ESTIMATION TECHNIQUES

As mentioned earlier, a variety of tools and techniques are available to assist in developing the life cycle cost estimate for a system [12]. In order to begin, an analyst must have a very good understanding of the system operations concept, the system functions, the system hardware (and software), the technology maturity, the quantities desired, and the system life cycle. All of these are developed in the course of applying the SDP in each life cycle stage. This information is necessary in order to develop a credible initial cost estimate.

In this section, we explore several techniques for developing life cycle cost estimates. First, we discuss the use of expert judgment and its role in establishing initial estimates. This technique is useful in developing initial estimates for comparison of alternatives early on the concept exploration phase. Second, we explore the use of cost estimating relationships as a vehicle to estimating the cost of a system, product, or process. These are used to provide more refined estimates of specific alternatives when selecting between alternatives and are often used to develop initial cost baselines for a system. Finally, we will discuss the use of learning curves in a production cost estimate. This tool is used to analyze and account for the effect quantity has on the cost of an item. This tool is often used in conjunction with cost estimating relationships (CERs) to build a life cycle cost estimate.

5.3.1 Estimating by Analogy Using Expert Judgment

Engineers are often asked to develop cost estimates for products and services that are in the system concept stage. The engineers may have nothing more than a preliminary set of functions and requirements and a rough description of a feasible system concept. Given this very preliminary information, the program manger, the cost analyst, the systems engineer, and the engineering design team are often asked to develop a life cycle cost estimate for the proposed system in order to obtain approval and funding for the system. Given the scarcity of information at this stage, many program managers and cost analysts will rely on their own experience and/or the experience of other stakeholders and experts to construct an initial cost estimate. The use of expert judgment to construct an initial estimate for a system is not uncommon and is often used for Class 4 and 5 estimates. This underscores yet another reason why a good deal of time and effort is dedicated to stakeholder analysis in the Problem Definition phase of the SDP.

Often, technological advances create the requirements and/or market opportunities for new systems. When this occurs, the existing system can serve as a reference point from which a baseline cost estimate for a new system may be constructed.

If historical cost and engineering data are available for a similar system, then that system may serve as a useful baseline from which modifications can be made based upon the complexity of the advances in technology and the increase in requirements for system performance.

Many times, an expert will be consulted to describe the increase in complexity by focusing on a single system element (e.g., the new processor is three times as complex). The cost analyst translates this into a cost factor by referencing past experience for example, "The last time we changed processors, it was two times as complex and it increased cost by 20% over the earlier generation." A possible first-order estimate may be to take the baseline cost, say $10,000, and create a cost factor based on the information elicited from the experts.

- Past cost factor: 2 × complexity = 20% increase
- Current estimate: 3 × complexity may increase cost 30%

These factors will be based on personal experience and historical precedent. In this example, the underlying assumption that the expert is making is that there is a linear relationship between the cost factor and the complexity factor. Given this assumption, a baseline estimate for the new technology might be ($10,000 × 1.3 = $13,000). Estimating by analogy can be accomplished at the meta system level as well when the new system has proposed characteristics in common with existing systems. For example, the cost of unmanned aeronautical vehicles (UAVs) could initially be estimated by drawing analogies between missiles and UAVs, because UAVs and missiles use similar technologies. By making appropriate adjustments for size, speed, payload, and other performance parameters, one could make an initial life cycle cost estimate based on historical missile data.

A major disadvantage associated with estimating by analogy is the significant dependency on the judgment of the expert and existing historical data. The credibility of the estimate rests largely upon the expertise and experience of the person constructing the analogy. Estimating by analogy requires significantly less effort and therefore is not as costly in terms of time and level of effort as other methods. Therefore, it is often used as a check of the more detailed estimates that are constructed as the system description evolves.

5.3.2   Parametric Estimation Using Cost Estimating Relationships

Parametric estimates are used to create estimates of life cycle costs using statistical analysis techniques. The use of parametric cost estimates was first introduced in the late 1950s by the RAND Corporation for use in predicting costs of military systems. These techniques rely heavily on historical data. In general, parametric cost estimates are preferred to expert judgment techniques. However, if sufficient historical data are not available or the product and its associated technology have changed significantly, then constructing a parametric cost estimate may not be possible.

The level at which parametric cost estimating is accomplished is largely dependent on the system life cycle stage. Parametric cost estimation is frequently used during early stages in the life cycle before detailed design information is available. Parametric techniques can also be constructed and/or revised using detailed design and production information. Because they are designed to forecast costs into the future, they are often used to estimate operation and support costs as well.

The end goal of this statistical approach is to develop a cost estimating relationship (mathematical relationship between one or more system physical and performance parameters and the total system cost estimate). For example, the cost of a satellite may be estimated as a function of weight, power, and orbit. The cost of a house may be estimated by forming a relationship between cost and the square footage, location, and number of levels in a house.

As mentioned previously, when constructing a system life cycle cost estimate, one should use the baseline work breakdown structure (WBS) for the system. This will help ensure that all the necessary elements of the system have been appropriately accounted for in the cost estimate. As an example, a three-level WBS for the air-vehicle portion of a missile system has been adapted from Mil-Hdbk-881a WBS Missiles [13] and is presented in Table 5.3. The missile system has many more level 2 components. For example, at WBS Level 2, one must also consider the costs of the command and launch components, the systems engineering and

TABLE 5.3   Levels of Cost Estimating [13]

Level 1	Level 2	Level 3
Missile system		
	Air vehicle	
		Propulsion system
		Payload
		Airframe
		Guidance and control
		Fuzing
		Integration and assembly
	Command and launch components	
	Systems engineering and program management	
	System test and evaluation	
	Training	
	Data	
	Support equipment	
	Site activation	
	Storage facilities	
	Initial spares	
	Operating and support costs	
	Retirement costs	

program management costs, the system test and evaluation costs, training costs, data costs, support equipment costs, site activation costs, facilities costs, initial spares costs, operational and support costs, and retirement costs. Each of these Level 2 elements can be further broken down into Level 3 WBS elements as has been done for the air vehicle.

A parametric cost estimating relationship can be developed at any of the three levels of the WBS depending on the technological maturity of the system components, available engineering and cost data, and amount of time available to create the estimate. In general, the further along in the life cycle, the more engineering data available and the lower the WBS level (higher level number) from which an estimate can be constructed.

In the next section, we present the basic approach for constructing a cost estimating relationship and provide guidance on how it can be used to develop a system estimate. We will use our simplified missile air vehicle system as an example.

Common Cost Estimating Relationship Forms A cost estimating relationship (CER) is a mathematical function that relates a specific cost category to one or more system variables. These variables must have some logical relationship to the system cost. One must be sure that the data used to develop the CER is relevant to the system and its specific technology. We will discuss the four basic forms for cost estimating relationships.

Linear CER with Fixed and Variable Cost. Many WBS elements can be modeled reasonably well by a simple linear relationship, $Y = aX$. For example, facility cost can be modeled as a function of area. Personnel costs can be modeled by multiplying labor rates by personnel hours. It is also possible to have a linear relationship that includes a fixed cost, denoted by b. For example, suppose the cost of the facility also includes the cost of the land purchase. Then it would have a fixed cost associated with the land purchase and a variable cost that is dependent on the size of the facility built on the land. The relationship is given by $Y = aX + b$. Both of these basic forms are shown below in Figure 5.2.

Power CER with Fixed and Variable Cost. Some systems may not have a linear relationship between cost and the selected estimating parameter. An economy of scale effect may occur; for example, as the size of a house increases, there will be a point at which the cost/ft^2 decreases. Similarly, one can also encounter situations were there are diseconomies of scale. For example, large gemstones often have higher costs per unit size than smaller gemstones. Figure 5.3 illustrates the various shapes that a power CER can take as well as the functional form of the various cost estimating relationships.

Exponential CER with Fixed and Variable Cost. Another functional form that is often used is the exponential cost estimating relationship. In this form, it is assumed

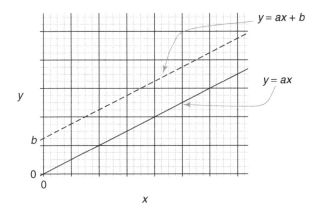

Figure 5.2 Linear CERs.

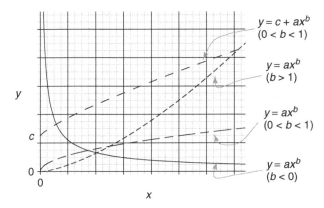

Figure 5.3 Power CERs.

that the relationship between the independent variable and cost is such that a unit change in the independent variable causes a relatively constant percentage change in cost. Figure 5.4 illustrates the shape for an exponential CER for a variety of functional forms.

Logarithm CERs. Finally, one other common form that may be useful for describing the relationship between cost and a particular independent variable is the logarithm CER. Figure 5.5 illustrates the shape for a logarithm CER for a variety of functional forms.

Constructing a CER   To construct a cost estimating relationship we need a sufficient amount of data to fit the curve. What is sufficient is a judgmental decision

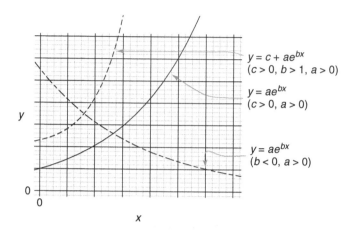

$$y = c + ae^{bx}$$
$$(c > 0, b > 1, a > 0)$$

$$y = ae^{bx}$$
$$(c > 0, a > 0)$$

$$y = ae^{bx}$$
$$(b < 0, a > 0)$$

Figure 5.4   Exponential CERs.

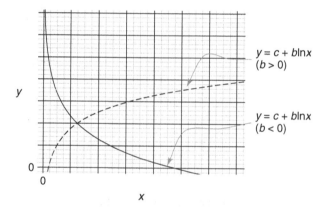

$$y = c + b\ln x$$
$$(b > 0)$$

$$y = c + b\ln x$$
$$(b < 0)$$

Figure 5.5   Logarithm CERs.

and is usually dependent on what is available. For most of these models, a minimum of three or four data points would be sufficient to construct a CER. A CER constructed from so few points is likely to have a significant amount of error. Ordinarily, linear regression is used to construct the cost estimating relationship. We can use linear regression on all of the functional forms discussed previously by performing a transformation on the data. Table 5.4, adapted from Stewart et al. [5], shows the relationship between the various CERs and their associated transformations.

Linear regression is used to estimate the parameters for the CERs once the data has been appropriately transformed. Linear regression is a statistical method used to fit a straight line through a set of data points. The goal is to determine the

TABLE 5.4 Linear Transformations for CERs

	Linear	Power	Exponential	Logarithmic
Equation form desired	$Y = a + bX$	$Y = ax^b$	$Y = ae^{bX}$	$Y = a + b \ln X$
Linear equation form	$Y = a + bX$	$\ln Y = \ln a + b \ln X$	$\ln Y = \ln a + bX$	$Y = a + b \ln X$
Required data transform	X, Y	$\ln X, \ln Y$	$X, \ln Y$	$\ln X, Y$
Regression coefficient obtained	a, b	$\ln a, b$	$\ln a, b$	a, b
Coefficient reverse transform required	None	$\exp(\ln a), b$	$\exp(\ln a), b$	None
Final coefficient	a, b	a, b	a, b	a, b

coefficient values for the parameters a and b of the linear equation. The parameters are determined by using the following formulas:

$$b = \frac{\sum_{i=1}^{n} x_i y_i - \left[\dfrac{\sum_{i=1}^{n} x_i}{n}\right] \sum_{i=1}^{n} y_i}{\sum_{i=1}^{n} x_i^2 - \left[\dfrac{\sum_{i=1}^{n} x_i}{n}\right] \sum_{i=1}^{n} x_i} \tag{5.1}$$

$$a = \frac{\sum_{i=1}^{n} y_i}{n} - b \left[\frac{\sum_{i=1}^{n} x_i}{n}\right] \tag{5.2}$$

Most of the time, especially when we have a reasonable size data set, a statistical analysis package such as Minitab [14], JMP [15], or even Excel will be used to perform the regression analysis on the cost data.

TABLE 5.5   Labor Hours and Costs for Highway Construction

X Labor hours	Y Cost [$]
940.87	252.87
5814.28	4708.28
302.31	137.31
292.44	303.44
149.46	149.46
2698.94	1385.94
680.64	362.64
1078.32	364.32
6961.21	5269.21
4174.96	1192.96
1277.78	813.78
1493.08	957.08
4731.84	2342.84

Example. Suppose we have collected the following data on labor hours and construction costs for highways in Table 5.5. We would like to establish a cost estimating relationship between labor hours and cost. Analyze the data using a linear model and a power model.

We will fit the data to a simple linear model. The first thing we should do is to construct a scatter plot of the data such as that shown in Figure 5.6. We can estimate the parameters for a line that minimizes the squared error between the line and the actual data points. If we summarize the data, we get the following:

$$\sum_{i=1}^{13} x_i = 30596.13$$

$$\sum_{i=1}^{13} y_i = 18240.13$$

$$\sum_{i=1}^{13} x_i y_i = 87361422.71$$

$$\sum_{i=1}^{13} x_i^2 = 135941716.08$$

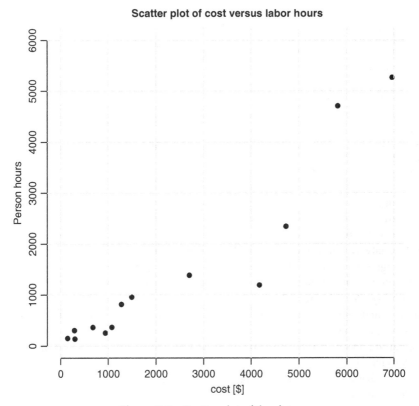

Figure 5.6 Scatter plot of the data.

Using the summary data we can calculate the following coefficients for the linear relationship:

$$b = \frac{87361422.71 - \left(\dfrac{30596.13}{13}\right)18240.13}{135941716.08 - \left(\dfrac{30596.13}{13}\right)30596.13} = 0.695$$

$$a = \frac{1824.13}{13} - 0.695\left(\frac{30596.13}{13}\right) = -232.61$$

If we enter the same data set into Minitab we obtain the following output:

```
The regression equation is
Cost = -233 + 0.695  Labor Hours
```

```
Predictor          Coef     SE Coef       T        P
Constant         -232.6       253.8   -0.92    0.379
Labor hours     0.69499     0.07849    8.85    0.000

S = 627.579 R-Sq = 87.7\% R-Sq(adj) = 86.6\%

Analysis of variance

Source              DF     SS         MS           F        P
Regression           1     30880140   30880140     78.40    0.000
Residual error      11     4332404    393855
Total               12     35212544
```

Examining the output, we see that the model is significant and that it accounts for approximately 87% of the total variation in the data. We note that the intercept term has a p-value of 0.379 and therefore could be eliminated from the model. As part of the analysis, one needs to check the underlying assumptions associated with the basic regression model. The underlying assumption is that the errors are normally distributed, with a mean of zero and a constant variance. If we examine the normal probability plot (Figure 5.7) and the associated residual plot (Figure 5.8), we see that our underlying assumptions may not be valid. We see that the residual data does not fall along a straight line and therefore is probably not normally distributed. Second, it appears that the variance is not constant. For larger values of man hours the variance increases.

Given that the underlying assumptions are not met for the basic linear regression model, one should consider some other type of cost estimating relationship. Let us consider a simple power cost estimating relationship. In this case, we need to transform our data according to Table 5.6, which contains the original data and the transformed data that we will use to fit a linear regression model. Note that

TABLE 5.6   Transformed Data

X	Y	$\ln(X)$	$\ln(Y)$	$(\ln(X))^2$	$(\ln(X))(\ln(Y))$
940.87	252.87	6.85	5.53	46.88	37.88
5814.28	4708.28	8.67	8.46	75.14	73.31
302.31	137.31	5.71	4.92	32.62	28.11
292.44	303.44	5.68	5.72	32.24	32.45
149.46	149.46	5.01	5.01	25.07	25.07
2698.94	1385.94	7.90	7.23	62.42	57.15
680.64	362.64	6.52	5.89	42.55	38.44
1078.32	364.32	6.98	5.90	48.76	41.19
6961.21	5269.21	8.85	8.57	78.29	75.83
4174.96	1192.96	8.34	7.08	69.5	59.06
1277.78	813.78	7.15	6.70	51.16	47.94
1493.08	957.08	7.31	6.86	53.42	50.17
4731.84	2342.84	8.46	7.75	71.60	65.65

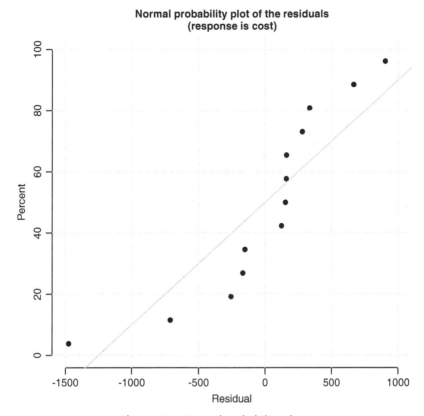

Figure 5.7   Normal probability plot.

we perform our transformation by taking the natural logarithm of the cost and the natural logarithm of the labor hours. Using this transformed data set, we can calculate our coefficients for our transformed linear model.

$$\sum_{i=1}^{13} \ln x_i = 93.43$$

$$\sum_{i=1}^{13} \ln y_i = 85.64$$

$$\sum_{i=1}^{13} (\ln x_i) \ln y_i = 632.25$$

$$\sum_{i=1}^{13} (\ln x_i)^2 = 689.66$$

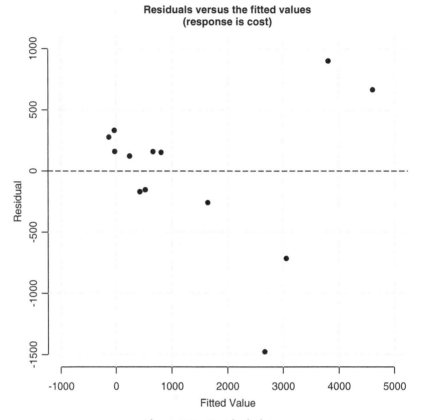

Figure 5.8   Residual plot.

Using the summary data, we can calculate the coefficients for the transformed linear relationship.

$$b = \frac{632.25 - \left(\dfrac{93.43}{13}\right)85.64}{689.66 - \left(\dfrac{93.43}{13}\right)93.43} = 0.9209$$

$$a = \frac{85.64}{13} - 0.921\left(\frac{93.43}{13}\right) = -0.0313$$

If we enter the same data set into Minitab we obtain the following output:

```
The regression equation is
LnY = -0.031 + 0.921 lnX
```

```
Predictor    Coef      SE Coef       T        P
Constant -0.0313      0.7613    -0.04    0.968
LnX       0.9209      0.1045     8.81    0.000

S = 0.446321 R-Sq = 87.6\% R-Sq(adj) = 86.5\%

Analysis of variance

Source              DF          SS       MS        F       P
Regression           1      15.461   15.461    77.62   0.000
Residual Error      11       2.191    0.199
Total               12      17.653
```

Examining the output, we see that the model is significant and that it accounts for approximately 87% of the total variation in the data. Again, we note that the intercept term has a p-value of 0.968 and therefore could be eliminated from the model at any level of significance less than or equal to $\alpha = 0.968$. Once again we must check the underlying assumptions associated with the basic regression model. The underlying assumption is that the errors are normally distributed, with a mean of zero and a constant variance. If we examine the normal probability plot and the associated residual plot (Figures 5.9 and 5.10) we see that our underlying assumptions appear to be valid. The residuals appear to be normal and have a constant variance with a mean of zero. Now we need to take the inverse transform to put our cost estimating relationship in its standard power form.

$$Y = e^{-0.031}X^{0.921}$$

$$Y = 0.97X^{0.921}$$

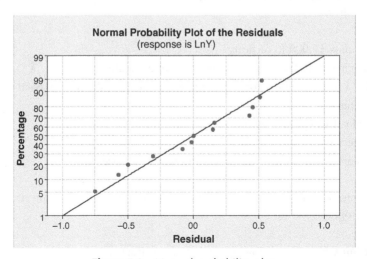

Figure 5.9 Normal probability plot.

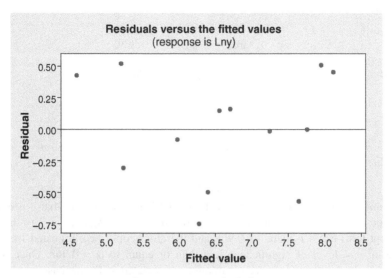

Figure 5.10   Residual plot.

Example. In this section, we provide several examples of hypothetical CERs that could be used to assemble a cost estimate for the air-vehicle component of the missile system [16, 17] described in the WBS given in Table 5.3. To estimate the unit production cost of the air-vehicle component, we sum the first unit costs for the propulsion system, the guidance and control system, the airframe, the payload, and the associated integration and assembly. Suppose the system that we are trying to estimate the first unit production cost for has the following engineering characteristics:

- Requires 15,000 lb of thrust
- Requires a 26-GHz guidance and control computer
- Has a 6-inch aperture on the antenna, operating in the narrow band
- Airframe weight of 300 lb
- Payload weight of 100 lb
- System uses electro-optics
- System checkout requires seven different test procedures

Suppose the following CERs have been developed using data from 20 different missile programs during the last 20 years.

Propulsion CER. The following CER was constructed using the propulsion costs from 15 of the 20 missile programs. Five of the programs were excluded because the technology used in those programs was not relevant for the missile system currently being estimated. The CER for the propulsion system is given by

$$\text{Mfg \$ (FY 2000)} = (\text{Thrust (lb)})e^{-0.1(\text{Yr}-1980)}$$
$$\text{Mfg \$ (FY 2000)} = (15{,}000)e^{-0.1(2000-1980)} = 1114.10$$

The manufacturing cost in dollars for the propulsion system is a function of thrust as well as the age of the motor technology (current year minus 1980).

Guidance and Control CER. The guidance and control CER was constructed using data from the three most recent missile programs. This technology has evolved rapidly and it is distinct from many of the early systems. Therefore, the cost analysts chose to use the reduced data set to come up with the following CER:

$$\text{Mfg \$ K (FY 2000)} \quad = \quad 7.43(\text{GHz})^{0.45}$$
$$\times \, (\text{Aper (inches)})^{0.35} e^{0.7(\text{Wide/Narrow})}$$
$$\text{Mfg \$ K (FY 2000)} \quad = \quad 7.43(26)^{0.45}(6)^{0.35} e^{0.7(1)} = 121.36$$

The manufacturing cost in thousands of dollars for the guidance and control system is a function of the operating rate of the computer, the diameter of the antenna for the seeker, and whether or not the system operates over a wide band (0) or narrow band (1).

Airframe CER. Suppose the following CER was constructed using the airframe cost data from the 20 missile programs. The CER for the airframe is given by

$$\text{Mfg \$ (FY 2000)} = 5.575(\text{Wt.lb})^{0.85}$$
$$\text{Mfg \$ (FY 2000)} = 5.575(300)^{0.85} = 710.88$$

Thus, the manufacturing costs in dollars for the airframe can be estimated if the analyst knows or has an estimate of the weight of the airframe.

Fuzing System. The following CER was established using data from five of the previous missile programs. The proximity fuse in the system being estimated is technologically similar to only five of the previous development efforts.

$$\text{Mfg \$ (FY 2000)} = 15(\text{payloadwt.lb})e^{0.3(\text{EO/RF})}$$
$$\text{Mfg \$ (FY 2000)} = 15(100)e^{0.3(0)} = 150$$

The manufacturing cost for the fuzing system in dollars is a function of the weight of the payload and the type of technology used. The term EO/RF is equal to 0 if it uses electro-optic technology and 1 if it uses radio-frequency technology.

Payload. The payload CER is given by the following relationship in dollars:

$$\text{Mfg \$ (FY 2000)} = 150(\text{payloadwt.lb})$$

$$\text{Mfg \$ (FY 2000)} = 150(100) = 15000$$

Integration and Assembly. This represents the costs in dollars associated with integrating all of the air-vehicle components, testing them as they are integrated, and performing final checkout once the air vehicle has been assembled. Letting *n* represent the number of system test procedures, we have the following.

$$\text{Mfg \$ (FY 2000)} = 1.25 \left(\sum \text{Hardware costs} \right) e^{-(n)}$$

$$\text{Mfg \$ (FY 2000)} = 1.25 \, (15{,}000 + 150 + 710.88 + 121.36 + 1114.10)$$

$$\times \, e^{-(7)}$$

$$\text{Mfg \$ (FY 2000)} = 157.68$$

Air-Vehicle Cost. Using this information, the first unit cost of the air-vehicle system is constructed below:

$$\text{Mfg \$ (FY 2000)} = 15{,}000 + 150 + 710.88 + 121.36$$

$$+ 1114.1 + 157.68$$

$$\text{Mfg \$ (FY 2000)} = 17254.02$$

This cost is in fiscal year 2000 dollars and it must be inflated to current year dollars (2006) using the methods discussed in Section 5.4.2. Once the cost has been inflated, the initial unit cost can be used to calculate the total cost for a purchase of 1000 missiles using an appropriate learning curve as discussed in the next section.

5.3.3 Learning Curves

Learning curves are an essential tool for adequately modeling the costs associated with the development and manufacture of large quantities of systems [8]. Many studies have shown that performance improves the more times a task is performed (supporting the old adage that practice makes perfect!). This "learning" effect was first noticed by Wright [18] when he analyzed aircraft data in the 1930s. Empirical evidence from a variety of other manufacturing industries has shown that human performance improves by some constant amount each time the production quantity is doubled [19]. This concept is especially applicable to labor-intensive products. Each time the production quantity is doubled, the labor requirements necessary to create a unit decrease to a fixed percentage of their previous value. This percentage is referred to as the *learning rate*.

Learning effects typically produce a cost and time savings of 5% to 30% each time the production quantity is doubled [10]. By convention, the 10% to 30% labor savings equates to a 90% to 70% learning rate. This learning rate is influenced by a variety of factors, including the amount of preproduction planning, the maturity of the design of the system being manufactured, training of the production force, the complexity of the manufacturing process, and the length of the production runs. Figure 5.11 shows a plot of a 90% learning rate and a 70% learning rate for a task that initially would take 100 hours. As evidenced by the plot, a 70% learning rate results in significant improvement of unit task times over a 90% curve. Delionback [5] defines learning rates by industry sector. For example:

- Aerospace—85%
- Repetitive electronics manufacturing—90% to 95%
- Repetitive machining—90% to 95%
- Construction operations—70% to 90%.

Unit Learning Curve Formula   The mathematical formula used to describe the learning effect shown in Figure 5.11 is given by

$$T_X = T_1 X^r \qquad (5.3)$$

where T_X = the cost or time required to build the Xth unit, T_1 = the cost or time required to build the initial unit, X = the number of units to be built, and r = negative numerical factor that is derived from the learning rate and is given by

$$r = \frac{\ln \ (\text{learning rate})}{\ln(2)} \qquad (5.4)$$

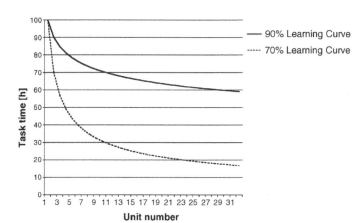

Figure 5.11   Plot of learning curves.

TABLE 5.7   Factors for Various Learning Rates

Learning Rate (%)	r
100	0
95	−0.074
90	−0.152
85	−0.2354
80	−0.322
75	−0.415
70	−0.515

Typical values for r are given in Table 5.7. For example, with a learning rate of 95%, the resulting factor is $r = \ln(0.95)/\ln(2) = 0.074$.

The total time required for all units of production run of size N is

$$\text{total time} = T_1 \sum_{X=1}^{N} X^r \tag{5.5}$$

Examining the above equation, and using the appropriate factor for a 90% learning rate, we can calculate the unit cost for the first eight items. Assuming an initial cost of \$100, Table 5.8 provides the unit cost for the first eight items as well as the cumulative average cost per unit required to build X units. Figure 5.12 plots the unit cost curve and the cumulative average cost curve for a 90% learning rate for 32 units.

Note that the cumulative average curve is above the unit cost curve. When using data constructed with a learning curve, the analyst must be careful to note whether

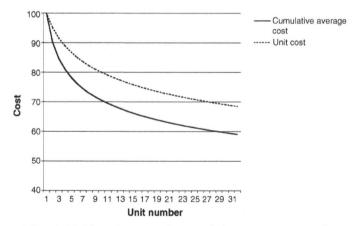

Figure 5.12   A 90% learning curve for cumulative average cost and unit cost.

TABLE 5.8   Unit Cost and Cumulative Average Cost

Total Units Produced	Cost to Produce (Xth) unit	Cumulative Cost	Cumulative Average Cost
1	100	100	100
2	90	190	95
3	84.6	274.6	91.53
4	81	355.6	88.9
5	78.3	433.9	86.78
6	76.2	510.1	85.02
7	74.4	584.5	83.5
8	72.9	657.4	82.175

they are using cumulative average data or unit cost data. It is easy to derive one from the other, but it is imperative to know what type of data one is working with to calculate the total system cost correctly.

Example 1.   Suppose it takes 40 minutes to assemble the fins for a rocket motor the first time and takes 36 minutes the second time it is attempted. How long will it take to assemble the eighth unit?

First, that task is said to have a 90% learning rate because the cost of the second unit is 90% of the cost of the first. If we double the output again, from two to four units, then we would expect the fourth unit to be assembled in (36 minutes) × (0.9) = 32.4 minutes. If we double again from four to eight units, the task time to assemble the eighth fin assembly would be (32.4) × (0.9) = 29.16 minutes.

Example 2.   Suppose we wish to identify the assembly time for 25th unit, assuming a 90% learning rate.

First, we need to define r:

$$r = \frac{\ln (0.9)}{\ln (2)} = -0.152$$

Given r, we can determine the assembly time for the 25th unit as follows:

$$T_X = T_1 X^r$$

$$T_{25} = (40 \text{ min}) (25)^{-0.152}$$

$$T_{25} = 24.52 \text{ min}$$

Figure 5.13 plots a 90% learning curve for the fin assembly.

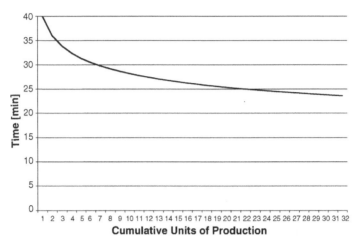

Figure 5.13   A 90% learning curve for fin assembly.

Example 3.   It is common for organizations to define a standard of performance based on the 100th or 1000th unit. Suppose your organization sets a target assembly time for the 100th unit of 50 h. Suppose that your company has historically operated at an 80% learning rate, what is the expected assembly time of the first unit?

$$T_X = T_1 X^r$$

$$T_X x^{-r} = T_1$$

$$T_1 = 50(100)^{-(-0.322)}$$

$$T_1 = 220.20\,\text{h}$$

Composite Learning Curves   Frequently, a new system will be constructed using a variety of processes, each of which may have its own learning rate. A single composite learning rate can be constructed that characterizes the learning rate for the entire system using the rates of the individual processes. One approach used to do this [5] weights each process in proportion to its individual dollar or time value. The composite learning curve is given by

$$r_c = \sum_p \left(\frac{T_p}{T} \right) r_p \tag{5.6}$$

where r_c = composite learning rate, r_p = learning rate for process p, T_p = time or cost for process p, and T = total time or cost for the system.

Example 4. Suppose our rocket has a final assembly cost of $50,000 and the final assembly task has a historic learning rate of 70%. Suppose that the rocket motor construction has a total cost of $100,000 and that it has a historic learning rate of 80%. Finally, the guidance section has total cost of $200,000 and a historic learning rate of 90%. Calculate the composite learning rate for the rocket?

$$r_c = \left[\frac{50000}{350000}\right](70\%) + \left[\frac{100000}{350000}\right](80\%) + \left[\frac{200000}{350000}\right](90\%)$$

$$= 84.29\%$$

Cumulative Average Formula The formula for calculating the approximate cumulative average cost or cumulative average number of labor hours required to produce X units is given by

$$T_c \approx \frac{T_1}{X(1+r)}\left[(X+0.5)^{(1+r)} - (0.5)^{(1+r)}\right] \tag{5.7}$$

This formula is accurate to within 5% when the quantity is greater than 10.

Example 5. Using the cumulative average formula, compute the cumulative average cost for eight units, assuming an initial cost of $100 and a 90% learning rate.

$$T_c \approx \frac{100}{8(1-0.152)}\left[(8.5)^{(1-0.152)} - (0.5)^{(1-0.152)}\right]$$

$$\approx 82.31$$

Note that this value is very close to the actual cost found in Table 5.8.

Constructing a Learning Curve from Historical Data The previous formulas are all dependent upon having a value for the learning rate. The learning rate can be derived for specific tasks in a specific organization by using historical cost and performance data. The basic data requirements for constructing a learning rate for an activity include the dates of labor expenditure, or cumulative task hours, and associated completed units. The learning rate is found by comparing the total hours expended at the end of a given date and the corresponding number of units completed.

By taking the natural logarithm of both sides of the learning curve function discussed in Section 5.3.3, one can construct a linear equation which can be used to find the learning rate.

$$T_x = T_1 X^r \tag{5.8}$$

$$\ln(T_X) = \ln(T_1) + r \ln(X) \tag{5.9}$$

The intercept for this linear equation is $\ln(T_1)$ and the slope of the line is given by r. Given r, the learning rate can be found using the following relation:

$$\text{learning rate\%} = 100(2^r) \tag{5.10}$$

This is best illustrated through an example.

Example 6.   Suppose the data in Table 5.9 is pulled from the company accounting system.

Transforming the data by taking the natural logarithm of the cumulative units and associated cumulative average hours yields Table 5.10. Figure 5.14 is a plot of the transformed data. Performing linear regression on the transformed data yields the following values for the slope and intercept of the linear equation.

TABLE 5.9   Accounting System Data

Week	Cumulative Hours Expended	Cumulative Units Complete	Cumulative Average Hours
1	040	0	
2	100	1	$100/1 = 100$
3	180	2	$180/2 = 90$
4	250	3	$250/3 = 83.33$
5	310	4	$310/4 = 77.5$
6	360	5	$360/5 = 72$
7	400	6	$400/6 = 66.67$

TABLE 5.10   Natural Logarithm of Cumulative Units Completed and Cumulative Average Hours

Cumulative Units Completed X	In X	Cumulative Average Hours TX	In TX
1	0	100	4.60517
2	0.693147	90	4.49981
3	1.098612	83.33	4.42281
4	1.386294	77.5	4.35028
5	1.609437	72	4.27667
6	1.791759	66.67	4.19975

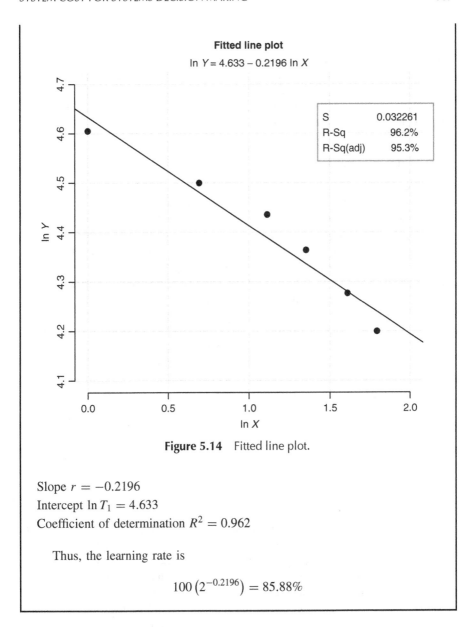

Figure 5.14 Fitted line plot.

Slope $r = -0.2196$

Intercept $\ln T_1 = 4.633$

Coefficient of determination $R^2 = 0.962$

Thus, the learning rate is

$$100\left(2^{-0.2196}\right) = 85.88\%$$

5.4 SYSTEM COST FOR SYSTEMS DECISION MAKING

In the previous section, we described how the system costs could be calculated using expert judgment, cost estimating relationships, and learning curves. Typically, this gives us a cost estimate in different year dollars. This section provides an overview of life cycle costing techniques that support the selection of economically feasible systems. A discussion of the time value of money is provided along with time equivalence formulas and a discussion of inflation impacts. These techniques are demonstrated in a system selection example, and a brief overview of additional life cycle costing techniques for system decision making is provided. In Chapter 12, we tie our development of life cycle cost estimates with system value so that a tradeoff decision can be made which takes into account all the candidate design solutions for a systems decision problem.

5.4.1 Time Value of Money

Life cycle costs must be identified in terms of timing as well as amount, because a dollar today is not the same as a dollar five years from now because of inflation and other economic effects. Two costs at different points in time are equivalent if they are equal to each other at some point in time for a given interest rate. For example, at a 10% annual interest rate, $500 today is equivalent to $605 two years from now. To compare costs occurring over the duration of a system's life cycle, we need to convert annual and future life cycle costs into their equivalent in present time. In addition, we may want to convert costs incurred today (present time) to future costs or analyze present and future costs as equal annual payments, depending on the specifics of the situation. Well-established equivalence formulas for converting costs in terms of time are shown in Table 5.11 along with an example for each conversion. These formulas are presented using the notation below:

i = Effective interest rate per period
N = Number of compounding periods
P = Present life cycle cost
F = Future life cycle cost
A = Annual life cycle cost

5.4.2 Inflation

It is possible that the data available to support system life cycle costing is collected from different years. In these cases, it is necessary to convert this cost data from actual dollars (actual dollars at the point of time in which they occurred) into constant dollars (adjusted dollars representing purchasing power at some base point in time). This conversion is referred to as an inflation adjustment where inflation refers to rising prices measured against a standard level of purchasing power. Annual inflation rates vary across different types of goods and services and over time. Life cycle cost analysts estimate future costs. Since this requires making economic assumptions about the future, inflation is one of the key economic assumptions the analysts need to make.

The consumer price index (CPI) is a measure of the average change in prices over time of goods and services purchased by households [4]. The CPI is commonly used in the conversion of actual to constant dollars. Table 5.12 provides the end of

TABLE 5.11 Equivalence Table [9]

Conversion	Formula	Example
Find F when given P	$F = P(1+i)^N$	A firm borrows $1000 for 10 years. How much must it repay in a lump sum at the end of the 10 year? $i = 5\%$ $F = 1000(1+0.05)^{10} = \$1628.89$
Find P when given F	$P = F\left[\dfrac{1}{(1+i)^N}\right]$	A company desires to have \$1,000 seven years from now. What amount is needed now to provide for it? ($i = 5\%$) $P = 1000\left[\dfrac{1}{(1+0.05)^7}\right] = \$710.68$
Find F when given A	$F = A\left[\dfrac{(1+i)^N - 1}{i}\right]$	If eight annual deposits of \$1000 each are placed in an account, how much money has accumulated immediately after the last deposit? ($i = 5\%$) $F = 1000\left[\dfrac{(1+0.05)^8 - 1}{(0.5)}\right] = \$9549.11$
Find A when given F	$A = F\left[\dfrac{i}{(1+i)^N - 1}\right]$	How much should be deposited each year in an account in order to accumulate \$5000 at the time of the sixth annual deposit? ($i = 5\%$), $A = 5000\left[\dfrac{0.05}{(1+0.05)^6 - 1}\right] = \$735.09$
Find P when given A	$P = A\left[\dfrac{(1+i)^N - 1}{i(1+i)^N}\right]$	How much should be deposited in a fund to provide for 10 annual withdrawals of \$500 each? ($i = 5\%$), $P = 500\left[\dfrac{(1+0.05)^{10} - 1}{0.05(1+0.05)^{10}}\right] = \$3860.87$
Find A when given P	$A = P\left[\dfrac{i(i+1)^N}{(1+i)^N - 1}\right]$	What is the size of 10 equal payments to repay a loan of \$10,000? $i = 5\%$, $A = 10{,}000\left[\dfrac{0.05(1+0.05)^{10}}{(1+0.05)^{10} - 1}\right] = \$1295.05$

TABLE 5.12 Consumer Price Index and Yearly Inflation Rate for 2000–2006

Year (i)	CPI (EOY)	Inflation rate % (f_i)
2000	174	—
2001	176.7	1.55
2002	180.9	2.38
2003	184.3	1.88
2004	190.3	3.26
2005	196.8	3.42
2006	201.8	2.54

year (EOY) CPI for years 2000 through 2006 and the corresponding year-by-year change rates as provided by the U.S. Bureau of Labor Statistics [4].

Constant dollars ($C\$$) at any time n of purchasing power as of any base time k can be converted to actual dollars ($A\$$) at any time n by using the equivalence formula for finding F given P.

$$A\$ = C\$_n^{(k)}(1+f)^{n-k} \tag{5.11}$$

where f = average inflation rate per period over the $n - k$ periods.

Example. $C\$10,000$ incurred in 2002 can be converted to 2005 dollars as follows:

$$f = \frac{f_{2003} + f_{2004} + f_{2005}}{3} = \frac{0.0188 + 0.0326 + 0.0342}{3} = 2.85\%$$

$$A\$ = \$10,000_{2005}^{(2002)}(1 + 0.0285)^{2005-2002}$$

$$A\$ = \$10,879.59$$

Example. Let us revisit the air-vehicle unit cost estimate developed in Section 5.3.2. Remember that the unit cost for the production of the first air vehicle was estimated to be $\$138,492.58$ in base year 2000. We can convert this to 2006 dollars as follows:

$$f = \frac{f_{2001} + f_{2002} + f_{2003} + f_{2004} + f_{2005} + f_{2006}}{6}$$

$$f = \frac{0.015 + 0.0238 + 0.0188 + 0.0326 + 0.0342 + 0.0254}{6} = 2.51\%$$

$$A\$ = \$138,492.58(1 + 0.0251)^{2006-2000} = \$160,702.9$$

5.4.3 Net Present Value

A system decision based on life cycle costs should take both time value of money and inflation into account. To conduct this analysis when selecting among multiple systems, the following conditions must exist [9]:

- Candidate system solutions must be mutually exclusive where the choice of one excludes the choice of any other.
- All systems must be considered over the same length of time. If the systems have different expected total life cycles, the time period should equal the lowest common multiples of their lives or the length of time the selected system will be used.

Example. An automated assembly system is being considered to assist in the production of our rockets. The initial purchase and installation cost is assumed to be $300,000. The life of the system is assumed to be seven years with annual operating and maintenance costs of $65,000. It is expected that an annual increase of $100,000 in revenue will be obtained from increased production, and the system can be salvaged for $175,000 at the end of its seven-year life. Our minimum attractive rate of return (interest rate) is 7%, and an inflation rate of 3% is assumed. Given these cost parameters, compute the net present value of the assembly system to determine if the system should be purchased. The first step is to adjust the annual net cash flows for inflation as shown in Table 5.13.

Once we have our adjusted annual net cash flows, the net present value (NPV) can be computed using the equivalence formulas provided in Table 5.11. The system will be selected as economically justified if the

TABLE 5.13   Inflation-Adjusted Annual Net Cash Flows

EOY	Cash Outflows [$]	Cash Inflows [$]	Net cash Flow [$]	Inflation Conversion Factor	Cash Flow in Actual [$]
0	$-300,000.00$		$-300,000.00$	$(1+0.03)^0$	$-300,000.00$
1	$-65,000.00$	$100,000.00$	$35,000.00$	$(1+0.03)^1$	$36,050.00$
2	$-65,000.00$	$100,000.00$	$35,000.00$	$(1+0.03)^2$	$37,132.00$
3	$-65,000.00$	$100,000.00$	$35,000.00$	$(1+0.03)^3$	$38,245.00$
4	$-65,000.00$	$100,000.00$	$35,000.00$	$(1+0.03)^4$	$39,393.00$
5	$-65,000.00$	$100,000.00$	$35,000.00$	$(1+0.03)^5$	$40,576.00$
6	$-65,000.00$	$100,000.00$	$35,000.00$	$(1+0.03)^6$	$41,794.00$
7	$-65,000.00$	$275,000.00$	$210,000.00$	$(1+0.03)^7$	$258,279.00$

NPV is greater than zero.

$$NPV = -\$300{,}000 + \$36{,}050 \left[\frac{1}{(1+0.07)^1}\right] + \$37{,}132 \left[\frac{1}{(1+0.07)^2}\right]$$

$$+ \$38{,}245 \left[\frac{1}{(1+0.07)^3}\right] + \$39{,}393 \left[\frac{1}{(1+0.07)^4}\right]$$

$$+ \$40{,}576 \left[\frac{1}{(1+0.07)^5}\right] + \$41{,}794 \left[\frac{1}{(1+0.07)^6}\right]$$

$$+ \$258{,}279 \left[\frac{1}{(1+0.07)^7}\right]$$

$$NPV = \$45{,}018.57$$

The resulting NPV of the assembly system is $45,018.57, and therefore we will recommend that the company approve the system for purchase and implementation.

5.4.4 Breakeven Analysis and Replacement Analysis

Two additional system selection techniques warrant discussion in this chapter due to their applicability to life cycle costing: breakeven analysis and replacement analysis [7]. In a breakeven analysis, the system output quantity required to earn a zero profit (breakeven) is determined as a function of the sales per output unit, variable cost per output unit, and total fixed cost. Once the required breakeven output quantity is determined, a judgment as to whether or not this level of output is reasonable determines if the system should be selected. Replacement analysis is generally performed using the equivalence formulas presented in Table 5.11. The primary decision is whether an existing system (defender) should be retired from use, continued in service, or replaced with a new system (challenger). Accurate replacement analysis is very important in the retirement stage of the life cycle.

5.5 RISK AND UNCERTAINTY IN COST ESTIMATION

As mentioned throughout this chapter, there is often considerable uncertainty associated with cost estimates, especially early in a systems life cycle. Cost risk has been defined in Chapter 3 as "the probability of exceeding the development, production, or operating budget in whole or in part." These probabilities are a function of the amount of uncertainty present in the cost estimate. All estimates have some level of uncertainty in them. Table 5.2 illustrates the relative accuracy for the different classes of cost estimates.

When presenting cost estimates to management, the analyst should attempt to quantify the uncertainty associated with the estimate. For example, when a linear regression model is used to develop a CER, the analyst can compute the two-sided confidence limits on the coefficients of the model. The confidence limits can be used to express a measure of uncertainty for the cost estimating relationship. If an analyst is using expert judgment and creates a cost estimate by analogy, the analyst can attempt to capture the level of certainty from the expert or could use the uncertainty expressed in Table 5.2 for Class 1 or Class 2 estimates.

For example, in the analogy estimate provided in Section 5.3.1, the expert stated that the technology was three times more complex than the previous generation of the system. The expert may really believe that it is most likely three times more complex than the previous technology but no less than two times more complex and is no more than 4.5 times as complex as the previous generation of technology. This information on the uncertainty associated with the expert's quantification of complexity can then bound the estimate:

$$\text{Cost}_{min} = (1.2) \times \$10,000 = \$12,000$$

$$\text{Cost}_{likely} = (1.3) \times \$10,000 = \$13000$$

$$\text{Cost}_{max} = (1.45) \times \$10,000 = \$14,500$$

5.5.1 Monte Carlo Simulation Analysis

Monte Carlo analysis is a useful tool for quantifying the uncertainty in a cost estimate. The Monte Carlo process creates a probability distribution for the cost estimate by rolling up all forms of uncertainty into a single distribution that represents the potential system costs. Once this distribution has been constructed, the analyst can provide management with meaningful insight about the probability that the cost exceeds a certain threshold, or that a schedule is longer than a specific target time. In Chapter 12 we will illustrate another use of Monte Carlo simulation analysis as a means of conducting sensitivity analysis on how uncertainty present in measure scores affects the total value for a system. Kerzner [10] provides five steps for conducting a Monte Carlo analysis for cost and schedule models. These steps are:

1. Identify the appropriate WBS level for modeling; the level will be a function of the stage in the system life cycle. In general, as the system definition matures, lower-level WBS elements can be modeled.
2. Construct an initial estimate for the cost or duration for each of the WBS elements in the model.
3. Identify those WBS elements that contain significant levels of uncertainty. Not all elements will have uncertainty associated with them. For example, if part of your system under study has off-the-shelf components and you have firm-fixed price quotes for the material, then there would be no uncertainty with the costs for those elements for the WBS.

4. Quantify the uncertainty for each of the WBS elements with an appropriate probability distribution.

5. Aggregate all of the lower-level WBS probability distributions into a single WBS Level 1 estimate by using a Monte Carlo simulation. This step will yield a cumulative probability distribution for the system cost. This distribution can be used to quantify the cost risk as well as identify the cost drivers in the system estimate.

Kerzner [10] emphasizes that caution should be taken when using Monte Carlo analysis. Like all models, the results are only as good as the data used to construct the model, the old adage that "garbage in, yields garbage out" applies to these situations. The specific distribution used to model the uncertainty in WBS elements depends on the information known about each estimate. Many cost analysts default to the use of a triangle probability distribution to express uncertainty. The choice of a triangle distribution is often a matter of convenience rather than the result of analysis. The probability distribution selected should fit some historical cost data for the WBS element being modeled. The triangle distribution will often be used for early life cycle estimates where minimal information is available (lower and upper bounds) and an expert is used to estimate the likeliest cost. When only the bounds on a WBS element are able to be reasonably estimated, a uniform distribution is frequently used in a Monte Carlo simulation to allow all values between the bounds to occur with equal likelihood. As the system definition matures and relevant cost data become available, other distributions should be considered and the cost estimate updated.

Example. Suppose you have been tasked to provide a cost estimate for the software nonrecurring costs for a Department of Defense (DoD) satellite system. The following CERs have been developed to estimate the cost of the ground control software, the system support software, and the mission embedded flight software during the conceptual design phase of the life cycle.

Application and system software: Person-months $= 4.3(\text{EKSLOC})^{1.2}(1.4)^{\text{DoD}}$

Support software: Person-months $= 5.3(\text{EKSLOC})^{0.95}$

Space mission embedded flight software: Person-months $= 8.7(\text{EKSLOC})^{1.6}$ $(1.6)^{\text{DoD}}$

The DoD parameter equals 1 if it is a DoD satellite, and 0 otherwise. EKSLOC is a measure of the size of the software coding effort. It is the estimated size measured in thousands of source lines of code, but the engineers still need to estimate these sizes for their project.

Suppose that it is early in the design process and the engineers are uncertain about how big the coding effort is. The cost analyst has chosen to use a triangle distribution to estimate the EKSLOC parameter. The analysts ask the expert to provide a most likely line of code estimate, m, which is the mode, and two other estimates, a pessimistic size estimate, b and an optimistic size estimate, a. The estimates of *a* and *b* should be selected such that the expert believes that the actual size of the source lines of code will never be less (greater) than *a*(*b*). These become the lower and upper bound estimates for the triangular distribution. Law and Kelton [20] provide computational formulas for a variety of continuous and discrete distributions. The expected value and variance of the triangle distribution are calculated as follows:

$$\text{Expected value} = \frac{a + m + b}{3} \tag{5.12}$$

$$\text{Variance} = \frac{a^2 + m^2 + b^2 - am - ab - mb}{18} \tag{5.13}$$

Suppose our expert defines the following values for EKSLOC for each of the software components.

Using these values and the associated CERs a Monte Carlo analysis is performed using Oracle® Crystal Ball [21] software designed for use with Excel. The probability density function (PDF) for the embedded flight software is shown in Figure 5.15.

The PDF for the estimated labor hours is given in Figure 5.16 below for 10,000 simulation runs. Finally, suppose that management is uncertain about the labor cost for software engineers. Management believes the labor cost is distributed as a normal random variable with a mean of \$20 per hour and standard deviation of \$5. The CDF for the total software development cost is given below. This estimate assumes that engineers work 36 hours in a week on coding and that there are 4 weeks in a month. Figure 5.17 shows the PDF for the software development costs. The primary observation to take away from Figure 5.17 is the spread in possible software development costs due to the uncertainty assumptions imposed on the WBS elements when the Monte Carlo simulation was constructed. For this example, while it is more likely that the actual software development costs will clump around \$12 million, it is possible for them to be up to four times as much or as little as one-tenth as much because of this uncertainty. In the former case, the project could be threatened; in the latter, the project would continue well within budget. What is the probability of these values occurring?

Once we have the PDF for the cost of development, we can construct the cumulative distribution function (CDF) to answer this question. Applications such as Crystal Ball accomplish this task easily. Figure 5.18 contains the CDF for

TABLE 5.14 EKSLOC Values by Software Component

Software type	Optimistic Size Estimate (KSLOC)	Most Likely Size Estimate (KSLOC)	Pessimistic Size Estimate (KSLOC)	Mean (KSLOC)	Variance (KSLOC)
Application	5	10	35	16.67	43.05
Support	70	150	300	173.33	2272.22
Embedded flight	7	25	80	37.33	241.06

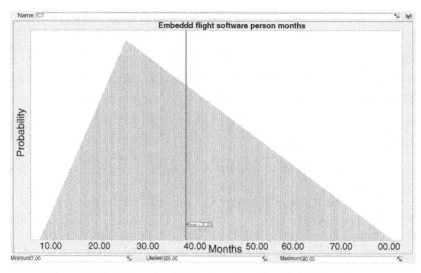

Figure 5.15   Triangular distribution for embedded flight software.

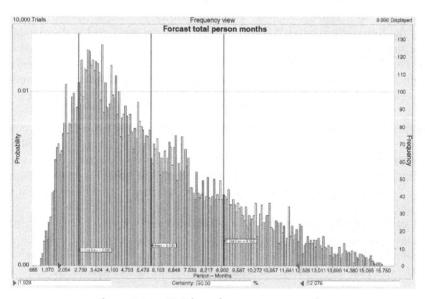

Figure 5.16   PDF for software person months.

software development costs. Using the CDF, we can make probability statements related to the software development cost. For example, we can state that there is a 50% probability that the software development costs will be less than $15.96 million; similarly, there is a 20% probability that the software development costs will exceed $26.59 million. This information is useful to senior-level management as they assess the cost risks associated with your program.

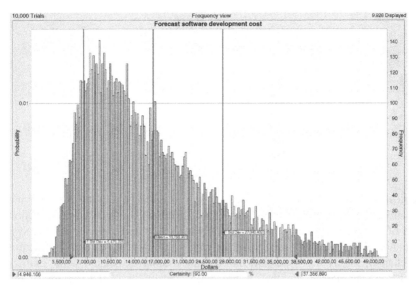

Figure 5.17   PDF for software development cost.

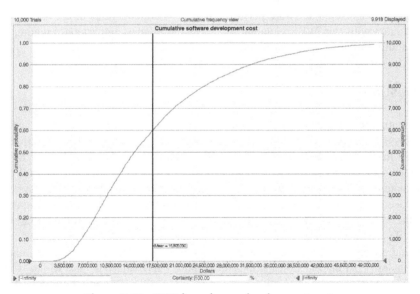

Figure 5.18   CDF for software development cost.

5.5.2   Sensitivity Analysis

The results can then be analyzed to identify those elements that are the most significant cost drivers. In our example, it is relatively easy because our total cost is only a function of three cost elements and one other factor. But realistic cost estimates may have on the order of 25 to 100 cost elements/factors, and choosing

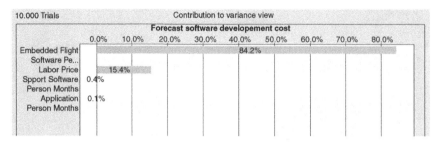

Figure 5.19   Sensitivity chart.

the cost drivers from this set is not so easy. Fortunately, Crystal Ball provides a tornado diagram (named for the shape) that analyzes the relative contribution of each of the uncertain components to the overall cost and variance for the system estimate. Figure 5.19 shows the sensitivity output for this example. Examining the chart in Figure 5.19, we find that the uncertainty associated with "embedded flight software person months" is the main contributor to the variability in the cost estimate, followed by the uncertainty in labor rate for software engineers. The cost analysis should consider spending more time getting a better estimate for the "embedded flight software person months," since reductions in the uncertainty associated with this WBS element will have the greatest impact on reducing the variability in the total cost estimate seen in Figure 5.17.

5.6   SUMMARY

Estimating the cost of a system is an essential component of the systems decision process. Since many decisions are based on cost, it is imperative that the estimates constructed be as accurate as possible. This chapter has introduced some of the fundamental concepts and techniques necessary for developing a meaningful cost estimate of a system. Life cycle cost estimating, like many things in the systems engineering process, is an iterative task. Costs estimates should undergo updates as the system is defined and as the system moves through the various phases of the life cycle. In this chapter, we have discussed the evolution of the cost estimating process. A variety of techniques have been introduced including estimating by analogy, parametric cost estimates, learning curve analysis, and Monte Carlo analysis. Systems engineers and engineering managers must know how to perform and interpret life cycle cost analysis.

5.7   EXERCISES

5.1. Describe how the system life cycle affects the life cycle costing process.

5.2. Think of three factors that influence the accuracy of a life cycle cost estimate and explain how each of the factors influences cost estimation accuracy.

5.3. Why is it important to consider the system uncertainty when developing a life cycle cost estimate?

5.4. To help manage production of a company's missile systems, the purchase of a production scheduling software is being considered. The software is expected to have a useful life of four years and can be purchased and installed for $150,000. Annual staff training and upgrades are expected to cost $40,000. The system engineers anticipate an operating cost savings of $75,000 due to an improved production schedule. It is predicted that the software will have no value at the end of its useful life. The company has a minimum attractive rate of return (interest rate) of 10% and uses an estimated annual inflation rate of 3%.

(a) Use net present value to determine if the software should be purchased.

(b) Suppose the company increases their minimum attractive rate of return to 15%, should they still purchase the software?

5.5. Open the daily newspaper to the local real estate section. Using the data available to you on the real estate pages of the newspaper, develop a CER for the cost of housing in a particular community. Using your CER, predict the cost of a particular house in another community and compare it to the list price. Explain the reasoning for any differences. Is this good practice?

5.6. Open the daily news paper to the automotive section and collect some cost data on at least 10 new automobiles. You will also need to collect some performance data and technical specifications for the vehicles. Using the information, construct a CER for the cost of a vehicle. Using your CER, predict the cost of another vehicle using its specifications and compare it to its list price. Explain the reason for any difference between the CER and the list price for the vehicle.

5.7. Suppose we have collected the following cost and performance data on rocket motors for missile propulsion systems. Establish a cost estimating relationship between cost and thrust,. Establish a CER between cost and length. Finally, establish a CER that utilizes both thrust and length.

Cost ($)	Thrust (pounds)	Length (inches)
33,003	1,000	36
34,712	1,200	40
27,253	500	24
39,767	2,000	42
51,093	5,000	62
62,073	10,000	98
36,818	1,500	40
79,466	25,000	124
68,628	15,000	78

5.8. One of Best Buys new Geek Squad members changed his first motherboard on a Thursday morning in 50 minutes. You know that the learning rate for this kind of task is 70%. If the chief Geek has him change 7 additional motherboards on Thursday morning, how long will it take him to change the last one?

5.9. Suppose we wish to identify the assembly time of the 100th unit for a disc brake system. Assume we know that the assembly time for the 1st unit is 120 minutes. Assuming a learning rate of 80% compute the assembly time for the 100th unit.

5.10. The automotive union has established a standard of performance for assembling the door panel based on the 100th unit. Suppose the standard is such that the target assembly time for the 100th unit is 150 minutes. Suppose the negotiated learning rate is 75%, what is the expected assembly time of the first unit?

5.11. Suppose you have collected the following data on time required to perform an overhaul on an aircraft system from the aircraft depot information system. Determine the learning rate for the overhaul process.

Week	Cumulative hours	Cumulative units completed
1	120	0
2	200	1
3	310	2
4	380	3
5	440	4
6	500	5

5.12. Consider problem nine again. Assume that you are uncertain of the actual learning rate, and suppose you believe it is most likely 80%, it would have a minimum value of 70% and a maximum value of 95%. Using a triangle distribution to represent your uncertainty, conduct a Monte Carlo analysis on the expected time of assembly for the 100th unit. Compute the probability that the assembly time of the 100th unit is longer than 90 minutes.

5.13. In problem 5.4, assume that you are uncertain about the interest rate as well as the inflation rate. Suppose the most likely value for the interest rate is 10% and the min value is 8% and the max value is 11%. Similarly, suppose the most likely value for the inflation rate is 3% and the min value is 1% while the max value is 6%. Assuming that both parameters can be modeled using a triangle distribution, conduct a Monte Carlo analysis to determine the distribution for the NPV. Based on the levels of uncertainty associated with the interest rate and inflation rate, would you purchase the software?

REFERENCES

1. International Committee for Systems Engineering (INCOSE). Retrieved 2007 March, from www.incose.org.
2. Society of Cost Estimating and Analysis. 2007. Glossary. Retrieved 2007 April, from http://www.sceaonline.org/prof_dev/glossary-l.cfm.
3. AACE, Inc. *AACE International Recommended Practice*. No. 17R-97. Cost Estimation Classification System. 1997.
4. U.S. Bureau of Labor Statistics, Division of Consumer Prices and Price Indexes. Retrieved April 2007, from http://www.bls.gov/CPI.
5. Stewart, R, Wyskida, R, Johannes, J. *Cost Estimator's Reference Manual*, 2nd ed. New York: John Wiley & Sons, 1995.
6. Canada, J, Sullivan, W, White, J, Kulonda, D. *Capital Investment Analysis for Engineering and Management*, 3rd ed. Upper Saddle River, NJ: Prentice-Hall, 2005.
7. Ostwald, P. *Cost Estimating*, 3rd ed. Englewood Cliffs, NJ: Prentice-Hall, 1992.
8. Lee, D. *The Cost Analyst's Companion*. McLean, VA: Logistics Management Institute, 1997.
9. Park, C. *Fundamentals of Engineering Economics*. Upper Saddle River, NJ: Pearson-Prentice-Hall, 2004.
10. Kerzner, H. *Project Management*, 9th ed. Hoboken, NJ: John Wiley & Sons, 2006.
11. GAO-07-406SP. Defense acquisitions: Assessment of selected weapon systems, United States Government Accountability Office. Retrived from http://www.gao.gov/new.items/d07406sp.pdf.
12. Fabrycky, W, Blanchard, B. *Life Cycle Cost and Economic Analysis*. Englewood Cliffs, NJ: Prentice-Hall, 1991.
13. Mil-Hdbk-881a. *Department of Defense Handbook*. Work Breakdown Structures for Defense Material Items, 30 July 2005. Retrieved from http://dcarc.pae.osd.mil/881handbook/881A.pdf.
14. Minitab Statistical Software. Available at http://www.minitab.com. Accessed 25 April 2010.
15. JMP Statistical Software. Available at http://www.jmp.com. Accessed 25 April 2010.
16. Brown, C, Horak, J, Waller, W, Lopez, B. *Users Manual for TBMD Missile Cost Model*. TR-9609-01. Santa Barbara, CA: Technomics, 1997.
17. Shafer, W, Golberg, M, Om, N, Robinson. M. *Strategic System Costs: Cost Estimating Relationships and Cost Progress Curves*. IDA Paper P-2702, November 1993.
18. Wright, T. Factors, affecting the cost of airplanes. *Journal of Aeronautical Sciences*, 1936; 3 (February), 122–128.
19. Thuesen, G, Fabrycky, W. *Engineering Economy*, 7th ed. Englewood Cliffs, NJ: Prentice-Hall, 1989.
20. Law, A, Kelton, W. *Simulation Modeling and Analysis*, 3rd ed. Boston, MA: McGraw-Hill, 2000.
21. ORACLE® Crystal Ball. 2005. Denver, CO: Decisioneering, Inc. Available at http://www. decisioneering.com; Internet.

Part II

Systems Engineering

Chapter 6

Introduction to Systems Engineering

GREGORY S. PARNELL, Ph.D.

Science determines what IS ... Component engineering determines what CAN BE ... Systems engineering determines what SHOULD BE and Engineering managers determine what WILL BE.

—Modified from Arunski et al. [1]

6.1 INTRODUCTION

This chapter introduces systems engineering. We begin by reviewing our definition of a system. We provide a brief history of the relatively new discipline of systems engineering. Then we identify the major trends in systems that provide challenges for systems engineers. Next, we review our definition of systems engineering and use it to identify the three fundamental tasks of systems engineers. Then we discuss the relationship of systems engineers to other engineering disciplines. Finally, we conclude with discussion of the education and training of systems engineers.

6.2 DEFINITION OF SYSTEM AND SYSTEMS THINKING

Understanding systems and using systems thinking are foundational for systems engineering. Recall our definitions of systems and systems thinking from Chapter 1.

Decision Making in Systems Engineering and Management, Second Edition
Edited by Gregory S. Parnell, Patrick J. Driscoll, Dale L. Henderson
Copyright © 2011 John Wiley & Sons, Inc.

A system is "an integrated set of elements that accomplish a defined objective. These elements include products (hardware, software, firmware), processes, people, information, techniques, facilities, services, and other support elements" [2].

Systems thinking is a holistic mental framework and world view that recognizes a system as an entity first, as a whole, with its fit and relationship with its environment being primary concerns.

6.3   BRIEF HISTORY OF SYSTEMS ENGINEERING

Systems engineering is a relatively new discipline. Of course, people have engineered large, complex systems since the Egyptians built the pyramids, Romans built their aqueducts, and fifteenth-century sea captains prepared their sailing ships to go around the world. However, until about a hundred years ago, system integration was generally in the hands of a craft specialist with a lifetime of experience in his craft, working mainly from a huge store of painfully learned rules of thumb. The need for disciplined and multidisciplinary engineering at the systems level was first widely recognized in the telephone industry in the 1920s and 1930s, where the approach became fairly common. Systems thinking and mathematical modeling got a further boost during World War II with the success of operations research, which used the scientific method and mathematical modeling to improve military operations. After the war, the military services discovered that systems engineering was essential to developing the complex weapon systems of the computer and missile age. The private sector followed this lead, using systems engineering for such projects as commercial aircraft, nuclear power plants, and petroleum refineries. The first book on systems engineering was published in 1957 [3]; and by the 1960s, degree programs in the discipline became widely available. The professional organization for systems engineers, now known as the International Council on Systems Engineering (INCOSE) [4], was founded in 1990.

The INCOSE mission is to enhance the state-of-the-art and practice of systems engineering in industry, academia, and government by promoting interdisciplinary, scalable approaches to produce technologically appropriate solutions to meet societal needs [4].

6.4   SYSTEMS TRENDS THAT CHALLENGE SYSTEMS ENGINEERS

There are several important systems trends that create significant systems engineering challenges. First, systems have become increasingly more complex, more dynamic, and more interconnected, involve more stakeholders than in the past, and face increasing security, and privacy challenges.

Increasing Complexity. Systems are more complex. Today's systems involve many science and engineering disciplines. New technologies (including information, biotechnology, and nanotechnology) create new opportunities and challenges. Interfaces are increasingly more complex and system integration is more difficult. To emphasize this complexity new terms have been used; for example, systems of systems, system architectures, and enterprise systems.

More Dynamic. Systems interact with their environment and the needs of stakeholders evolve in concert with this interaction. Rapid changes in the environment require systems to be dynamic to continue to provide value to consumers of products and services.

Increasing Interconnectedness. The Internet and advances in information technology have led to business-to-business collaboration and a global economy enabled by pervasive interconnectedness. Anyone can start a global business by establishing a website. We now have an international supply chain for electronic components; and, increasingly, hardware development, software development, component production, and services are being done globally.

Many Stakeholders. The increasing complexity and interconnectedness contribute to the increase in the number of stakeholders involved in the system life cycle. In addition to considering the perspectives of scientists, engineers, and engineering managers, system engineers must consider the perspectives of functional managers (production, sales, marketing, finance, etc.), regulators, professional organizations, legal, environmentalists, government, community groups, and international groups to name just a few of the many stakeholders with vested interests in the system.

Increasing Security Concerns. Many systems face increasing security challenges due to threats from malicious adversaries ranging from hackers to terrorists. Information assurance, which is the activity of protecting data and its flow across communication networks, is a major concern of system developers and users. In a similar fashion, physical security is an important design criteria for many systems as well.

Increasing Privacy Concerns. As systems become more complex, more interconnected, and face more security challenges the potential for privacy violations increases. The protection of personal information in systems is now a major system challenge.

Complexity is a challenging concept when it comes to systems decision problems. Table 6.1 illustrates a modified and expanded spectrum of complexity for low, medium, and high complexity problems across 10 problem dimensions of systems decision problems based on the complexity research of Clemens [5]. The third category, called a "Wicked Problem," is a recently characterized phenomenon typifying a growing number of systems in existence today. These descriptions are particularly helpful for the Problem Definition phase of the SDP during which the systems

TABLE 6.1 Dimensions of Problem Complexity [6]

Problem Dimension	Low (Technical Problem)	Medium (Complex Problem)	High (Wicked Problem)
Boundary Type	Isolated, defined Similar to solved problems	Interconnected, defined Several unique features and new constraints will occur over time	No defined boundary Unique or unprecedented
Stakeholders	Few homogeneous stakeholders	Multiple with different and/or conflicting views and interests	Hostile or alienated stakeholders with mutually exclusive interests
Challenges	Technology application and natural environment requirements	New technology development, natural environment, adaptive adversaries	No known technology, hostile natural environment, constant threats
Parameters	Stable and predictable	Parameter prediction difficult or unknown	Unstable or unpredictable
Use of Experiments	Multiple low-risk experiments possible	Modeling and simulation can be used to perform experiments	Multiple experiments not possible
Alternative Solutions	Limited set	Large number are possible	No bounded set
Solutions	Single optimal and testable solution	Good solutions can be identified and evaluated objectively and subjectively	No optimal or objectively testable solution
Resources	Reasonable and predictable	Large and dynamic	Not sustainable within existing constraints
End State	Optimal solution clearly defined	Good solutions can be implemented but additional needs arise from dynamic needs	No clear stopping point

team is dedicated to identifying the structure, characteristics, and challenges of the system under study.

Second, the risks associated with systems have increased dramatically. As systems become more complex, dynamic, and interconnected and face security challenges from determined adversaries, the consequences of system failures increase dramatically. System failures can become system catastrophes. Examples of system catastrophes include the Exxon Valdez oil spills, the terrorist attacks on 11 September 2001, the loss of the Challenger space shuttle, the loss of the Columbia space shuttle, Hurricane Katrina the financial crisis that began in 2007, and the British Petroleum (BP) Gulf oil spill of 2010. These catastrophic events have resulted in significant direct and indirect consequences including loss of life, economic, social, environmental, and political consequences, precisely the type of events that the risk management approach introduced in Chapter 3 is intended to address. Each of these events fall into at least one of the risk categories identified: environmental risk (hurricanes), technical risk (space shuttles), operational risk (deep water oil drilling), organizational risk (space shuttles), and systemic risk (financial crisis of 2007).

6.5 THREE FUNDAMENTAL TASKS OF SYSTEMS ENGINEERS

In Part I, we identified the need for systems thinking and introduced a host of tools for applying systems thinking to systems decision problems. All participants in the system life cycle should use systems thinking. However, systems engineers are the primary system thinkers. We define systems engineering using the INCOSE definition [2]:

> Systems engineering is an interdisciplinary approach and means to enable the realization of successful systems. It focuses on defining customer needs and required functionality early in the development cycle, documenting requirements, then proceeding with design synthesis and system validation while considering the complete problem.

From this definition, we can derive the three most fundamental tasks of systems engineers and the key questions that systems engineers must answer for each task.

Task 1: Use an interdisciplinary systems thinking approach to consider the complete problem in every systems decision in every stage of the system life cycle.

The problems change in each stage of the life cycle. The initial problem statement from one decision maker or stakeholder is never the total problem. In each stage, an interdisciplinary approach to systems thinking and problem definition is required by the system trends described in the previous section.

As they perform the first fundamental task, they must answer the following key questions modified from [7]:

- What is the system under consideration?
- What is the system boundary?
- What is the actual problem we are trying to solve?
- Who are the decision makers and stakeholders?
- What are the influencing factors and constraints of the system environment?
- How will we know when we have adequately defined the problem?
- What value can the system provide to decision makers and stakeholders including clients, system owners, system users, and consumers of products and services?
- How much time do we have to solve the problem?

Chapter 10 describes the first phase in our systems decision process (SDP), including some useful techniques for problem definition based on research and stakeholder analysis.

Task 2: Convert customer needs to system functions, requirements, and performance measures.

The opening quote of this chapter clearly defined the difference between scientists, component (discipline) engineers, systems engineers, and engineering managers. Systems engineers determine what should be. One of the most surprising facts of systems engineering is that it is not always easy to identify the future users or consumers of the system that will be designed as a solution to a problem. Many times the organization funding the system is not the user or the consumer of the product or service. This is especially true when the users do not directly pay for the service or product the system provides. This happens in many government systems engineering efforts. Working with customers and users to determine the functions that the system must perform is a daunting task when dealing with complex, dynamic, interdependent systems involving many stakeholders and facing significant security and privacy challenges. Once the functions have been determined, the system requirements must be specified and assigned to system elements (components) so the component engineers can begin design.

The following are some of the key questions for this task:

- Who are the stakeholders (clients, system owners, system users, and consumers of product and services) holding a vested interest in the system?
- What is our methodology for implementing the systems engineering process to define system functions and requirements?
- How do we involve decision makers and stakeholders in our systems engineering process?
- What are the functions the system needs to perform to create value for stakeholders?
- What are the design objectives for each function?
- How will we measure the ability of a design solution to meet the design objectives?
- What are the requirements for each function?
- How will we allocate system functions to system elements?
- How, when, and why do system elements interact?
- What are the design, operational, and maintenance constraints?
- How will we verify that elements meet their requirements and interfaces?

Chapter 10 describes the tasks and some useful techniques for functional and requirements analyses and value modeling. The screening criteria (the requirements that any potential solution must be able to meet) and the value model (which defines the minimum acceptable levels on each value measures) define the major system requirements.

Task 3: Lead the requirements analysis, design synthesis, and system validation to achieve successful system realization.

After identifying system functions, requirements, and performance measures, the systems engineer must lead the requirements analysis and design synthesis and validate that the design solution solves the defined problem. The basis for system design and validation is usually an iterative sequence of functional and requirements analyses modeling, simulation, development, test, production, and evaluation. For complex systems, a spiral development approach may be used to develop system prototypes with increasing capabilities until the system requirements and feasibility are established. Once the system design is validated, the systems engineer must continue to work on the successful system realization.

One of the most essential systems engineering tasks is to lead the resolution of requirements, configuration control, design integration, interface management, and test issues that will occur during the life cycle stages. The chief (or lead) systems engineer creates a technical environment that encourages the early identification, multidisciplinary assessment, creative solution development, timely decision making, and integrated resolution of engineering issues. To achieve the value the system was designed to obtain, the following are some of the key questions for this task:

- How will we know when we have adequately solved the problem?
- How do we ensure that the design will meet the requirements?
- How do we resolve conflicting requirements, interfaces or design issues?
- How can we allocate system performance to system elements?
- How can we identify and validate component and system interfaces?
- Can we trade off one performance measure versus another measure?
- How will we verify that system performance has been achieved?
- How do we identify, assess, and manage risk during the system life cycle?
- How do we trade off system performance with life cycle cost to ensure affordability?

Several chapters of this book provide information on performance measurement:

- Chapters 4 and 8 provide mathematical techniques for modeling system performance and system suitability.
- Cost is almost always a critical systems performance measure. Chapter 5 provides techniques for life cycle costing.
- Chapter 10 describes the multiple objective decision analysis value model that identifies the value measures (often system performance measures) for the objectives of each system function. The value model captures the value added for each increment of the range of the value measures. It also captures the relative importance of the value measure range to the system design.
- Chapter 4 describes modeling and simulation techniques to assess system performance, screen alternatives, and enhance candidate solutions.

- Chapter 12 describes the use of multiple objective decision analysis to assess the value of candidate solutions, improve the value of candidate solutions with Value-Focused Thinking (Chapter 9), and perform tradeoff analysis among value measures and between system value and system cost.

Systems thinking is critical to each of the three fundamental systems engineering tasks. Systems engineer must help everyone think about the problem and the solution that is being designed for the problem. Systems engineers must continue to be the consumer and user advocates as they help convert owner, user, and consumer needs to systems functions, requirements, and value measures. In addition, they must define, validate, and test the element interfaces. These three fundamental tasks are made more challenging by the increasing system complexity, dynamic changes, interconnectedness, number of stakeholders, and increasing security and privacy concerns.

6.6   RELATIONSHIP OF SYSTEMS ENGINEERS TO OTHER ENGINEERING DISCIPLINES

As noted in our opening quote of this chapter, systems engineering is not the only engineering discipline involved in the system life cycle. Section 1.6, Table 1.1 provides a comparison of systems engineering and other engineering disciplines' views on several dimensions. Scientists play a critical role by developing the fundamental theory that supports each of the engineering disciplines. Component (or discipline) engineers develop the technology that determines what can be. By performing the tasks described above, systems engineers determine what should be. Finally, engineering managers determine what will be. The manager approves all the products developed by the systems engineers and obtains the organizational approvals and resources to design, produce, and operate the system.

Systems engineers help provide design synthesis. To do this job, they need to understand the system component engineering disciplines, work effectively with their engineering colleagues, and know when to bring interdisciplinary teams together to solve requirement, design, test, or operational problems.

6.7   EDUCATION, TRAINING, AND KNOWLEDGE OF SYSTEMS ENGINEERS

The education and training of systems engineers includes undergraduate, graduate, and continuing education, and certification programs. Most of these education programs introduce and reinforce the systems thinking and systems engineering tools needed to prepare systems engineers for work throughout the system life cycle of modern systems. Training programs are sometimes tailored to a stage in the system life cycle and the roles of the systems engineer in that stage.

Since it is a relatively new discipline, systems engineering programs are still being developed at several colleges and universities. Systems engineering programs are accredited by ABET Inc., which accredits all engineering disciplines [8]. INCOSE is currently working with ABET to become the professional society to establish systems engineering centric accrediting standards for systems engineering programs [8]. In addition to undergraduate systems engineering degree programs, several universities offer masters and Ph.D. programs in systems engineering. These programs have different names. Some of the common names are systems engineering, industrial and systems engineering, and information and systems engineering. A list of these programs can be found at [9]. Many engineers, including the editors, have undergraduate degrees in another engineering discipline before obtaining a graduate degree in systems engineering.

There are also several systems engineering continuing education programs offered. INCOSE established a systems engineering certification program in 2004 which has continued to expand [10].

As an evolving and growing field, systems engineering knowledge is available from many sources including textbooks and technical books. Specific series have been established by various publishers to highlight a select offering of references considered foundational to understanding the principles and practices of systems engineering. Two examples are the Wiley Series in Systems Engineering and Management [11] and the CRC Press Complex and Enterprise Systems Engineering series [12]. For a broader list of systems engineering books, readers can refer to Wikipedia. A second source of useful reference material is contained in primers, handbooks, and bodies of knowledge. INCOSE versions of these type of reference materials are available on their website. A third source of systems engineering knowledge is found in published standards. And in this, INCOSE committees continue to work to develop and improve worldwide SE standards [13].

6.7.1 Next Two Chapters

This completes our brief introduction to systems engineering. The next two chapters provide more information on roles and tools of systems engineers. Chapter 7 discusses systems engineering in professional practice. Key topics include the roles of systems engineering, the tasks and responsibilities of systems engineers in each stage of the system life cycle, and how they relate to other professionals in the system life cycle. Two of the key responsibilities of systems engineers are to assess system effectiveness and system risk. Chapter 8 discusses analysis techniques used by systems engineers to model systems effectiveness and systems suitability. Key topics include reliability modeling, maintenance modeling, and availability modeling.

6.8 EXERCISES

6.1. Write a simple one sentence definition of systems engineering that you can use to define your major and/or department to family and friends.

6.2. Who were the first system engineers? What were the factors that caused the discipline of systems engineering to be established?

6.3. What is the name of the professional society for systems engineering? Visit the website and find the purpose of the society. What resources are available on the website that might be useful to you in your academic program or professional career?

6.4. List and describe some system trends that create challenges for systems engineers. Illustrate each of these trends using the events of 9/11.

6.5. List and describe the three fundamental tasks of systems engineers.

6.6. Explain why an interdisciplinary approach is needed for systems engineering.

6.7. Explain why system performance measurement is important. Who develops performance measures and what are they used for? List three system performance measures for an automobile.

6.8. What is the difference between a system function and a system requirement? How are the functions and requirements identified for a new system? What is the role of the systems engineer?

6.9. What is the difference among the following three programs: an undergraduate systems engineer, a graduate systems engineer, and a systems engineering certification program?

6.10. Find an ABET accredited undergraduate systems engineering program at another college or university. How does their program compare to your program? What are the similarities and differences?

ACKNOWLEDGMENT

Dr. Roger Burk, United States Military Academy, provided the initial draft of the section on the history of systems engineering.

REFERENCES

1. Arunski, JM, Brown, P, Buede, D. Systems engineering overview. Presentation to the Texas Board of Professional Engineers, 1999.
2. What Is Systems Engineering? www.incose.org/practice/whatissystemseng.[aspx]. Accessed January 26, 2006.
3. Goode, H, Machol, R. *Systems Engineering: An Introduction to the Design of Large-Scale Systems*. New York: McGraw-Hill, 1957.
4. International Committee for Systems Engineering (INCOSE), www.incose.org. Accessed January 26, 2006.
5. Clemens, M. The Art of Complex Problem Solving, http://www.idiagram.com/CP/cpprocess.html. Accessed May 10, 2010.
6. Parnell, G. Evaluation of risks in complex problems. In: Williams, T, Sunnevåg, K, and Samset, K, editors. *Making Essential Choices with Scant Information: Front-End Decision-Making in Major Projects*. Basingstoke, UK: Palgrave Macmillan, 2009.

7. *Systems Engineering Primer*. International Committee on Systems Engineering and American Institute of Aeronautics and Astronautics, Systems Engineering Technical Committee, 1997.

8. http://www.abet.org/http://www.abet.org/. Accessed June 26, 2006.

9. http://www.incose.org/educationcareers/academicprogramdirectory.aspx

10. http://www.incose.org/educationcareers/certification/index.aspx. Accessed June 25, 2006.

11. Wiley Series in Systems Engineering and Management. Editor: Andrew P. Sage. http://www.wiley.com/WileyCDA/Section/id397384.html. Accessed April 20, 2010.

12. *Complex and Enterprise Systems Engineering*, CRC Press, Taylor & Francis Publishing. http://www.crcpress.com/ecommerce_product/book_series.jsf?series_id=2159. Assessed April 20, 2010.

13. INCOSE Standards Update, http://www.incose.org/practice/standardsupdate.aspx. Accessed March 10, 2010.

Chapter 7

Systems Engineering in Professional Practice

ROGER C. BURK, Ph.D.

I have been involved with many milestone reviews as the principal for the Defense Acquisition Board. I see examples of well-run and poorly run projects. The difference becomes clear in the first few minutes of the review. What is also clear is the important role that systems engineering plays in making a project run smoothly, effectively, and efficiently, as well as the contrary, where the lack of systems engineering and the discipline that comes with proper implementation can cause tremendous problems [1].
— Michael W. Wynne, Secretary of the Air Force

7.1 THE SYSTEMS ENGINEER IN THE ENGINEERING ORGANIZATION

> The typical job for a professional systems engineer is technical integrator supporting a Program Manager who is developing a complex system.

Systems Engineering as a discipline was introduced in Section 1.7, with Chapter 6 describing it in more detail. This chapter is focused on the practitioner of the discipline: his or her place in the organization, responsibilities, and specific activities and tasks, so as to convey what it is like to be a systems engineer. I also want to give some references for important aspects of the systems

Decision Making in Systems Engineering and Management, Second Edition
Edited by Gregory S. Parnell, Patrick J. Driscoll, Dale L. Henderson
Copyright © 2011 John Wiley & Sons, Inc.

engineer's job that are beyond the scope of this book. This section discusses the job in general, including the title and the organizational placement. Section 7.2 describes the activities during each part of the system life cycle. Section 7.3 describes relationships with other practitioners involved in system development. Section 7.4 goes into team formation and building, since systems engineers often lead interdisciplinary teams, either formally or informally. Section 7.5 gives more detail on typical responsibilities of a systems engineer—for instance, which documents he or she would be expected to prepare. The next two sections reflect on the nature of the systems engineering job: 7.6 describes the various roles an SE can play, and 7.7 gives the personality characteristics conducive to being a good systems engineer. Section 7.8 summarizes the chapter.

Figure 7.1 is a map of the concepts presented in this chapter. Typically, a professional systems engineer works for a Chief Systems Engineer, who in turn works for a Program Manager. The systems engineer is the technical interface with clients, users, and consumers and is often the one responsible for building a team. This team will adopt an attitude concerning the challenge (good or bad) and usually lasts throughout the team's life cycle. The systems engineer has roles and responsibilities, and ideally it has certain personal characteristics that contribute significantly to success. The specific activities that a systems engineer performs are distributed (nonuniformly) over the system life cycle. Throughout this chapter, the acronym SE will be used to describe both a systems engineer and the discipline of systems engineering. The distinction will be clear from the context within which the acronym is used.

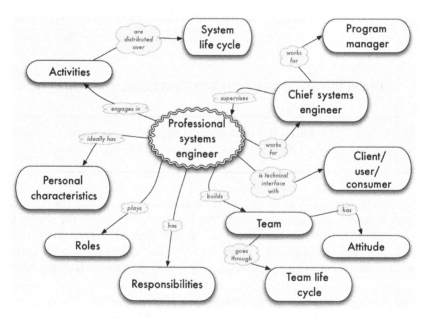

Figure 7.1   Concept map for Chapter 7.

The Systems Engineering Job

Anyone in any walk of life can use systems thinking to find good solutions to complex problems in a technological setting. To some extent, any professional engineer will use systems considerations to (a) determine the requirements for a system he or she is designing, (b) define its interfaces with other systems, and (c) evaluate how well it is performing. However, in professional practice some people are given specific big-picture engineering responsibility for a system or group of systems. These people have "systems engineer" in their job title or job description, and they are responsible for making sure that the technical efforts of everyone involved work together to produce an operational system that meets all requirements. For instance, a team developing a new helicopter will include mechanical engineers to make sure all the moving parts work together, aeronautical engineers to make sure the vehicle flies, electrical engineers to make sure the control systems work, software engineers to make sure the onboard computer systems operate properly, and many other discipline or specialty engineers: reliability engineers, test engineers, human factors engineers, and so on. It is the systems engineer who takes the overall view and is responsible for ensuring that all these engineers work together at the technical level to produce a system that meets the customer's needs. This chapter is about what these professional systems engineers actually do.

The International Council on Systems Engineering (INCOSE) has produced a Systems Engineering Handbook [2] that covers much of the material of this chapter in more detail, but in condensed handbook form for practitioners.

Three Systems Engineering Perspectives

There are three perspectives on the practice of systems engineering: the organization's, the system's, and the individual's. The organization is interested in (a) the process and what people do and (b) meeting customer needs and expectations. However, we also have to remember the point of view of the system: the product and its development throughout its life cycle. This is the long-term point of view, concerned with the design, manufacture, test, deployment, and operation of the system, regardless of the developing organization. Finally, the individual SE has a different perspective, concerned with the here and now, tasks and responsibilities, and getting the job done. This chapter stresses the organizational perspective and also describes the system and individual perspectives as they relate to the engineering organization.

Organizational Placement of Systems Engineers

A civil or mechanical engineer sometimes works on his or her own, doing small design projects for individual customers. Such an engineer can also work for a small engineering company that takes on contracts to build bridges, levees, industrial facilities, and so on. In contrast, most systems engineers work for larger companies, because those are the companies that take on complex projects that cannot be done without a systems approach. The systems engineer will be a part of a team developing a complex system such as an aircraft or a telecommunications network, or perhaps he or she will be in a systems engineering staff overseeing a number of smaller projects. If a systems engineer works for a small company, it is generally a consulting firm whose clients are large companies or government agencies seeking advice on how to develop or acquire major complex systems.

The organizational placement of systems engineers can vary widely based on the scale and complexity of the system involved and on the technical practices of the organization. For a major system such as an enterprise software system, an aircraft, or a spacecraft, where the development cost is hundreds of millions of dollars or more, there will often be a dedicated chief systems engineer who reports to the program manager. This chief SE may have a staff of systems engineers working for him or her, especially early in system development when key decisions about system architecture and system requirements are made. In this case, "Systems Engineering" (SE) can refer to a functional organization. The chief SE may also be given responsibility for technical specialists, such as configuration management personnel. For a smaller system, there may be a chief systems engineer working with a smaller staff, or even working alone as an advisor to the program manager on overall technical integration of the program. Sometimes the program manager him- or herself is also designated as the program systems engineer. If the organization is responsible for a family of relatively small systems, such as a set of communications or electronic devices, there may be a single SE with technical oversight for all of them.

7.2   SYSTEMS ENGINEERING ACTIVITIES

> An SE coordinates technical efforts over the lifetime of the system, but he or she is usually busiest toward the beginning, when he or she defines the system requirements, develops the system concept, and coordinates and integrates the efforts of the other design engineers.

Whether a system is large or small, its development will go through a series of system life cycle stages (described in Chapter 3). The execution of the fundamental SE tasks (described in Chapter 6) will be distributed in a sequence of activities during these stages. The distribution is not uniform. Task 1 ("Use an interdisciplinary approach . . . ") applies constantly throughout the system life cycle, but Task 2 ("Convert customer needs . . . ") is predominantly in the early stages, and Task 3 ("Lead the requirements analysis . . . ") has peaks of activity both early, when system requirements are established, and later, during systems test, when proper function at the system level is verified. The systems engineer's changing set of activities during these life cycle stages can be described as follows. Table 7.1 summarizes them.

TABLE 7.1 Summary of Systems Engineering Activities During a Systems Life Cycle

Life Cycle Stage	Major SE Activities
Establish system need	Stakeholder interaction (especially users)
	Requirements definition
	Functional analysis
Develop system concept	Stakeholder interaction (especially users)
	Team building
	Requirements analysis and management
	Architectural tradeoff studies
	Definition of system architecture, elements, boundaries, and interfaces
	Creation of systems engineering master plan
Design and develop the system	Stakeholder interaction (especially users)
	Interface control
	Overall design coordination
	Requirements analysis and management
	Configuration control
	Specialty engineering
	Coordinating major design reviews
	System development testing
Produce system	Coordination with production engineers
	System acceptance testing
Deploy system	Coordinating deployment with users and other stakeholders
Operate system	Gathering data on system performance and user feedback
	Coordinating problem resolutions and system upgrades
Retire system	Coordinating hand-off of mission to replacement system
	Coordinating system disposal

Establish System Need

Stakeholder interaction is a key role for SEs during this stage. The SE will talk to the stakeholders, paying special attention to consumers of the system's products and services, the system users, and the system owner, in order to understand and document the needs that are driving the system development. For instance, if the system is a passenger aircraft, the consumer is the airline passenger, the user is the aircrew, and the owner is the airline. The SE will use the needs to identify the system functions and define the specific technical requirements for the system. The SE often acts as an interface between the consumer or user and the designer. He or she translates between "consumer-speak" or "user-speak" and "engineer-speak" and makes sure they understand each other correctly. The result is usually a formal statement of the system's form, fit, and function, agreed to in writing by all parties. After this stage, the SE continues to manage the requirements, ensuring no change without sufficient cause and agreement by all parties.

Develop System Concept

At this point, the need is established and a project initialized, and the SEs play perhaps their most important role. In the organizational dimension, this is a time of team building, organization, and planning. This is when people are recruited, background research in the problem area is done, key tasks identified, and a preliminary schedule created. At the technical level, the SE continues to manage and refine the system requirements. The SE also examines different candidate system architectures and helps make the selection of the one to be pursued. This crucial process is described in more detail in Section 7.5; it includes identifying system functions, system boundaries, and major system elements and defining their interfaces. A Systems Engineering Management Plan may also be written during this stage, to plan SE activities for the rest of the system's life cycle. A Test and Evaluation Master Plan may be written to describe how testers will ensure that the finished system meets requirements.

Design and Develop the System

Detailed design is primarily the responsibility of discipline-specific engineers, but the SE also has a role to ensure system-level requirements are met. The SE maintains technical cognizance over the whole program to ensure that all elements will work together as they should. The SE often coordinates interfaces between system elements and between the system and its environment. Often these interfaces will be defined in a set of interface control documents (ICDs), which will be under configuration control, meaning that they can be changed only after review by the SE and approval by a Configuration Control Board (CCB). The ICDs will define the interfaces exactly, often in mind-numbing detail. For instance, the ICD for a data exchange will identify each data field, the type of data in it, the units of measurement, the number of characters, and so forth. An SE is also often involved in specialty engineering such as reliability, maintainability, and usability (see Section 7.5 and Chapter 8 for more on specialty engineering and its role in SE). When it is time to integrate the various elements and test them at the system level, the SE is usually in charge. He or she will be involved in planning, executing, and evaluating operational tests, in which test personnel not involved in system development use the system under field conditions. This is the second peak of activity for fundamental systems engineering Task 3 as described in Chapter 6: the time when the SE ensures system validation and successful system realization.

Produce System

The SE may be involved in the production plan and in monitoring performance to ensure that the system is built as designed. During this stage, the SE plays a key role in the analysis and approval of engineering change proposals and block upgrades.

Deploy System

The SE will help plan the deployment to ensure that all operability needs are met.

Operate System

The SE examines how well the system meets the consumers' needs, users' needs, and the client's expectations. This often involves gathering data and feedback from the consumers and users after the system is deployed. These provide input for deciding on modifications for future systems, problem fixes or upgrades for systems already in use, and improvements on systems engineering practices.

Retire System

The SE ensures that system retirement and disposal are properly planned for and executed.

7.3 WORKING WITH THE SYSTEMS DEVELOPMENT TEAM

Because of their wide responsibilities, SEs often find themselves working with a wide variety of other professionals, who may have widely varying engineering expertise. The following sections sketch out how an individual SE typically works with other professionals. Of course, people do not always fill these roles as a full-time job, though the larger the program, the more likely that they will. These relationships are summarized in Table 7.2.

The SE and the Program Manager

Broadly speaking, every project or program will have a manager who has overall responsibility for both technical and business success. In private organizations, the program manager (PM) is responsible for making sure that the program makes money for the company. In public organizations, the program manager is responsible for making sure the program provides the product or service for a given budget. This is the most responsible job in the program. The chief SE is the PM's chief advisor on overall technical aspects of the program, and the SE staff provides system integration for the PM. Table 7.3 compares the responsibilities of the PM and the SE.

The SE and the Client, the User, and the Consumer

The client pays to develop the system, the user operates it, and the consumer receives products or services from it. These may not be the same individuals, or even in the same organization. For instance, in Army acquisition the customer is a program manager in Army Material Command stationed at a place like Redstone Arsenal in Alabama, whereas the user and the consumer may be soldiers in the

TABLE 7.2 Summary of Program Roles in Relation to the Systems Engineer

Individual	Basic Responsibility	Provides to SE	Receives from SE
Program manager (PM)	Business and technical management of program	Program direction	Technical advice and leadership
Customer	Acquire the best system for his or her organization	System requirements	Technical information and recommendations
User	Operate the system	System requirements	Technical information
CTO or CIO	Coordinate a company's technology policy and/or information systems	Guidance and cooperation	Technical information and recommendations
Operations researcher or system analyst	Mathematical modeling to support decision making	Well-supported technical recommendation	Tasking for trade studies and analyses
Configuration manager	Ensure no changes occur without agreed level of review	Assurance	Cooperation
Life cycle cost estimator	Estimate system costs	Cost information	Technical information
Engineering manager	Appropriate engineering processes	Well-managed discipline engineering	Cooperation
Discipline engineer	Detailed design	Sound design and interface requirements	System architecture
Test engineer	Ensure materials, components, elements, or systems meet specifications and requirements	Test results	Requirements
Specialty engineer	Ensure system meets requirements in area of specialization	Specialized expertise	Requirements
Industrial engineer	Technical operation of industrial plant	Efficient plant	Coordination
Quality assurance	Ensure that manufacture and test is performed as intended	Assurance	Cooperation

TABLE 7.3   Comparison of the Program Manager and the Systems Engineer [5]

Domain	Program Manager	Systems Engineer
Risk	Manages risk	Develops risk management process
	Sets guidelines	Analyzes risk
Changes	Controls change	Analyzes changes
		Manages configuration
Outside interfaces	Primary customer interface	Primary user interface
Internal interfaces	Primary internal management interface	Primary internal technical interface
Resources	Provides and manages resources	Delineates needs
		Uses resources

field. For an information technology (IT) system, the client may be an IT organization within a company, whereas the consumers may be the clients of the company and the users may be distributed throughout the company's other divisions. Clients, users, and consumers are all critically important, but by definition the client has control over the acquisition process. The program manager is responsible for relations with the client and the SE is the PM's primary advisor on technical issues in that relationship. The SE is also responsible for coordinating relationships with consumers and users, especially developing system requirements that take into account consumer and user needs and desires. This process is much easier if the client has good relations with the consumers and users.

The SE and the CTO or CIO

A company oriented toward research, technology, or systems development may designate a high-level executive to have responsibility for technical issues such as research and development and strategic technical direction. This person is often called the chief technology officer (CTO). The CTO is responsible for using technology to achieve the organization's mission. The exact scope of responsibilities varies widely from company to company, and some high-tech companies do not use this title. The CTO's role is somewhat like that of a systems engineer for the entire company as a single enterprise. Other companies designate an executive as chief information officer (CIO), who will be responsible for the company's information strategy and for information technology systems to achieve the organization's mission.

The SE and the Operations Researcher or System Analyst

These specialties involve studying existing or proposed systems and environments and evaluating system performance. They are sometimes regarded as part of systems engineering, though their emphasis is on mathematical and computer-based modeling and simulation (see Chapter 4) and not on such SE activities as managing user interfaces, requirements definition and allocation, and system performance verification. These individuals specialize in answering quantitative questions about complex systems, and they are often invaluable in making a sound and defensible decision when there is a lot at stake and the best course of action is not clear.

The SE and the Configuration Manager

Configuration management (CM) is the process of ensuring that things do not change without due review and approval and without all stakeholders being aware of the change and its implications. This includes key documents, such as those that describe system requirements, design, and interfaces, as well as the actual physical configuration of the system being built. Experience has shown that without strong CM discipline, people will have good ideas that cause them to introduce small changes into the system, and these changes will accumulate and cause chaos later when everyone has a slightly different version of the system. Typically, detailed configuration is under the control of the design engineer early in the design process. At some point the design is brought under configuration control; after that, any change requires paperwork, reviews, signatures, and approval by a configuration control board that is often chaired by the PM. The configuration manager administers this paperwork process. It is not a romantic job, but it is absolutely essential in developing a system of any significant complexity. In some organizations, CM is part of the SE shop; in others it reports independently to the PM.

The SE and the Life Cycle Cost Estimator

The program manager is deeply concerned with system costs in both the near and far terms, because costs help determine system viability and profitability. The total cost of a system over its entire life cycle is especially important. This includes costs to design, build, test, deploy, operate, service, repair, upgrade, and finally dispose of the system. Cost estimators have their own methods, models, and databases that they use to help estimate total life cycle cost (see Chapter 5). Some of these methods are empirical or economic in nature; others are technical, and the SE can expect to be involved in them. Since life cycle cost is a system-level criterion of great interest to the PM, the SE will want to to use life cycle cost as a key consideration in all systems decisions.

The SE and the Engineering Manager

An engineering manager (EM) is in charge of a group of engineers. He or she is concerned with ensuring that sound engineering methods are used, as well as performing the usual personnel management functions. To the extent that sound engineering always involves some element of a systems perspective, an EM will also be involved in promoting systems engineering. However, an EM's basic responsibility is sound discipline-specific engineering. When a functional or matrix organization is used, the EM may be in charge of engineers working on many different programs and systems, so his or her system perspective may be weaker than the cognizant systems engineer. In contrast, the SE is primarily a technology leader, integrator, coordinator, and advisor.

The SE and the Discipline Engineer

By discipline engineer, we mean mechanical engineer, civil engineer, electrical engineer, aerospace engineer, software engineer, chemical engineer, environmental engineer, information security engineer, and so forth. These engineers design things, and they are responsible for every detail of what they design. In contrast, a systems engineer is responsible for the high-level structure of a system, and at some level of detail he turns it over to the appropriate discipline engineer for completion. Frequently an SE starts out professionally as a discipline engineer and moves into systems work as his or her career progresses. This provides a useful background that enables the SE to work with other engineers, particularly in his or her original discipline. However, the SE must have (or develop) the knack for dealing with experts in fields other than his or her own. He must convince them that he can grasp the essentials of a sound argument in the expert's field. Also, the SE often spends a great deal of time translating what the user or consumer says into precise language that the design engineer can use, as well as translating what the design engineer says into language that the user or consumer can understand.

The SE and the Test Engineer

Testing can occur at the material, component, element, system, or architecture level. System test is usually considered a systems engineering responsibility, and elements that are complex enough to be treated as systems in themselves may also be given to systems engineers to test. An engineer who specializes in system test is a specialized SE; other test engineers specialize in material or component testing, and they are considered specialty engineers.

The SE and the Specialty Engineer

Specialty engineers are those who concentrate on one aspect of design engineering, such as human factors, reliability, maintainability, or information assurance (see Section 7.5 below for a longer list). Sometimes these specialties are collectively referred to as *systems effectiveness* or "the ilities." These specialties require a systems outlook, though they are narrower in focus than general systems engineering. An SE should have a basic understanding and appreciation of these specialty engineering disciplines. In a large program, there may be one or more engineers specializing in each of these areas, and they may be organized separately or within a systems engineering office.

The SE and the Industrial Engineer

An industrial engineer (IE) can be regarded as a systems engineer for an industrial operation, such as a manufacturing plant. An IE's responsibilities might include facility layout, operation scheduling, and materials ordering policies. Other IEs design efficient processes for service industries; yet others deal with human factors in commercial operations. A program SE can expect to interact with a plant IE when working out producibility and related issues. Industrial Engineering as an academic discipline predates the emergence of SE, and many universities teach both in the same department because of the related history and substantial overlap in material.

The SE and Quality Assurance

Quality assurance (QA) means making sure an item is produced exactly as it was designed. In many organizations, there is a separate quality assurance organization that reports directly to a high-level manager. QA personnel are not concerned directly with engineering, but with process. They ensure that all checks are made, that all necessary reviews are completed, and that all required steps are executed. QA personnel provide an important independent check to ensure that the fabrication and test process was executed exactly as the engineers intended. A QA person may carry a personal stamp to apply to paperwork to verify that it has been reviewed and that all necessary signatures are on it. That QA stamp will be required before work can proceed to the next step.

7.4   BUILDING AN INTERDISCIPLINARY TEAM

> Systems engineers often form and lead interdisciplinary teams to tackle particular problems.

Because of their role as both leader and integrator, systems engineers often find themselves in the organizational role of assembling teams of people from various backgrounds to work on a particular task or project. The systems engineer must identify early on the people with the best mix of skills to work the given problem. The team membership will vary by the nature of the problem. The team may include electrical, mechanical, and civil engineers; it can also include architects, computer or political scientists, lawyers, doctors, and economists. Technical skills are only a part of the mix.

Team Fundamentals

Figure 7.2 shows the key ingredients and products of a successful team. The vertices of the triangle show the products of a successful team. The sides and inner triangles describe what it takes to make the results happen. In *The Wisdom of Teams* [3],

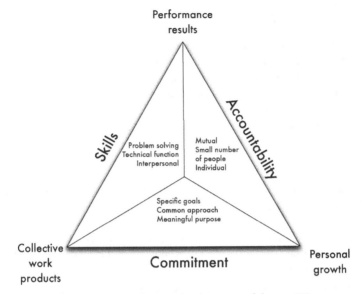

Figure 7.2 Fundamentals of a successful team [3].

Katzenbach and Smith stress that the performance ethic of the team, comprising accountability and commitment, is essential for team success. They build on this to create a definition that distinguishes a team from "a mere group of people with a common assignment."

A team is a small number of people with complementary skills who are committed to a common purpose, performance goals, and approach for which they hold themselves mutually accountable.

Each member of the team must understand and be committed to the answers to three fundamental questions:

1. Why are we here?
2. What are we to accomplish?
3. What does success look like, and how will we know when we get there?

The answers to these questions will differ based on the type of team assembled.

Team Attitude

It is vital that team members have an attitude that "only the team can succeed or fail." This is difficult to foster in an ad hoc team drawn from many sources.

Members not fully assimilated into the team may have a loyalty to their home organization and seek what is best for their constituency. A key to building an effective SE team is to assemble the complete team early, rather than adding members over the life of the project. This encourages cooperation, buy-in, and a sense of ownership early by everyone. A common occurrence is the addition of expertise such as finance or marketing during later stages, when their contribution is more directly related. However, this often leads to a feeling of outsider-ship that can result in lackluster enthusiasm, if not outright sabotage.

Team Selection

Building and managing a successful team requires several up-front decisions by the lead systems engineer [4]:

- What is the right mix of skills and power level? Should members represent a spectrum of rank and authority to promote varied viewpoints, or should the power level be the same to avoid undue influence by superiors?
- What attitude toward collaboration and problem solving is required?
- How much time is available?
- How well-defined must the problem be to fully engage team members?

Team Life Cycle

The systems engineer must evaluate the impact of the team as it works through the problem. Katzenbach and Smith [3] designed the team performance curve shown in Figure 7.3 to illustrate how well various teams achieve their goals.

A working group relies on the sum of the individual "bests" for their performance. Members share information, best practices, or perspectives to make

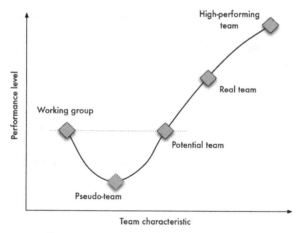

Figure 7.3 Team performance curve [3].

decisions to help each individual perform within his or her area of responsibility. Pseudo-teams are teams in name only and are not focused on the collective performance. Katzenbach and Smith rate their performance impact below that of work groups because their interactions detract from other individual performances without delivering any joint benefit. Potential teams have significant incremental performance and are trying to improve their impact. However, they have not yet established collective accountability. Their performance impact is about the same as a working group. Real teams are those that meet the definition of a small number of people with complementary skills who are equally committed to a common purpose, goals, and a working approach for which they hold themselves mutually accountable. Teams with the highest performance impact, high-performance teams, are those that are *real teams*, and their members are also deeply committed to one another's personal growth and success. That commitment usually transcends teams.

Cross-Cultural Teams

A word is in order about the special problems of teams with members from different ethnic, national, or religious cultures. Such teams are becoming more and more common as professionals from all countries become more mobile, as more international projects are undertaken, and as "virtual teams" that interact only electronically become more common. Even men and women from the same culture can sometimes have different behavioral expectations. If team members are not used to working with people from the other cultures on the team, misunderstandings can arise that interfere with team formation. Different cultures have different customs governing how business is done, including norms on very basic things, such as:

- How close to each other should two people stand when speaking together?
- What physical contact is appropriate between colleagues of the same or of the opposite sex?
- How should one dress in the workplace?
- How much privacy should one expect?
- How much organizational structure is needed?
- How many pleasantries must be exchanged before getting to business?
- How frank or tactful should one be when expressing dissatisfaction?
- How much deference should be shown to those in authority?
- How much individual ambition is appropriate to express?
- What kind of humor is acceptable?
- How diligent must one be when on the job?
- How scrupulously honest must one be in matters small and great?

Cultural norms like these are taken in with one's mother's milk. At first many people are hardly aware that they are only the customs of the people they grew up with and not the universal rules of decent behavior. They can be unaware of how behavior that seems to them entirely normal and ordinary can be considered offensive according to other cultural norms. This is certainly true of some Americans, and it is equally true of some from Asia, Africa, Europe, Latin America, and every other place in the world.

Fortunately, if one is aware of the potential for misunderstanding and has a little bit of goodwill, these problems are not too hard to avoid, especially among the well-educated people likely to be on engineering teams. The important point here is to be aware of the potential problems due to cultural differences, to maintain a spirit of forbearance and understanding, and to be a good listener. Also, keep in mind that there is a lot of variation between individuals from the same culture, so not everyone will behave as one might expect based solely on their background. Harris and Moran [5] provide a text for those who want to better understand cultural issues and develop cross-cultural teamwork skills.

7.5   SYSTEMS ENGINEERING RESPONSIBILITIES

> Common specific assigned responsibilities of SEs include writing a systems engineering management plan, external technical interface, requirements analysis, requirements management, system architecting, interface control, writing a test and evaluation master plan, configuration management, specialty engineering, coordinating technical reviews, and system integration and test.

From the point of view of the individual, the following are the specific responsibilities that are often assigned to an SE. From the point of view of the organization, these are the tasks that the systems engineering office will accomplish.

Systems Engineering Management Plan (SEMP)

In a major project, a systems engineering management plan (SEMP)(often pronounced as one syllable: "semp") should be written when the system concept is defined but before design starts. This document describes how the ideal systems engineering process is going to be tailored for the problem at hand, and it is the basis for all technical planning. It relates technical management to program management, provides a technical management framework, sets coordination requirements and methods, and establishes control processes. It is a communication vehicle that lets the client, users, consumers, and everyone on the project know how systems engineering will be carried out. The SEMP is a living document that can be modified from time to time as circumstances change. A typical SEMP might contain the following:

- Description of the envisioned development process and system life cycle
- SE activities in each envisioned phase

- Participants and involved organizations
- Planned major system reviews, audits, and other control points, including success criteria
- Products and documentation to be produced by the SE
- Risk management plan
- System requirements, including method of testing
- Identification of key measures of technical progress
- Plan for managing internal and external interfaces, both physical and functional
- Description of any trade studies planned
- Integration plan for configuration management, quality assurance, system effectiveness engineering, and other specialties, as required

Technical Interface with Users and Consumers

Systems engineers will usually be charged with user and consumer relationships as described above. The SE will meet with the user and consumer, travel to their locations, conduct interviews, focus groups, or surveys, and do whatever else is required to ensure that the users' and consumers' needs and desires are captured. The SE will be responsible for ensuring that the written system requirements truly describe what the user and consumer need with sufficient precision to design the system, and with sufficient accuracy that a system that meets the requirements will also meet the needs. The SE also has to interpret engineering constraints for nontechnical stakeholders, so that they can understand when a particular requirement should perhaps be relaxed because of its disproportionate effect on system cost, reliability, or other criterion.

Analysis and Management of Systems Requirements

In any major system development, the SE should be in charge of the system requirements, and those requirements should be written down in a document that only the CCB can change. These requirements define what the system needs to do and be in order to succeed. They can determine whether or not a billion-dollar contract has been properly executed and how much (if anything) the contractor will be paid. There is an important tradeoff in determining requirements. Typically, if the performance requirements are set at the high level that users and customers would like, the resulting system will be too expensive, unreliable, or both. If the requirements are set too low, the system may not perform well enough to be worth building. Ultimately, it is the PM's responsibility to make a requirements tradeoff between cost and performance while remaining cognizant of schedule. It is the SE's responsibility to make sure the PM understands the consequences and risks of the decision. A common approach is to establish both *threshold* requirements, which must be met to have a worthwhile system, and *goal* (or *objective*) requirements, which represent a challenging but feasible level of higher performance.

Requirements development should start with *customer requirements* that describe what is expected from the system by the operator, the client, and other stakeholders. These requirements should cover where and how the system will be used, in what environments, what its minimum performance should be, how long the system life cycle should last, and so forth. Requirements can start with a high-level objective such as, "We're going to land a man on the moon and return him safely to Earth." Analysis of such high-level requirements will produce *functional requirements* for such things as rocket propulsion, temperature control, breathable atmosphere, and communications. *Nonfunctional requirements* may specify criteria that will be used to judge the system but are not related to specific functional behaviors—for instance, color, finish, or packaging. *Performance requirements* will specify how well or to what level functional requirements have to be performed—for instance, speed in knots, availability in percent, or reliability as a probability.

Further analysis of a moon mission requirement will produce requirements for a launch vehicle, a command module, a lunar lander, and so forth. At the lowest level there will be *design requirements* describing exactly what must be built, coded, or bought. High-level requirements should not vary with the implementation, but as detail increases, the requirements become more and more dependent on the particular technical design chosen. For instance, a lunar mission going directly to the Moon's surface without a rendezvous in lunar orbit would not have separate requirements for a command module and a lander. The value measures identified in the systems decision process (see Chapter 10) are natural bases for system-level requirements, if they are direct and natural measures suitable for formal testing.

Requirements that come from an analysis of other requirements, rather than directly from an analysis of what the system must do and be, are called *derived requirements*. One kind of derived requirement is an *allocated requirement*, which is laid upon a subsystem to partially fulfill a higher-level requirement. For instance, two subsystems may have allocated reliability requirements of 0.9 and 0.8 in order to meet a system reliability requirement of 0.72.

The SE has the job of documenting both high-level and derived system requirements, at least down to a certain level of detail. One common way to visualize the process is the systems engineering "V," an example of which is shown in Figure 7.4. The highest-level requirements should be written in language that the important stakeholders can readily understand, and they may be tested in an acceptance test. From these the SE may derive system requirements that are in engineering language and which will be tested at the system level. Further analysis may derive subsystem requirements, or may allocate system requirements directly to subsystems, and the subsystems will be tested in integration tests. At the lowest level are component requirements and testing. Thus a requirement to go to the Moon leads to a requirement for a booster with a certain performance, which leads to a requirement for a rocket motor of a certain power, which leads to a requirement for a rocket nozzle with certain characteristics. At the lower levels, the design and requirements allocation will be in the hands of discipline engineers, but the SE will have oversight of the integrity of requirements traceability.

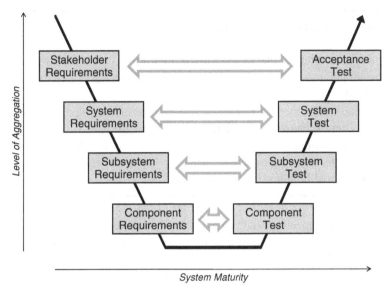

Figure 7.4 A systems engineering "V" [6].

Requirements can be of different types (sometimes overlapping):

- Customer requirements, which describe what the client expects
- Functional requirements, which define what the system has to do
- Nonfunctional requirements, which specify criteria not related to system behavior
- Performance requirements, which specify in engineering units how well the functions have to be performed
- Constraint requirements, which describe the constraints under which the performance needs to be demonstrated
- Design requirements, which specify in detail what is to be built, coded, or bought
- Derived requirements, which are developed from higher-level requirements
- Allocated requirements, which are derived from a higher-level requirement and assigned to a subsystem
- Physical requirements, which give the form, fit, and finish of the system

The SE should ensure that all requirements are:

- Unique, meaning that no two requirements overlap or duplicate each other
- Unambiguous, so that there can be no misunderstanding of exactly what is required
- Testable, so that there is a practical way of determining whether the delivered system actually meets the requirement
- Traceable, so that every high-level requirement is met by one or more low-level testable requirements, and every low-level requirement supports a documented high-level requirement

Once set, requirements should be changed only with extreme reluctance. Fluctuating requirements typically waste effort by requiring redesigns; they also promote integration problems when the full implications of the change are not grasped at first. It is also common for stakeholders to come late to the process with new requirements, leading to "requirements creep" that accumulates until the system becomes unaffordable or unworkable. The resulting schedule and cost impacts have led to the demise of many programs.

Later chapters in this book present techniques for requirements analysis and trade studies. These techniques include functional analysis (for functional requirements), screening criteria (for performance requirements), and value modeling (for goal requirements). These are described in Chapters 10 and 11 and their use is illustrated in Chapter 12. For the most complicated systems, requirements engineering can become a discipline of its own. See Hull, Jackson, and Dick [6] and Laplante [7] for in-depth treatments of requirements analysis.

System Architecting

This is the first stage in establishing the design for a complex system. For example, in the Apollo program, the decision to use one launch vehicle to put an entire mission into Earth orbit, rather than two launches with an Earth orbit rendezvous, was an architectural decision. So was the decision to have a separate lander to go from lunar orbit to the surface and back, rather than landing the entire spacecraft. These crucial decisions have to be made early in the design process. Making them wisely requires the participation of the best systems engineers, supported by the best discipline engineers and other specialists to provide expertise in their particular areas.

The first step in deciding on a system architecture is a functional analysis (see Chapter 10). A function is a task, action, or activity that the system must perform in order to accomplish its purpose. For a complex system, the SEs will be tasked to identify all system functions, analyze them to identify all required subfunctions, and develop a hierarchical functional architecture that documents them all. A function is identified using a verb and object. Adequately defining and describing the requirements may require block diagrams, data flow diagrams, state transition diagrams, and so forth, depending on the nature and complexity of the system.

The second step is to define the major elements or subsystems of the system. This is the point where a particular solution starts to be defined. The decisions made here will have a fundamental effect on the outcome of the design, so they should be made very carefully. The SE defines the system boundary (see Chapter 2)—that is, what is to be regarded as part of the system being designed and what is part of the environment. For instance, in designing a cargo ship, the freight handling in ports could be defined as part of the system and subject to redesign, or as part of the environment that constrains the ship design. The SE defines the interface between the system and its environment, for instance, by defining the port facilities a ship must be able to use.

The SEs help conceptualize and analyze various system concepts to meet the need. Tools for conceptual systems design are described in Chapter 10. Systems engineers, along with relevant discipline engineers, can expect to put much effort into architectural design trade studies during this step. In these studies, they will develop models of different system architectures and evaluate how well they will be able to meet system requirements. A complex system may consist of many different physical elements in different locations, as a spacecraft system might include ground stations, communications relay stations, a mission control center, a satellite, and a launch vehicle. In other cases, a system may contain subsystems or elements that are primarily the domain of one engineering discipline, as a helicopter might have a power system, flight control system, electrical system, and data processing system. The SE is responsible for identifying such major elements and defining their relationships. To the greatest extent possible, the architecture should be selected such that each system requirement can be allocated to exactly one element. The final architecture is often documented in a hierarchical diagram called a work breakdown structure (WBS), which provides the framework for subsequent breakouts of tasks on the project.

The third and final step in system architecting is functional allocation. The system requirements are allocated to the architecture elements. Element interactions and interfaces are defined. External interfaces (between system elements and the environment) are also defined. System elements are turned over to discipline engineers for further design work, or to systems engineers for elements that themselves are unusually complex or require the integration of different engineering disciplines.

Architecting has become an subdiscipline of its own. There are several good texts that give extended advice on how to do it, including Maier and Rechtin [8].

Systems Engineering Tools and Formal Models

Some projects use standard description and documentation models, sometimes software-based. A systems engineer will usually be responsible for creating and maintaining such models. Some projects use a standard format, such as the Department of Defense Architecture Framework, or DoDAF [9]. Such a standard framework eases communication and defines several standard ways to view the system description—for example, from the point of view of capabilities, operations, functions, or schedules. Many projects use a commercial software product to manage requirements and system models; examples are CORE® [10] from Vitech Corporation and Rational DOORS® [11] from IBM. These tools can be complicated to use and require a good deal of training and experience, but they automate much of the tedious record keeping and consistency checking that the SE would otherwise have to do by hand. They keep track of dependencies and interaction, making it easier to determine the effect of a change in one part of the system. They ensure that a requirements or configuration change made in one place is also reflected everywhere else. They can produce a draft of a System Requirements Document and other standard documents while ensuring that they are all consistent with each other.

Interface Control Documents (ICDs)

The interactions between system elements and between the system and its environment are recorded in ICDs, which are the responsibilities of systems engineers and configuration managers to write and to update as required, as described in Section 7.2. This is often one of the SE's major tasks during much of the system development period, after system architecting is complete and before system test starts, when much of the technical effort is in the hands of discipline engineers.

Test and Evaluation Master Plan (TEMP)

This document is sometimes assigned to the SE to write, and sometimes to a separate test organization. It describes all testing to be done on the project, from part and component test, to development testing of new element designs to evaluate how well they work, to systems testing of the entire system under field conditions to demonstrate that it meets user needs (see Figure 7.4). The more complex the system, the earlier it should be written.

Configuration Management (CM)

The role of CM has been described previously in Section 7.3. CM gains control of the major documents described above, the SEMP, the TEMP, and the ICDs, after the SE has written the initial versions and they are accepted by the configuration control board. The CCB will control many other documents, some of which will be written by systems engineers and some of which will only be reviewed by systems engineers.

Specialty Engineering

It is common for specialists in some or all of following areas to be part of the systems engineering organization, reporting to the chief SE. They are responsible for reviewing the entire design and development process for the impact on their areas of responsibility and for recommending areas for improvement.

- *Risk Management*. Systems that have a significant chance of total failure commonly have one or more engineers dedicated to risk management. Such systems include spacecraft, high-tech military systems, and complex systems using cutting-edge technology. Risk cannot be completely eliminated from systems like these. Risk management involves identifying and tracking the sources of risk (e.g., piece part failure or extreme environmental conditions), classifying them by likelihood of occurrence and severity of effect, and guiding risk-reduction efforts into areas with the highest expected payoff. Risk management is discussed in Chapter 3 and in Chapter 12.

- *Reliability, Maintainability, Availability* (RMA). Engineers who specialize in RMA are focused on producing a system that is working when the user needs it. Reliability refers to the likelihood of malfunction, maintainability to the ease with which the system can be serviced and repaired, and availability to the overall level of readiness for use (the result of reliability, maintainability, and spare parts availability). RMA engineers use models to calculate expected availability and recommend efforts to improve it. Chapter 8 discusses reliability models in some detail.

- *Producibility*. This is the attribute of being relatively easy and cheap to manufacture. Producibility engineering involves selecting the right design, materials, components, piece parts, and industrial processes for manufacture.

- *Quality*. The role of the quality assurance function was described earlier (Section 7.3). Quality engineers design engineering processes that produce high-quality output (i.e., items having few defects), and they design systems so that they tend to develop few defects.

- *Integrated Logistics Support* (ILS). This function encompasses the unified planning and execution of system operational support, including training, maintenance, repairs, field engineering, spares, supply, and transportation. ILS is particularly important in military systems, which often have dedicated ILS engineers from the very beginning of system development, when they are responsible for establishing logistics-related requirements.

- *Human Factors*. This specialty focuses on how the system interacts with people, particularly with users and consumers. It includes both cognitive and perceptual factors (man–machine interface, situational awareness) and ergonomics (fit, comfort, controls, etc.). Other areas of concern are workload, fatigue, human reliability, and the impact of stress. Human factors engineers specialize in the human elements of the system and their interfaces.

- *Safety*. Safety engineers are concerned with preventing not only injury and death from accidents during system manufacture and test, but also mishaps that damage high-value equipment or result in long schedule delays. Because of the importance of safety and the natural human tendency to take more and more risks when behind schedule, it is common to have a separate safety office reporting directly to the program manager or to another high-level officer.

- *Security and Information Assurance*. Engineers in these areas are responsible for ensuring that information about the program does not get to people the customer does not want to have it, for either commercial or national security reasons. They also provide systems and procedures to protect privacy data and to protect computer systems and data from attack. Finally, they assist in designing a system so that it can operate without having information about it obtained or tampered with by others. This is a particularly important function for financial and for military systems.

- *Environmental Impact.* Major government projects often cannot be done without an environmental impact statement, and that statement is often a major hurdle. Environmental engineers will help write it and then ensure that the system is developed, built, and operated in accordance with it so that environmental impact can be kept as low as possible.

- *Independent Verification and Validation* (IV&V). Verification is ensuring that the system was built as designed; validation is ensuring that the system as designed and built meets the user's needs. An IV&V engineer is a disinterested authority (not involved in the original development) responsible for ensuring that the system is correctly designed and built. The function is especially important in software system testing. The military services also have IV&V organizations to ensure new systems meet requirements.

Major Program Technical Reviews

Chapter 3 described the various life cycle models that are used in system development. Regardless of the model, usual practice is to hold a formal review as a control point (or gate) when moving from one stage to the next. The purpose of these reviews is to allow inspection and assessment of the work, to gain concurrence and approval, and to educate the staff, the management, the customer, and the user. These reviews go by such names as system requirements review, preliminary design review, critical design review, design readiness review, and full rate production readiness review, depending on the life cycle model used and the stage of the project. Design engineers, specialty engineers, testers, quality assurance personnel, and others present the status of their work and any important open issues, as appropriate to the project stage.

A systems engineer will often be tasked with organizing and emceeing the review, as well as presenting such SE topics as requirements, risk management, and systems test. For a major system, these reviews can be lengthy affairs. An auditorium full of people will look at slide after slide of Microsoft PowerPoint, for several days. The major program decision makers (PM, chief SE, etc.) will sit through the whole thing; others may come and go based on their involvement in each topic. People often find these reviews to be of compelling interest when the topic is in one's own area of responsibility, and crushingly boring at other times.

The review will result in a list of action items that identify areas that need clarification or further work. These can range from minor points like small discrepancies between two presentations of the same data to "show-stoppers" that threaten the development. Successful completion of the review (as judged by the PM or customer) is required to enter the next stage.

System Integration and Test

When the element development work is done, it is time to put the system together and test it to make sure that all the elements work together as intended, that the system interacts with its environment as it should, and that it meets client, user, and consumer needs. These are the activities on the upper right-hand side of the systems engineering "V" (Figure 7.4). It is normally an SE's responsibility to coordinate these efforts. In acceptance testing, the customer or user should be closely involved, and sometimes runs the testing.

7.6   ROLES OF THE SYSTEMS ENGINEER

> Systems engineers can play a number of roles that may or may not align closely with their formally assigned responsibilities.
> *—adapted from Sheard* [12]

These are short statements of the roles often played by a designated SE, whether or not they are really part of systems engineering and whether or not they are formally assigned. This is a more subjective account of the different roles an SE as an individual may play.

Technical Client Interface. The PM often relies on the SE for dealing with the client on technical issues, when no business matters are at stake.

User and Consumer Interface. This is a primary SE job; it is part of translating possibly inchoate needs into engineering requirements.

Requirements Owner. The SE investigates the requirements, writes them down, analyzes them, and coordinates any required changes for the lifetime of the project.

System Analyst. The SE builds models and simulations (Chapter 4) and uses them to predict the performance of candidate system designs.

System Architect. The SE defines system boundaries, system interfaces, system elements, and their interactions, and assigns functions to them.

Glue among Elements. The SE is responsible for integrating the system, identifying risks, and seeking out issues that "fall through the cracks." He or she is the technical conscience of the program, a proactive troubleshooter looking out for problems and arranging to prevent them. Since many problems happen at interfaces, the SE carefully scrutinizes them to ensure that the elements do not interfere with each other.

Technical Leader. SEs frequently end up as the planners, schedulers, and trackers of technical work; sometimes the role is formally assigned by the PM.

Coordinator. SEs coordinate the efforts of the different discipline engineers, take charge of resolving system issues, chair *integrated product/process teams* (IPTs) assembled to provide cross-disciplinary oversight of particular areas, and head "tiger teams" assembled to resolve serious problems.

System Effectiveness Manager. SEs oversee reliability, availability, human factors, and the other specialty engineering areas that can make the difference between a usable and a worthless system.

Life Cycle Planner. SEs provide for such necessities as users' manuals, training, deployment, logistics, field support, operational evaluation, system upgrades, and eventual system disposal.

Test Engineer. SEs are usually in charge of overall test planning and evaluation, and of execution of system-level tests.

Information Manager. SEs often write key program documents, review all important ones, control document change, and manage system data and metrics.

7.7   CHARACTERISTICS OF THE IDEAL SYSTEMS ENGINEER

> A good SE has a systems outlook, user orientation, inquisitiveness, common sense, professional discipline, good communication skills, a desire to cooperate, and a willingness to stand up for what's technically right.
> *—adapted from SAIC* [13]

As a final look at systems engineering practice, we will describe the personality traits that make an individual a good SE. Like any other job, some people fit more naturally into it than others. While anyone with the necessary technical skills and discipline can become a good SE, people with the following characteristics will find themselves easily falling into the role, liking the job, and doing well.

Systems Outlook. A natural SE tends to take a holistic, systems-level view on problems. Other engineers may gravitate toward looking at the details and making sure that all the crucial little things are done right; many people tend to be most concerned with organizational relationships and personalities. A good SE looks at the system as a whole, considering both technical and human factors, and is comfortable leaving element details to other experts.

Client, User, and Consumer Orientation. The ideal SE has field experience relevant to the system being worked on, or at least can readily identify with the user's and customer's perspectives. The SE should feel or develop a strong affinity with the user and customer, since one of the SE's key jobs is facilitating the consumer–user–designer interfaces.

Inquisitiveness. An SE should have a natural curiosity, and should indulge it by inquiring into areas that "just don't look right." He or she wants to know as much about the system as can be absorbed by one person, and also about the design and development process. When the SE comes across something that does not seem to make sense, he or she presses the inquiry until the doubt is resolved. In this way, the SE gains a better systems-level understanding and also often uncovers problems that had escaped notice by others with more narrow responsibilities.

Intuition. A good SE has the ability to quickly grasp essentials of an unfamiliar field, and it has a good feel for what level of detail he or she should be able to understand. He or she has good judgment on when it is necessary to press a question and when it is safe to hold off and leave it to other experts.

Discipline. A good SE adheres to engineering processes, knowing that they are essential for imposing structure on the formless and that they enable both understanding of the state of progress and control of the development. This includes objectivity: The SE maintains an objective and systems-level view of the project, and it does not let him or herself become identified with any other group working on the project. A good SE will be accepted as an honest broker when there are internal disagreements.

Communication. This is essential to the SE's role as glue among the elements. It has three parts. The SE is ready to listen to everyone involved in the project, especially the users and others not in the same chain of command. The SE is also ready to talk to everyone, to make sure everyone has a common understanding of the big picture. Finally, the SE is ready to act on what he or she finds out, bringing problems to the attention of the appropriate people and getting them working together to find a solution.

Cooperation, but not Capitulation. A natural SE is cooperative and eager to get everybody working together toward a common goal. The SE works to get buy-in from all parties. However, he or she knows when not to give in to resistance. If the issue seems important enough, the SE will insist on an appropriate explanation or investigation and is willing to take the problem to the program manager (or perhaps to the Chief Technology Officer) if necessary. That is what the SE is paid for.

7.8 SUMMARY

This chapter has focused on the realities of those who have "systems engineer" in their job title. These SEs can have a great variety of jobs, but perhaps the most typical is as the technical leader and integrator supporting a program manager who is building a complex system like an aircraft or a telecommunications system. The SE will be responsible for coordinating technical efforts over the lifetime of the system, but he or she will probably be busiest toward the beginning, when he or she is in charge of defining the system requirements, developing the system concept, and coordinating and integrating the efforts of the other design engineers.

The coordinating role of SEs means that they will work with a wide variety of other professionals and specialists, including discipline and specialty engineers, analysts, testers, inspectors, managers, executives, and so on. The SE will often have the task of forming and leading interdisciplinary teams to tackle particular problems. Other specific tasks assigned to SEs will vary with the organization and the project; they often include defining the top-level system architecture, performing risk analysis and other specialty engineering, coordinating major technical reviews, and analyzing and maintaining system requirements, system and element interfaces, and the system test plan.

Besides accomplishing these tasks, SEs may find themselves playing many different roles during system development, such as external and internal technical interface, requirements owner, system analyst and architect, system effectiveness manager, and overall technical coordinator and planner, again depending on the organization and what the program manager desires (or allows). The person most likely to enjoy and succeed at this kind of professional systems engineering is a person who naturally has a systems outlook, user orientation, inquisitiveness, common sense, professional discipline, good communication skills, a desire to cooperate, and a willingness to stand up for what is technically right. The SE is the one responsible to the program manager for making sure that all the elements work together in a system that meets the needs of the client, the user, and the consumer.

7.9 EXERCISES

7.1. Is the SE more important at the beginning of a project or at the end? Why?

7.2. Identify the client, the user, and the consumer in each of the following situations:

(a) You buy a car

(b) A taxi driver buys a taxicab

(c) A taxi company buys a fleet of taxicabs

7.3. Develop system requirements for a clock. Try to make them implementation-independent. Develop a list of functions for the clock.

7.4. Make a list of alternative system concepts for getting across a river.

7.5. Identify the functions and elements of an automobile. Draw a matrix that assigns each function to one or more elements.

7.6. Write an ICD for information exchange between a baseball pitcher and a catcher.

7.7. Explain why the performance of a newly formed task team often goes down at first, but then improves.

7.8. What factors allow a team to perform exceptionally well?

7.9. Which of the roles of a systems engineer seem to be most important to you? Justify your choice. Which is second? Third?

7.10. Research a major historical technical failure and write a three-page paper on how (or whether) better systems engineering could have averted or ameliorated it. Some examples follow; a little online research will provide a rich supply of engineering disasters.

(a) The San Francisco fire following the 1906 earthquake

(b) The sinking of the Titanic (1912)

(c) The crash of the airship Hindenburg (1937)

(d) The collapse of the Tacoma Narrows bridge (1940)

(e) The meltdown at the Three Mile Island nuclear power plant (1979)

(f) The loss of the Space Shuttle orbiter Challenger (1986) or the Columbia (2003)

(g) The flooding of New Orleans by Hurricane Katrina (2005)

7.11. Select a well-known individual from history or fiction and make a case why that person would or would not make a good systems engineer, based on his or her personality traits.

ACKNOWLEDGMENT

Dr. Paul West, United States Military Academy, provided the initial draft of the section on team building.

REFERENCES

1. Wynne, MW, Shaeffer, MD. Revitalization of systems engineering in DoD. *Defense AT&L*, 2005;XXXIV(2):14–17.

2. Haskins, C, editor. *Systems Engineering Handbook: A Guide for System Life Cycle Processes and Activities*, version 3, INCOSE-TP-2003-002-03. June 2006.

3. Katzenbach, JR, Smith, DK. *The Wisdom of Teams*. New York: HarperCollins, 1993.

4. Aranda, L, Conlon, K. Teams: *Structure, Process, Culture, and Politics*. Upper Saddle River, NJ: Prentice-Hall, 1998.

5. Harris, P, Moran, R. *Managing Cultural Differences*, 4th ed. Houston: Gulf Publishing, 1996.

6. Hull, E, Jackson, K, Dick, J. *Requirements Engineering*, 2nd ed. London: Springer, 2005.

7. Laplante, PA. *Requirements Engineering for Software and Systems*. Boca Raton, FL: CRC Press, 2009.

8. Maier, MW, Rechtin, E. *The Art of Systems Architecting*, 3rd ed. Boca Raton, FL: CRC Press, 2009.

9. DoDAF, http://cio-nii.defense.gov/sites/dodaf20/index.html. Accessed April 22, 2010.

10. CORE software, http://www.vitechcorp.com/products/Index.html. Accessed April 22, 2010.

11. DOORS software, http://www-01.ibm.com/software/awdtools/doors. Accessed April 22, 2010.

12. Sheard, SA. Systems engineering roles revisited. Software Productivity Consortium, NFP, 2000.

13. SAIC (Science Applications International Corporation). Systems Engineering Definitions & Scene Setting [briefing], 1995.

Chapter **8**

System Reliability

EDWARD POHL, Ph.D.

Failure should be our teacher, not our undertaker. Failure is delay, not defeat. It is a temporary detour, not a dead end. Failure is something we can avoid only by saying nothing, doing nothing, and being nothing.

—Denis Waitley

Would you like me to give you a formula for success? It's quite simple, really. Double your rate of failure. You are thinking of failure as the enemy of success. But it isn't at all. You can be discouraged by failure—or you can learn from it. So go ahead and make mistakes. Make all you can. Because, remember, that's where you will find success.

—Thomas J. Watson

8.1 INTRODUCTION TO SYSTEM EFFECTIVENESS

Modeling and measuring the effectiveness of a proposed solution to a problem is a necessary component of the systems decision process. When analyzing potential solutions to system problems, we generally speak about system effectiveness. To understand a system's effectiveness, we break apart the term and investigate its components. There are many definitions of what constitutes a system across a variety of fields and disciplines. In the first chapter we defined a system using the definition adopted by INCOSE:

Decision Making in Systems Engineering and Management, Second Edition
Edited by Gregory S. Parnell, Patrick J. Driscoll, Dale L. Henderson
Copyright © 2011 John Wiley & Sons, Inc.

> A system is an integrated set of elements that accomplish a defined objective. These elements include products (hardware, software, firmware), processes, people, information, techniques, facilities, services, and other support elements [1].

Webster's Dictionary defines "effective" as "producing a decided, decisive, or desired effect" [2].

Soban and Marvis [3] combine these two components to form a definition that captures the need to quantify the intended effects of the system under study. They define *system effectiveness* as follows:

> System effectiveness is a quantification, represented by system level measure, of the intended or expected effect of the system achieved through functional analysis [3].

When studying complex systems, we are concerned with how well the system, performs its "mission." Blanchard [4] specifies several of the common system level measures that Soban and Marvis [3] allude to in their definition. The key system level measures that appear in value hierarchies for most complex systems are reliability, availability, and capability. Capability is a system specific measure that captures the overall performance objectives associated with the system. Reliability and availability are measures that apply to all types and levels of system analysis. Because of their importance in analyzing system effectiveness, we spend the remainder of the chapter discussing how to model and analyze the reliability and availability of a system.

8.2   RELIABILITY MODELING

Reliability is one of the core performance measures for any system under study. Today, more than ever, reliability is not only expected, but is in demand in the market. The current global economic environment is forcing system designers to find creative ways to make cost-effective, reliable systems that meet or exceed the performance expectations of their consumers and users. To maintain their competitiveness, system designers must design for reliability and make those system level trades early in the designing process. Attempting to improve reliability after the system has been designed is a costly approach as illustrated in Figure 8.1.

Reliability is the *probability* that an item (component, subsystem, or system) or process operates properly for a *specified amount of time* (design life) under *stated use conditions* (both environmental and operational conditions) without failure. What constitutes failure for a component, subsystem, system, or process must be clearly defined as the item or process is developed. In addition, proper operating and environmental conditions must be adequately defined so that the designers and operators have a common understanding of how, when, and where the item or process should be used. In simple terms, reliability is nothing more than the

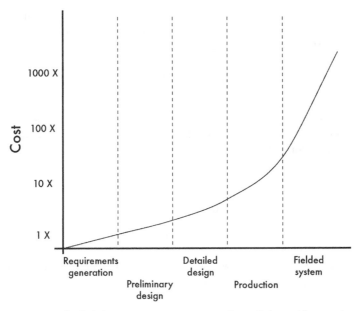

Figure 8.1   Cost of reliability improvement versus time. (Adapted from reference 5.)

probability that the system under study operates properly for a specified period of time.

> Reliability is the *probability* that an item (component, subsystem, or system) or process operates properly for a *specified amount of time* (design life) under *stated use conditions* (both environmental and operational conditions) without failure.

8.3   MATHEMATICAL MODELS IN RELIABILITY

As discussed previously, the definition of reliability has to be precisely constructed. In this section we are going to develop basic mathematical models of reliability that focus on items that can be in one of the two states:

- working ($X(t) = 1$) and
- not working ($X(t) = 0$).

Basic mathematical models of reliability are often constructed by analyzing test data and using these data in conjunction with the probability theory to characterize the items' reliability. Suppose we put N_0 identical items on test at time $t = 0$. We will assume that each item that is put on test at time $t = 0$ is functional (i.e., $X_i(0) = 1, i = 1$ to N_0).

Let $N_s(t)$ be the number of items that have survived to time t. Let $N_f(t)$ be a random variable representing the number of items that have failed by time t, where $N_f(t) = N_0 - N_s(t)$. Thus, the reliability at time t can be expressed as

$$R(t) = \frac{E[N_s(t)]}{N_0} \tag{8.1}$$

Remember that reliability is a probability and it represents the probability that the item is working at time t. If we let T be a random variable that represents the time to failure of an item, then the reliability can be written as

$$R(t) = P(T > t) \tag{8.2}$$

and the unreliability of an item $F(t)$ can be expressed as

$$F(t) = P(T \leq t) \tag{8.3}$$

$F(t)$ is commonly referred to as the *cumulative distribution function* (CDF) of failure. $F(t)$ can also be expressed as

$$F(t) = \frac{E[N_f(t)]}{N_0} \tag{8.4}$$

Using this information, we can establish the following relationship:

$$R(t) = \frac{E[N_s(t)]}{N_0} = \frac{E[N_0 - N_f(t)]}{N_0} = \frac{N_0 - E[N_f(t)]}{N_0} = 1 - F(t). \tag{8.5}$$

Given the cumulative distribution function of failure $F(t)$, the probability density function (PDF) of failure $f(t)$ is given by

$$f(t) = \frac{d}{dt}F(t) \tag{8.6}$$

Thus

$$f(t) = \frac{d}{dt}(1 - R(t)) = -\frac{d}{dt}R(t) \tag{8.7}$$

Another important measure is the hazard function. The hazard function represents the probability that a unit fails in the next increment of time and is represented mathematically by the equation below:

$$h(t) = \lim_{\Delta t \to 0} \left\{ \frac{1}{\Delta t} \Pr[T \leq t + \Delta t | T > t] \right\} \tag{8.8}$$

$$= \lim_{\Delta t \to 0} \left\{ \frac{1}{\Delta t} \frac{F(t + \Delta t) - F(t)}{R(t)} \right\} \tag{8.9}$$

$$= \frac{f(t)}{R(t)} \tag{8.10}$$

The hazard function provides another way of characterizing the failure distribution of an item. Hazard functions are often classified as increasing failure rate (IFR), decreasing failure rate (DFR), or constant failure rate (CFR), depending on the behavior of the hazard function over time. Items that have an increasing failure rate exhibit "wear-out" behavior. Items that have a CFR exhibit behavior characteristic of random failures.

A general relationship between the reliability function $R(t)$ and the hazard function $h(t)$ can be established as follows:

$$h(t) = \frac{1}{R(t)} \left[\frac{-dR(t)}{dt} \right] \tag{8.11a}$$

$$h(t)\, dt = \frac{-dR(t)}{R(t)} \tag{8.11b}$$

Integrating both sides, we obtain

$$\int_0^t h(u)\, du = \int_{R(0)}^{R(t)} \left[\frac{-dR(t)}{dt} \right] \tag{8.12a}$$

$$\int_0^t h(u)\, du = \int_1^{R(t)} \frac{-dR(t)}{dt} \tag{8.12b}$$

$$-\int_0^t h(u)\, du = \ln(R(t)) \tag{8.12c}$$

$$R(t) = \exp\left[-\int_0^t h(u)\, du \right] \tag{8.12d}$$

and therefore

$$F(t) = 1 - \exp\left[-\int_0^t h(z)\, dz \right] \tag{8.13}$$

The most common form of the hazard function is the bathtub curve (see Figure 8.2). The bathtub curve is most often used as a conceptual model of a population of items rather than a mathematical model of a specific item. Early on, during the development of an item, initial items produced will oftentimes be subject to manufacturing defects. Over time, the manufacturing process is improved and these defective units are fewer in number, so the overall hazard function for the remaining population decreases. This portion of the bathtub curve is referred to as the "infant mortality" period. Once the manufacturing system matures, fewer items will fail early in their lifetime and items in the field will exhibit a constant hazard function. This period is known as the "useful life" period. During this time, failures are purely random and usually caused by some form of unexpected or random stress placed on the item. At the end of its useful life, items in the field may begin to "wear out" and as a result the hazard function for the population of items remaining in the field will exhibit a rapidly increasing rate of failure.

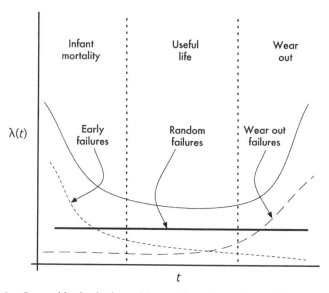

Figure 8.2 General bathtub-shaped hazard function. (Adapted from reference 6.)

To remain competitive, most product developers attempt to reduce or eliminate the "infant mortality" period and the associated "quality" related failures by using a variety of quality control tools and initiatives. For high-reliability items, many manufacturers may use environmental stress screening (ESS) or "burn-in" to enhance their quality initiatives and further reduce or eliminate this infant mortality period. ESS subjects the items to a variety of stresses (i.e., temperature, vibration, voltage, humidity, etc.) to cause the "weak" components to fail before they enter the field. Burn-in is used to filter out the defective items by having each item operate for some predefined period of time, often at an increased temperature. Ideally, to maximize reliability, we would like items to operate in the useful life period. Many organizations develop maintenance strategies that allow their products to remain in this period. The goal of these maintenance strategies is to replace the items in the field before they enter the wear-out period and fail in the field.

One final measure that is often used to characterize the reliability of an item is the items' mean time to failure (MTTF). The MTTF is nothing more than the expected value of the failure time of an item. Mathematically, the MTTF is calculated as follows:

$$\text{MTTF} = E\,[T] = \int_0^\infty tf(t)\,dt \qquad (8.14)$$

or

$$\text{MTTF} = E\,[T] = \int_0^\infty R(t)\,dt \qquad (8.15)$$

The variance for the time to failure T is given by the following relationship:

$$\text{Var}\,[T] = E\left[T^2\right] - E\,[T]^2 = \int_0^\infty t^2 f(t)\,dt - \left[\int_0^\infty t f(t)\,dt\right]^2 \qquad (8.16)$$

8.3.1 Common Continuous Reliability Distributions

As was mentioned previously, reliability is a function of time, and time is a continuous variable. Thus, most items that operate over continuous periods should be modeled using a continuous time-to-failure distribution. There are many continuous time-to-failure distributions that can be used to model the reliability of an item. Some of the well-known distributions include those listed in Table 8.1.

Most reliability books (see references 7–11) cover in detail many of these distributions. Leemis [12] has done an exceptional job by describing each of the above distributions and their relationships with each other. Despite this large number of possible failure distributions, the two most widely used continuous failure distributions are the exponential and Weibull distributions. We will discuss each in detail.

Exponential Failure Distribution The exponential distribution is probably the most used and often abused failure distribution. It is a single parameter distribution that is easily estimated for a variety of data collection methods. Its mathematical tractability for modeling complex combinations of components, subsystems, and systems make it attractive for modeling large-scale systems and system of systems. Empirical evidence has shown that systems made up of large numbers of components exhibit exponential behavior at the system level. The exponential distribution is a CFR distribution. It is most useful for modeling the "useful life" period for an item. The exponential distribution possesses a unique property called the memoryless property. It is the possession of this property that often results in this distribution being used in inappropriate situations. The memoryless property states that if an item has survived until a specific time t, the probability that it will survive for the next time period $t + s$ is independent of t and only dependent

TABLE 8.1 Common Continuous Reliability Distributions

• Exponential	• Extreme value
• Weibull	• Logistic
• Normal	• Log logistic
• Lognormal	• Pareto
• Beta	• Inverse Gaussian
• Gamma	• Makeham
• Rayleigh	• Hyperexponential
• Uniform	• Muth

on s. Therefore, this distribution should not be used to model components that have wear-out failure mechanisms. The PDF, CDF, reliability function, and hazard function for the exponential distribution are

$$f(t) = \lambda e^{-\lambda t} \tag{8.17}$$

$$F(t) = 1 - e^{-\lambda t} \tag{8.18}$$

$$R(t) = e^{-\lambda t} \tag{8.19}$$

$$h(t) = \lambda \tag{8.20}$$

Figures 8.3–8.5 illustrate the PDF, reliability function, and hazard function for an exponential distribution with $\lambda = 0.0001$. The MTTF for this distribution is

$$\text{MTTF} = E[T] = \int_0^\infty t\lambda e^{-\lambda t} = \frac{1}{\lambda} \tag{8.21}$$

There are several methods available to estimate the hazard rate of an exponential distribution. Techniques include the method of moments, maximum likelihood, and rank regression. We will explore the use of the first two techniques for the exponential distribution. The method of moments estimator is found by equating the population moments with the sample moments. Let $T_1, T_2, \ldots, T_n$ denote a

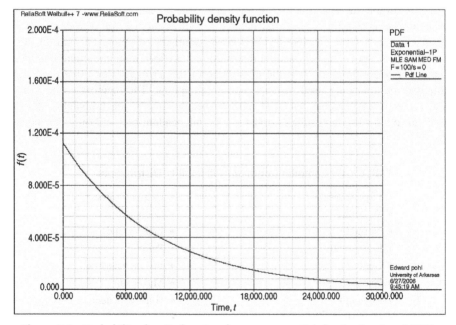

Figure 8.3 Probability density function for an exponential distribution $\lambda = 0.0001$.

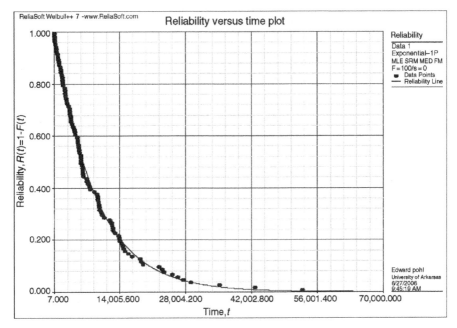

Figure 8.4   Reliability function for an exponential distribution $\lambda = 0.0001$.

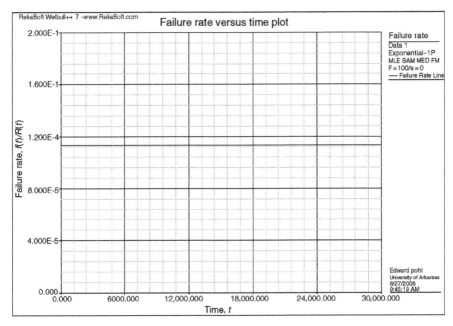

Figure 8.5   Hazard function for the exponential distribution $\lambda = 0.0001$.

random sample of failure times for n items from the total population of items. We can construct the sample mean, and sample variance as follows:

$$\bar{T} = \sum_{i=1}^{n} \frac{T_i}{n} \tag{8.22}$$

$$S^2 = \sum_{i=1}^{n} \frac{(T_i - \bar{T})^2}{n-1} \tag{8.23}$$

We can then equate the population mean, the MTTF, to the sample mean and solve for the parameter. Thus,

$$\text{MTTF} = \frac{1}{\lambda} = \bar{T} = \sum_{i=1}^{n} \frac{T_i}{n} \tag{8.24}$$

$$\hat{\lambda} = \frac{1}{\bar{T}} = \frac{n}{\sum_{i=1}^{n} T_i} \tag{8.25}$$

A second approach and the one that is most often used, especially for medium to large samples, is the method of maximum likelihood. Caution should be exercised for small samples as the maximum likelihood estimation (MLE) has been shown to be biased for small samples. Maximum likelihood estimators are found by maximizing the likelihood function. The likelihood function is derived by observing the status of all the items in the sample. This technique allows us to account for censored data. Censoring occurs when the exact failure time of an item on test is unknown. The most common situation occurs when n items have been put on test for s hours, p items have failed, and the remaining $n - p$ items have not failed, by time s. These $n - p$ items have been censored (i.e., the test has been terminated before they where allowed to fail). The general form of the likelihood function is given as

$$L(T; \lambda) = \prod_{i \in \text{failed}} f(T; \lambda) \prod_{j \in \text{censored}} R(s; \lambda) \tag{8.26}$$

$$L(T; \lambda) = \prod_{i=1}^{p} \lambda e^{-\lambda T_i} \prod_{j=1}^{n-p} e^{-\lambda s} \tag{8.27}$$

For the case where no censoring has occurred (i.e., every item was run to failure in the sample), the following likelihood function is obtained:

$$L(T; \lambda) = \prod_{i=1}^{n} \lambda e^{-\lambda T_i} \tag{8.28}$$

It is often easier to maximize the log-likelihood function. This is accomplished by setting the first partial derivatives equal to zero and solving for the parameter.

$$\ln L(T; \lambda) = n \ln \lambda - \lambda \sum_{i=1}^{n} T_i \tag{8.29}$$

$$\frac{\partial}{\partial \lambda} \ln L(T; \lambda) = \frac{n}{\lambda} - \sum_{i=1}^{n} T_i = 0 \tag{8.30}$$

$$\hat{\lambda} = \frac{n}{\sum_{i=1}^{n} T_i} \tag{8.31}$$

Example. The next generation over the horizon radar system is currently under development. As part of the development process, seven systems have been tested for 2016 hours. Two systems failed, one at 1700 hours and the other at 2000 hours. The remaining five systems were still operating at the end of the test period. Given your test data, what is the probability that the system will operate 24 hours a day, seven days a week for 30 days? Assume that the time to failure is adequately modeled by the exponential failure distribution.

Solution. First, we need to estimate the parameter for the exponential distribution. We will first derive the likelihood function and then construct the MLE.

$$L(T; \lambda) = \prod_{i=1}^{2} \lambda e^{-\lambda T_i} \prod_{j=1}^{5} e^{-\lambda s} = \lambda^2 e^{\sum_{i=1}^{2} T_i} e^{-\lambda \sum_{j=1}^{5} s}$$

$$\ln L(T; \lambda) = 2 \ln \lambda - \lambda \sum_{i=1}^{2} T_i - 5\lambda s$$

$$\frac{\partial \ln L(T; \lambda)}{\partial \lambda} = \frac{2}{\lambda} - \left[\sum_{i=1}^{2} T_i + 5s \right]$$

$$= \frac{2}{\lambda} - [1700 + 2000 + 5(2016)] = 0,$$

$$\hat{\lambda} = \frac{n}{\sum_{i=1}^{2} T_i + 5s} = \frac{2}{13780} = 1.4514 \times 10^{-4}$$

Once we know the parameter we can calculate the probability the system can work continuously for 30 days without failure.

$$R(t) = e^{-\hat{\lambda} t},$$

$$R(5040) = e^{-(1.4514 \times 10^{-4})5040} = 0.6937.$$

Weibull Failure Distribution The Weibull distribution is commonly used because of its flexibility in modeling a variety of situations. It has also been shown to fit a wide variety of empirical data sets, especially for mechanical systems. The most common form of the Weibull distribution is a two-parameter distribution. The two parameters are the scale parameter, η, and the shape parameter, β. When $\beta < 1$, the hazard function for the distribution is DFR. When $\beta > 1$, the hazard function for the distribution is IFR. Finally, when $\beta = 1$, the hazard function is CFR. The PDF, CDF, reliability function, and hazard function for the Weibull distribution are given by the following relationships:

$$f(t) = \frac{\beta}{\eta} \left(\frac{t}{\eta}\right)^{\beta-1} e^{-\left(\frac{t}{\eta}\right)^{\beta}} \tag{8.32}$$

$$F(t) = 1 - e^{-\left(\frac{t}{\eta}\right)^{\beta}} \tag{8.33}$$

$$R(t) = e^{-\left(\frac{t}{\eta}\right)^{\beta}} \tag{8.34}$$

$$h(t) = \frac{\beta}{\eta} \left(\frac{t}{\eta}\right)^{\beta-1} \tag{8.35}$$

Figures 8.6–8.8 illustrate the various shapes the PDF and hazard function take when the shape parameter is varied. Notice that when $\beta \cong 3$ the PDF takes a shape similar to a normal distribution. The MTTF for this distribution is

$$\text{MTTF} = E[T] = \int_0^{\infty} t \frac{\beta}{\eta} \left(\frac{t}{\eta}\right)^{\beta-1} e^{\frac{t^{\beta}}{\eta}} = \eta \Gamma \left(1 + \frac{1}{\beta}\right) \tag{8.36}$$

where the Γ function is

$$\Gamma(n) = \int_0^{\infty} e^{-x} x^{n-1} dx \tag{8.37}$$

A three-parameter version of the Weibull distribution is sometimes used and has the following PDF, CDF, reliability function, and hazard function:

$$f(t) = \frac{\beta}{\eta} \left(\frac{t-\gamma}{\eta}\right)^{\beta-1} e^{-\left(\frac{t-\gamma}{\eta}\right)^{\beta}}$$

$$F(t) = 1 - e^{-\left(\frac{t-\gamma}{\eta}\right)^{\beta}}$$

$$R(t) = e^{-\left(\frac{t-\gamma}{\eta}\right)^{\beta}}$$

$$h(t) = \frac{\beta}{\eta} \left(\frac{t-\gamma}{\eta}\right)^{\beta-1}$$

where γ is the location parameter for the distribution, and is sometimes called the "guaranteed life" parameter because it implies that if $\gamma > 0$, then there is zero

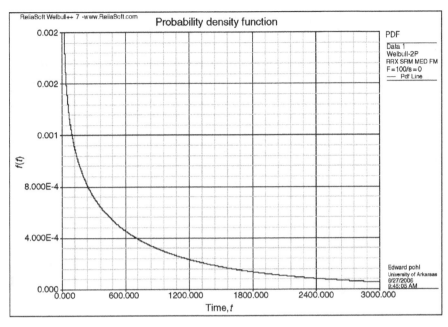

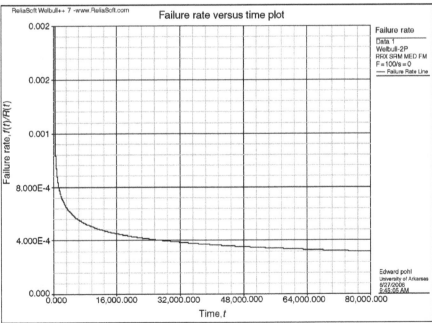

Figure 8.6 Weibull PDF and hazard function, $\beta = 0.77$.

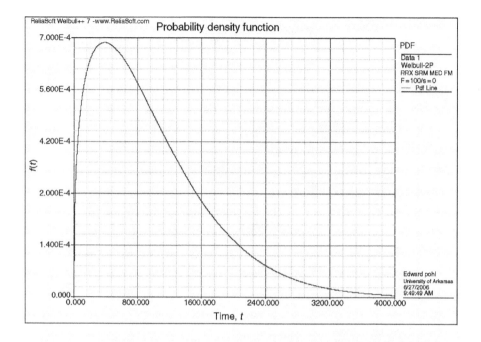

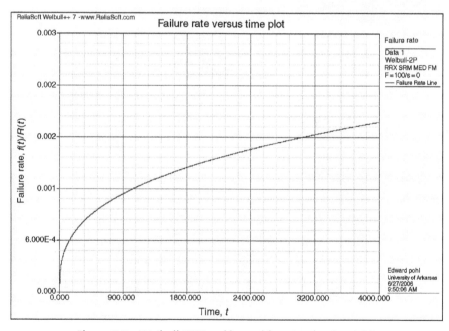

Figure 8.7 Weibull PDF and hazard function for $\beta = 1.36$.

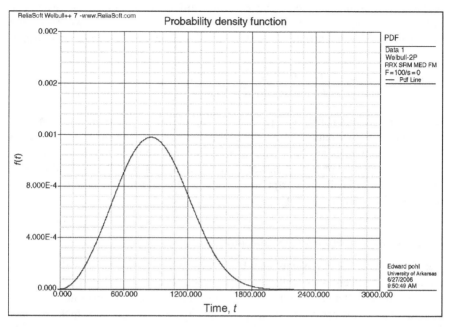

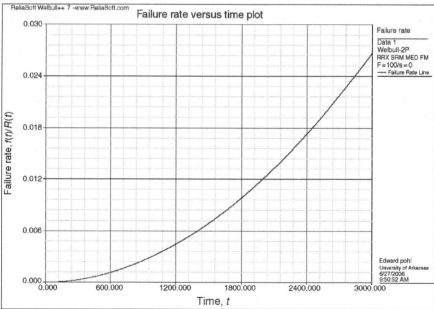

Figure 8.8 Weibull PDF and hazard function, $\beta = 2.96$.

probability of failure prior to γ. This is often a difficult assumption to prove and one of the reasons that the two-parameter model is used more often. The MTTF for the three-parameter Weibull failure distribution is given by the following relationship:

$$\text{MTTF} = E[T] = \int_0^\infty t \frac{\beta}{\eta} \left(\frac{t-\gamma}{\eta}\right)^{\beta-1} e^{-\left(\frac{t-\gamma}{\eta}\right)^\beta} = \gamma + \eta \Gamma \left(1 + \frac{1}{\beta}\right) \qquad (8.38)$$

Like the exponential distribution, there are a variety of techniques available for estimating the distribution parameters. The method of moments and maximum likelihood techniques are both reasonable techniques and constructed in the same manner as was demonstrated earlier on the exponential distribution. Kececioglu [13] provides a detailed description of each of these techniques for the Weibull distribution as well as several others.

8.3.2 Common Discrete Distributions

Certain components or systems may have performance characteristics that require them to be modeled using a discrete distribution. For example, a switch's performance may be better characterized by the number of cycles (on/off) rather than the amount of time it is operated. Another example is a satellite launch vehicle. It either launches successfully or does not. Time to failure is not an adequate measure to describe the performance of the launch vehicle. We will explore three of the common discrete distributions utilized in measuring system performance. Method of moments and maximum likelihood methods are common approaches for estimating the parameters for these distributions.

Binomial Distribution   The binomial distribution is a distribution that characterizes the sum of n independent Bernoulli trials. A Bernoulli trial occurs when an item's performance is a random variable that has one of two outcomes; it either works (success) or fails (failure) when needed. The probability of success, p, is constant for each trial. Mathematically, the PDF for a Bernoulli random variable is

$$f(x) = p^x (1-p)^{1-x} \qquad (8.39)$$

The mean and variance of the distribution are

$$E[x] = p \qquad (8.40)$$

$$\text{Var}\,[x] = p(1-p) \qquad (8.41)$$

The PDF for the binomial distribution is given by

$$f(x) = \binom{n}{x} p^x (1-p)^{n-x} \qquad (8.42)$$

where

$$\binom{n}{x} = \frac{n!}{(n-x)!x!} \tag{8.43}$$

The mean and variance for the binomial distribution are

$$E[x] = np \tag{8.44}$$

$$\text{Var } [x] = np(1-p) \tag{8.45}$$

Example. Suppose the next bomber is designed such that it has three engines. Assume that for the plane to complete its mission, one engine must operate. If the probability that an engine fails during flight is $(1-p)$ and each engine is assumed to fail independently, then what is the probability that a plane returns safely.

Solution. Since each engine is assumed to fail independently, then the number of engines remaining operative can be modeled as a binomial random variable. Hence, the probability that the three-engine next-generation bomber makes a successful flight is

$$= \binom{3}{1} p(1-p)^2 + \binom{3}{2} p^2(1-p)^1 + \binom{3}{3} p^3$$

$$= \frac{3!}{2!1!} p^{(1}-p)^2 + \frac{3!}{1!2!} p^2(1-p) + \frac{3!}{0!3!} p^3$$

$$= 3p(1-p)^2 + 3p^2(1-p) + p^3$$

Geometric Distribution   The geometric distribution is commonly used to model the number of cycles to failure for items that have a fixed probability of failure, p, associated with each cycle. In the testing arena this distribution has been used to model the distribution of the number of trials until the first success. The PDF for this distribution is given by

$$f(x) = (1-p)^{x-1}p. \tag{8.46}$$

The mean and variance for the geometric distribution are given by

$$E[x] = \frac{1}{p} \tag{8.47}$$

$$\text{Var} = \frac{1-p}{p^2} \tag{8.48}$$

Example. A manufacturer of a new dipole light switch has tested 10 switches to failure. The number of on/off cycles until failure for each switch is given below:

Switch	# Cycles	Switch	# Cycles
1	30,000	6	75,000
2	35,000	7	80,000
3	40,000	8	82,000
4	56,000	9	83,000
5	70,000	10	84,500

The marketing department believes they can improve market share if they advertise their switch as a high-reliability switch. They want to advertise that their product has reliability greater than 98% for a 5-year period. They assume that the switch is cycled three times a day, 365 days a year. Using the estimated value of p, calculate the point estimate for the reliability of the switch for a 5-year period.

Solution.

$$\hat{p} = \frac{n}{\sum_{i=1}^{10} C_i} = \frac{10}{635500} = 1.5736\text{E} - 05$$

$$E[C] = \frac{1}{\hat{p}} = 63550$$

$$P[C > (3 \times 365 \times 5)] = P[C > 5475] = 1 - P[C \leq 5475]$$

$$= 1 - \sum_{x=1}^{5475} (1-p)^{x-1} p$$

$$= 1 - 0.0826 = 0.9174$$

8.4 BASIC SYSTEM MODELS

Most systems are composed of many subsystems, each of which can be composed of hundreds or thousands of components. In general, reliability analysis is performed at the lowest levels and the results then aggregated into a system level estimate. This is done to save time and money during the development process. System level testing cannot be accomplished until the entire system is designed and assembled. Waiting to test components and subsystems until the entire system is assembled is not time- or cost-effective. Usually, a system's functional and physical decompositions are used to help construct a system level reliability block diagram. We utilize this structure to compute system level reliability performance in terms of the component and subsystem reliabilities. In the next couple of sections, we will use basic probability concepts to explore some of the common system structures utilized by design engineers. We will begin with the two basic structures, series and parallel systems.

8.4.1 Series System

Assume that a system consists of N functionally independent components, each with individual measures of reliability $R_1, R_2, \ldots, R_N$ for some specified period of performance in a specified environment. The set of components constitute a series system if the success of the system depends on the success of all of the components. If a single component fails, then the system fails. The reliability block diagram for this situation is given in Figure 8.9. The series system reliability, which is the probability of system success, is given by

$$R_s(t) = R_1(t) \cdot R_2(t) \cdot \ldots \cdot R_n(t) = \prod_{i=1}^{N} R_i(t) \tag{8.49}$$

It should be noted that the system reliability is always smaller than the worst component. Thus, the worst component in a series system provides a loose upper bound on the system reliability.

Example. Four identical components form a series system. The system reliability is supposed to be greater than 0.90 for a specified period. What should the minimum reliability be for each of the components to achieve the desired system reliability for that specified time?

Solution.

$$R_s = R_1 \cdot R_2 \cdot R_3 \cdot R_4 = R^4 \geq 0.90$$

$$R \geq (0.90)^{1/4} \geq 0.974$$

8.4.2 Parallel System

Assume that a system consists of N functionally independent components, each with individual measures of reliability $R_1, R_2, \ldots, R_N$ for some specified period of performance in a specified environment. The set of components constitute a parallel

Figure 8.9 Series system.

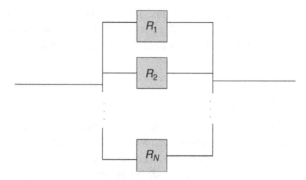

Figure 8.10   Parallel system.

system if the success of the system depends on one or more of the components operating successfully. If a single component survives, then the system succeeds. The reliability block diagram for this situation is given in Figure 8.10. The reliability for a parallel system can be expressed as the probability that at least one component in the system survives. Mathematically, this can be expressed as one minus the probability that all of the components fail. The system reliability for a parallel set of components is given by

$$R_s(t) = 1 - [(1 - R_1(t))(1 - R_2(t)) \dots (1 - R_N(t))] \qquad (8.50)$$

It should be noted that the system reliability for a parallel system is always larger than the reliability of the best component. Thus, the reliability of the best component in a parallel system provides a loose lower bound on the system reliability.

Example.   Suppose a system consists of three time-dependent components arranged in parallel, one component has a Weibull failure distribution with a scale parameter of 1000 h, and a shape parameter of 2, the other two components have exponential failure distributions with $\lambda_1 = 0.005$ and $\lambda_2 = 0.0001$. Derive an expression for the system reliability and compute the reliability of the system for the first 1000 h.

Solution.

$$R_s(t) = 1 - \left[\left(1 - e^{-\left(\frac{t}{\eta}\right)^\beta}\right) \left(1 - e^{-\lambda_1 t}\right) \left(1 - e^{-\lambda_2 t}\right) \right],$$

$$R_s(1000) = 1 - \left[\left(1 - e^{-\left(\frac{1000}{1000}\right)^2}\right) \left(1 - e^{-0.005(1000)}\right) \left(1 - e^{-0.0001(1000)}\right) \right]$$

$$= 0.94025$$

8.4.3 *K*-out-of-*N* Systems

K-out-of-*N* systems provide a very general modeling structure. It includes both series systems and parallel systems as special cases. In this structure, we assume that a system consists of *N* functionally independent components each with identical measures of reliability, for some specified period of performance in a specified environment. The set of components constitute a *K*-out-of-*N* structure if the success of the system depends on having *K* or more of the components operating successfully. If less than *K* components are operating, then the system has failed. The reliability for a *K*-out-of-*N* system can be expressed as the probability that at least *K* components in the system survive. Mathematically, this can be modeled as an application of the binomial distribution. The system reliability for a *K*-out-of-*N* system is given by

$$R_s = \sum_{i=k}^{N} \binom{N}{j} R^j (1 - R)^{N-j} \tag{8.51}$$

An *N*-out-of-*N* system is equivalent to a series system and a 1-out-of-*N* system is equivalent to a parallel system.

8.4.4 Complex Systems

Most systems are complex combinations of series and parallel system structures of components and subsystems. Consider the following bridge network of functionally independent components with individual reliabilities R_1, R_2, R_3, R_4, R_5 for a specified period of performance in a specified environment. The components are arranged in what is commonly called a bridge network (Figure 8.11). The system reliability for this structure can be constructed by the method of system decomposition. Decomposing the system around component 3 and using conditional probability, we get the following expression where C_3 is the event that component 3 is working, $\bar{C}_3$ is the event that component 3 has failed, and S is the event that the system is working:

$$R_s = P(S|C_3)P(C_3) + P(S|\bar{C}_3)P(\bar{C}_3) \tag{8.52}$$

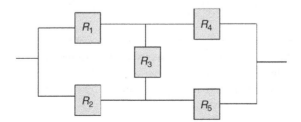

Figure 8.11 Bridge network.

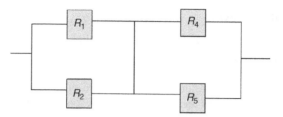

Figure 8.12 System structure when C_3 is working.

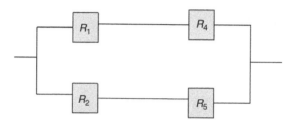

Figure 8.13 System structure when C_3 has failed.

If component 3 is working, then the system structure reduces to the series arrangement of parallel systems as shown in Figure 8.12. The reliability for the system is given by the following:

$$R_s(t)|C_3 = \{1-[(1-R_1(t))(1-R_2(t))]\} \cdot \{1-[(1-R_4(t))(1-R_5(t))]\} \quad (8.53)$$

If component 3 fails, then the system structure reduces to a parallel arrangement of series components as shown in Figure 8.13. The reliability for this system is given by the following:

$$R_s(t)|\bar{C}_3 = 1 - [(1 - R_1(t)R_4(t))(1 - R_2(t)R_5(t))] \quad (8.54)$$

Thus, the unconditional system reliability is given by the following relationship:

$$R_s(t) = \{1-[(1-R_1(t))(1-R_2(t))]\} \cdot \{1 - [(1 - R_4(t))(1 - R_5(t))]\} \cdot R_3(t)$$
$$+ (1-[(1-R_1(t)R_4(t))(1-R_2(t)R_5(t))]) (1 - R_3(t)) \quad (8.55)$$

Example. Suppose all five components in the bridge structure are identical. Assume that the components have a Weibull distribution with a scale parameter of 10,000 h and a shape parameter of 3. What is the 5000-h reliability for the bridge system?

Solution. The first step is to calculate the component reliability for a 5000-h period.

$$R_1(5000) = e^{-\left(\frac{5000}{10000}\right)^3} = 0.8825$$

Substituting this probability into the system reliability equation, we get the following:

$$
\begin{aligned}
R_s(5000) = &\{1 - [(1 - 0.8825)(1 - 0.8825)]\} \\
&\cdot \{1 - [(1 - 0.8825)(1 - 0.8825)]\}(0.8825) \\
&+ (1 - [(1 - (0.8825)(0.8825))(1 - (0.8825)(0.8825))]) \\
&\times (1 - 0.8825) = 0.970
\end{aligned}
$$

8.5 COMPONENT RELIABILITY IMPORTANCE MEASURES

One of the general concerns associated with a complex system of components is which component is the most important for the success of the system. This question often arises when system designers are trying to decide which component should be improved first if resources become available [14]. To determine which component to improve first, we take the partial derivative of the system reliability function, $R_s(t)$, with respect to each of the component reliabilities, $R_i(t), i = 1, 2, \ldots, N$.

8.5.1 Importance Measure for Series System

Assume that a system consists of N functionally independent components, each with individual measures of reliability $R_1, R_2, \ldots, R_N$ for some specified period of performance in a specified environment, and these components are functionally arranged as a series system as shown in Figure 8.9. The component importance measure is calculated as follows:

$$\frac{\partial R_s(t)}{\partial R_i(t)} = \frac{\partial}{\partial R_i(t)}[R_1(t)R_2(t)\ldots R_i(t)\ldots R_N(t)] = \prod_{\substack{j=1 \\ j \neq i}}^{N} R_j(t) = \frac{R_s(t)}{R_i(t)} \qquad (8.56)$$

This suggests that the most important component R_{i*} is the component that has the largest importance measure. Thus,

$$\frac{R_s(t)}{R_{i*}} = \max \frac{R_s(t)}{R_i(t)} \qquad (8.57)$$

$$R_{i*} = \min R_i(t) \qquad (8.58)$$

This tells us that for a series system we should improve the design of the least reliable component.

8.5.2 Importance Measure for Parallel System

Assume that a system consists of N functionally independent components, each with individual measures of reliability $R_1, R_2, \ldots, R_N$ for some specified period of performance in a specified environment, and these components are functionally arranged as a parallel system as shown in Figure 8.10. The component importance measure is calculated as follows:

$$\frac{\partial R_s(t)}{\partial R_i(t)} = \frac{\partial}{\partial_i(t)} \left[1 - \prod_{i-1}^{N} [1 - R_i(t)] \right] = \prod_{\substack{j=1 \\ j \neq i}}^{N} (1 - R_j(t)) = \frac{1 - R_s(t)}{1 - R_i(t)} \tag{8.59}$$

The most important component, R_{i*}, is the component that has the largest importance measure. Thus,

$$\frac{(1 - R_s(t))}{(1 - R_{i*})} = \max \left(\frac{1 - R_s(t)}{1 - R_i(t)} \right) \tag{8.60}$$

$$R_{i*} = \max R_i(t) \tag{8.61}$$

This tells us that for a parallel system we should improve the design of the most reliable component.

8.6 RELIABILITY ALLOCATION AND IMPROVEMENT [5]

In Section 8.5, we saw that the reliability importance measure can be used to identify which component should be improved to maximize the system reliability. Unfortunately, system designers generally do not operate in an unconstrained environment. Often, depending on the system, there are costs associated with improving a component's reliability. Costs can take a variety of forms such as dollars, weight, volume, quantity, and so on. If we let the cost per unit reliability of the ith component be C_i, then the incremental cost to improve the reliability of component i is given by

$$R_i(t) + \Delta_i = C_i \Delta_i \tag{8.62}$$

Now, if we assume that the system under study is a series system, then the improvement in system reliability as a result of the improvement in component i is given by

$$R_s^*(t) = \prod_{\substack{j=1 \\ j \neq i}}^{N} R_j(t)[R_i(t) + \Delta_i] = R_s(t) + \frac{R_s(t)}{R_i(t)} \Delta_i \tag{8.63}$$

If we assume that $R_s^*(t)$ can also be obtained by increasing the reliability of one of the other components by an incremental amount Δ_j at a cost of $C_j \Delta_j$, then the reliability of the improved system is given by

$$R_s^*(t) = R_s(t) + \frac{R_s(t)}{R_j(t)}\Delta_j = R_s(t) + \frac{R_s(t)}{R_i(t)}\Delta_i \qquad (8.64)$$

Therefore,

$$\frac{\Delta_j}{R_j(t)} = \frac{\Delta_i}{R_i(t)} \qquad (8.65)$$

Multiplying both sides by the associated costs for the incremental improvement in reliability and rearranging terms, we get

$$C_i \Delta_i = \frac{C_i\, R_i(t)}{C_j\, R_j(t)} C_j \Delta_j \qquad (8.66)$$

Now, for $C_i \Delta_i < C_j \Delta_j$ to be true, the following relationship must hold:

$$\frac{C_i\, R_i(t)}{C_j\, R_j(t)} < 1 \qquad (8.67)$$

$$C_i R_i(t) < C_j R_j(t) \qquad (8.68)$$

Therefore, to improve the system reliability for a series system to $R_s^*(t)$ at minimum cost, the component that should be improved is the component that satisfies the following relationship:

$$C_{i^*} R_{i^*}(t) = \min\, C_i R_i(t) \qquad (8.69)$$

Example. An advanced optical package has been designed for the next generation weather satellite for the National Weather Service. The basic optical package can be modeled functionally as a three-component series system. Each component has a 5-year mission reliability of 0.99, 0.995, and 0.98, respectively. Due to the design constraints, there is a weight constraint for the optical package of 1000 lb. Suppose that the reliability of the system can be improved by adding redundant components to the optical package. Your task is to determine the optimal combination of components that maximize reliability at minimal cost subject to the 1000-lb weight constraint. Assume the weights of the components are 150, 200, and 300 lb, respectively. The initial reliability of the system is given by

$$R_s = (0.99)(0.995)(0.98) = 0.9653$$

Solution. Let us assume that the effectiveness of the satellite is measured by its reliability. Let n_1, n_2, and n_3 represent the number of each type of component used in the recommended optical package. The initial weight for the system is given by $(150 + 200 + 300) = 650$ lb. To proceed, we should investigate the cost in terms of weight per unit of reliability improvement for each of the components. We will start with component 1 by investigating the improvement in reliability for the function associated with component 1 if a redundant component is added:

$$R_1^* = 1 - (1 - 0.99)(1 - 0.99) = 0.9999$$

Thus, the improvement in the contribution to the reliability for component 1 is 0.0099 at a cost of 150 lb. Therefore,

$$C_1 = \frac{150}{0.0099} = \frac{15151.51 \text{ lb}}{\text{unit of reliability}}$$

Similarly, we can calculate the same costs for components 2 and 3.

$$C_2 = \frac{40201 \text{ lb}}{\text{unit of reliability}}$$

$$C_3 = \frac{15306.12 \text{ lb}}{\text{unit of reliability}}$$

Thus, we should add a redundant component 1. Adding an additional component 1 increases the weight of the optical package to 800 lb. Now, we can compare the cost of adding a third component 1 or a second component 2. Since we only have 200 lb available, we cannot consider adding an additional component 3.

$$R_i^{**} = 1 - (1 - 0.99)(1 - 0.99)(1 - 0.99) = 0.999999$$

$$C_1^{**} = \frac{150}{0.000099} \frac{1515151.51 \text{ lb}}{\text{unit of reliability}}$$

Given this enormous cost associated with adding a third component 1, the best solution is to add an additional component 2. Thus, the reliability for the final configuration is given by

$$R_s^* = (0.9999)(0.999975)(0.98) = 0.979877$$

8.7 MARKOV MODELS OF REPAIRABLE SYSTEMS [15]

In this section we focus our modeling efforts on using continuous-time Markov chains (CTMCs) to model *repairable systems*. A *repairable system* (RS) is a system that, after failure, can be restored to a functioning condition by some maintenance action other than replacement of the entire system [16]. A CTMC is a stochastic process that moves from state to state in accordance with a discrete-time Markov chain (DTMC). It differs from a DTMC in that the amount of time it spends in each state before it transitions to another state is exponentially distributed [17]. Like a DTMC, it has the Markovian property whereby the "future is independent of the past, given the present." In this section, we assume that the CTMC has stationary (homogeneous) transition probabilities (i.e., $P[X(t+s) = j|X(s) = i]$ is independent of s). Ross [18] formally defines a CTMC as a stochastic process where each time it enters state i,

1. The amount of time it spends in state i before it transitions into a different state is exponentially distributed with a rate v_i.
2. When the process leaves state i, it will enter state j with some probability p_{ij}, where $\sum_{j \neq i} P_{ij} = 1$.

8.7.1 Kolmogorov Differential Equations

In the discrete-time case, $p_{ij}(n)$ represents the probability of going from state i to j in n transitions. In the continuous case we are interested in $p_{ij}(t)$, which represents the probability that a process currently in state i will be in state j in t time units from the present. Mathematically, we denote this by

$$p_{ij}(t) = P[X(t+s) = j|X(s) = i] \qquad (8.70)$$

In the continuous-time case, we can define the intensity at which transitions occur by examining the *infinitesimal transition rates*:

$$-\frac{d}{dt}p_{ij}(0) = \lim_{t \to 0} \frac{1 - p_{ij}(t)}{t} = v_i \qquad (8.71)$$

$$-\frac{d}{dt}p_{ij}(0) = \lim_{t \to 0} \frac{p_{ij}(t)}{t} = q_{ij} \qquad (8.72)$$

where v_i represents the rate at which we leave state i, and q_{ij} represents the rate at which we move from state i to state j. However, for small Δt, $q_{ij}\Delta t$ can be interpreted as the probability of going from state i to state j in some small increment of time Δt, given we started in state i. Using the transition intensities, as well as making use of the Markovian property, one can derive the Kolmogorov differential equations for $p_{ij}(t)$. The backward and forward Kolmogorov equations are given by Equations (8.73) and (8.74). These equations can be used to derive the transient probabilities of a CTMC. This is best illustrated through the use of an example.

$$\frac{d}{dt}p_{ij}(t) = \sum_{k \neq i} q_{ik}p_{kj}(t) - v_i p_{ij}(t) \tag{8.73}$$

$$\frac{d}{dt}p_{ij}(t) = \sum_{k \neq j} q_{kj}p_{ik}(t) - v_i P_{ij}(t) \tag{8.74}$$

8.7.2 Transient Analysis

Consider a single-component system that fails according to an exponential failure distribution with rate λ and whose repair time is exponentially distributed with rate μ. This system can be in one of the two states. It can be working (state 0) or can fail and be undergoing repair (state 1). A state transition diagram for this system is given in Figure 8.14. This diagram shows the states and the associated transition rates between the states. Using the state transition diagram and the Kolmogorov forward equation, Equation (8.74), we can derive the transition probabilities for the CTMC.

$$\frac{d}{dt}p_{ij}(t) = \sum_{k \neq j} q_{ik}p_{ik}(t) - v_i p_{ij}(t) \tag{8.75}$$

$$\frac{d}{dt}p_{00}(t) = \sum_{k \neq j} q_{10}p_{01}(t) - v_0 p_{00}(t) \tag{8.76}$$

$$\frac{d}{dt}p_{00}(t) = \mu p_{01}(t) - \lambda p_{00}(t) \tag{8.77}$$

$$\frac{d}{dt}p_{00}(t) = \mu[1 - p_{00}(t)] - \lambda p_{00}(t) \tag{8.78}$$

$$\frac{d}{dt}p_{00}(t) = \mu - (\mu + \lambda)p_{00}(t) \tag{8.79}$$

$$\frac{d}{dt}p_{00}(t) + (\mu + \lambda)p_{00}(t) = \mu \tag{8.80}$$

Solving this differential equation, we obtain

$$e^{(\lambda+\mu)t}\left[\frac{d}{dt}p_{00}(t) + (\mu + \lambda)p_{00}(t)\right] = \mu e^{(\lambda+\mu)t} \tag{8.81}$$

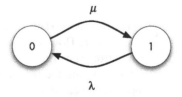

Figure 8.14 Single-component state transition diagram.

$$\frac{d}{dt}\left[e^{(\lambda+\mu)t}p_{00}(t)\right] = \mu e^{(\lambda+\mu)t} \tag{8.82}$$

$$e^{(\lambda+\mu)t}p_{00}(t) = \frac{\mu}{\lambda+\mu}e^{(\lambda+\mu)t} + c \tag{8.83}$$

$$\text{since } p_{00}(t) = 1, \quad c = \frac{\mu}{\lambda+\mu} \tag{8.84}$$

Therefore, $p_{ij}(t)$ for $i = j$ are given below:

$$p_{00}(t) = \frac{\lambda}{\lambda+\mu}e^{(\lambda+\mu)t} + \frac{\mu}{\lambda+\mu} \tag{8.85}$$

$$p_{11}(t) = \frac{\mu}{\lambda+\mu}e^{(\lambda+\mu)t} + \frac{\lambda}{\lambda+\mu} \tag{8.86}$$

Note that $p_{00}(t)$ represents the probability that the system is operating at time t. This is also known as the *system availability* $A(t)$. If we take the limit of $p_{00}(t)$ as t goes to infinity, we get the limiting or steady state availability. The limiting availability is given below:

$$\lim_{t\to\infty} A(t) = \lim_{t\to\infty} p_{00}(t) = \frac{\mu}{\lambda+\mu} \tag{8.87}$$

In general, we can establish a set of N first-order differential equations that characterize the probability of being in each state in terms of the transition probabilities to and from each state. Mathematically, the set of N first-order differential equations is summarized in matrix form in Equation (8.88), and the general form of the solution to this set of differential equations is given by Equation (8.89) [19].

$$\frac{d\underline{P}(t)}{dt} = [T_R]\underline{P}(t) \tag{8.88}$$

$$\underline{P} = \exp[T_R]t \cdot \underline{P}(0) \tag{8.89}$$

In Equation (8.89), T_R is the rate matrix. For our simple single-system example, using Figure 8.14, we get the following rate matrix.

$$T_R = \begin{pmatrix} -\lambda & \mu \\ \lambda & -\mu \end{pmatrix} \tag{8.90}$$

To solve the set of differential equations one must compute the matrix exponential. There are several different approaches to computing the matrix exponential. Two such methods include the infinite series method and the eigenvalue/eigenvector approach. Such routines are readily available in many of the commercially available mathematical analysis packages (Maple™, Mathematica®, and MATLAB®). In many instances, as the problem complexity increases, the Kolmogorov differential equations cannot be solved explicitly for the transition probabilities. In such cases, we will use numerical solution techniques; we might use simulation (see reference 20), or for a variety of reasons, focus our attention on the steady-state performance of the system.

8.7.3   Steady-State Analysis

For many systems, it is the limiting availability (aka, steady-state availability), $A(\infty)$, that is of interest. Another common name for the steady state availability is the uptime ratio. For example, the uptime ratio is of critical importance in a production facility. Similarly, for a communication system, the average message transfer rate will be the design transfer rate times the uptime ratio. So knowing the uptime ratio is essential for analyzing the performance of many systems.

We can compute the steady-state probabilities by making use of the following:

$$\text{let } \rho_j = \lim_{t \to \infty} p_{ij}(t) \tag{8.91}$$

We can then state the following:

$$v_j \rho_j = \sum_j p_i q_{ij} \; \forall j = 0, 1, 2, \dots, N \tag{8.92}$$

$$\sum_j \rho_j = 1 \tag{8.93}$$

Expression 8.92 is called the "balance" equations. The balance equations state that the rate into each state must be equal to the rate out of each state for the system to be in equilibrium. Equation (8.93) states that we must be in some state, and the sum of the probabilities associated with each state must be equal to 1. Using $(N - 1)$ of the balance equations and Equation (8.93), we can easily derive the steady-state probabilities for each state.

8.7.4   CTMC Models of Repairable Systems

In this section we illustrate how to model and analyze a variety of repairable systems using continuous-time Markov chains. We focus specifically on the single machine cases. Consider a single repairable machine. Let T_i denote the duration of the ith interval of machine function, and assume that $T_1, T_2, \dots$ is a sequence of independent identically distributed exponential random variables having failure rate λ. Upon failure, the machine is repaired. Let D_i denote the duration of the ith machine repair, and assume $D_1, D_2, \dots$ is a sequence of independent identically distributed exponential random variables having repair rate μ. Assume that no preventive maintenance is performed on the machine.

Recall that $X(t)$ denotes the state of the machine at time t. Under these assumptions, $\{X(t), t \geq 0\}$ transitions among two states, and the time between transitions is exponentially distributed. Thus, $\{X(t), t \geq 0\}$ is a CTMC having the rate diagram shown in Figure 8.15. We can easily analyze the "steady-state" behavior of the CTMC. Let ρ_j denote the long-run probability that the CTMC is in state j. We use balance equations to identify these probabilities. Each state of the CTMC has

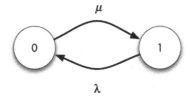

Figure 8.15 Single-machine rate diagram.

a balance equation that corresponds to the identity "rate in" = "rate out." For the rate diagram in Figure 8.15, the balance equations are

$$\text{state } 0: \quad \lambda\rho_1 = \mu\rho_0 \tag{8.94}$$

$$\text{state } 1: \quad \mu\rho_0 = \lambda\rho_1 \tag{8.95}$$

These balance equations are equivalent, so we need an additional equation to solve for ρ_0 and ρ_1. We use the fact that the steady-state probabilities must sum to 1.

We then use the two equations to solve for the two unknowns.

$$\rho_1 = \frac{\mu}{\lambda + \mu} \tag{8.96}$$

$$\rho_2 = \frac{\lambda}{\lambda + \mu} \tag{8.97}$$

Note that ρ_1 is equivalent to the steady-state availability found from taking the limit of the transient probabilities in Equation (8.87).

Let us consider another single-machine example. Just like the first example, let T_i denote the duration of the ith interval of machine function, and assume $T_1, T_2, \ldots$ is a sequence of independent identically distributed exponential random variables having failure rate λ. Upon failure, the machine is repaired. But this time, each repair requires two distinct repair operations, A and B. Assume that the duration of repair is exponentially distributed with rate μ_j where $j = (A, B)$. For this example, assume that there are enough resources available so that the repairs can be done concurrently.

This problem differs significantly from the first in that we now have four different states. State 0 is when the machine is operating; State 1 is when the machine is down and we are awaiting the completion of repair process A; State 2 is when the machine is down and we are awaiting the completion of repair process B; and State 3 is when the machine is down and we are awaiting the completion of both repair processes. The rate diagram for this model is shown in Figure 8.16. Using the rate diagram, the set of balance equations can be written as

$$\mu_A\rho_1 + \mu_B\rho_2 = \lambda\rho_0 \tag{8.98}$$

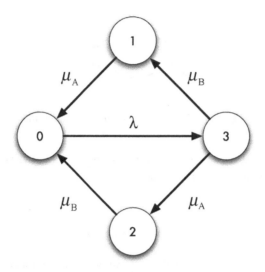

Figure 8.16 Rate diagram for multiple repair process.

$$\mu_B \rho_3 = \mu_A \rho_1 \tag{8.99}$$

$$\mu_A \rho_3 = \mu_B \rho_2 \tag{8.100}$$

$$\lambda \rho_0 = (\mu_A + \mu_B)\rho_3 \tag{8.101}$$

Using the balance equations in conjunction with the total probability equation, we can solve for the individual steady-state values for each of the states. The state of interest is state 0, as it represents the system steady-state availability. Suppose the system described above has a mean time between failure of 100 h, and the mean repair time for process A is 10 hours and for process B it is 5 h. Determine the system steady-state availability. Using the balance equations, we can derive Equation (8.102) and determine that the system has a steady state availability 0.96.

$$\rho_0 = \frac{\mu_A + \mu_B}{\lambda}\left(\frac{\mu_A + \mu_B}{\lambda} + \frac{\mu_A}{\mu_B} + \frac{\mu_B}{\mu_A}\right)^{-1} = \frac{0.1 + 0.2}{0.01}(32.5)^{-1} = 0.9231 \tag{8.102}$$

8.7.5 Modeling Multiple Machine Problems

Suppose the repairable "system" of interest actually consists of m identical machines that correspond to the assumptions of the previous section. To model this situation using a CTMC, we must first modify our definition of the system state $X(t)$. Let $X(t)$ now represent the number of machines functioning at time t. However, $\{X(t), t_0\}$ is still a CTMC because the number of states is discrete and transition times are exponentially distributed.

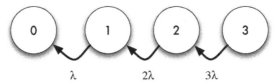

Figure 8.17 Partial rate diagram when $m = 3$.

A partial rate diagram for the case in which $m = 3$ is constructed in Figure 8.17. Note that the repair rates on the diagram depend on s, the number of maintenance technicians in the system. Note that we assume each repair requires exactly one technician.

Suppose $m = 3, s = 2, \lambda = 2$ failures per day, and $\mu = 10$ repairs per day. The completed rate diagram for the resulting CTMC is given in Figure 8.18. Note that the transition rate from state 3 to state 2 is 6. This is because three machines are functioning; each has a failure rate of 2, so the total failure rate is 6. Note that the transition rate from state 1 to state 2 is 20. This is because two machines have failed; this implies that both technicians are repairing at a rate of 10, so the total repair rate is 20. The corresponding balance equations are

$$\text{state 0:} \quad 2\rho_1 = 20\rho_0 \rightarrow \rho_1 = 10\rho_3 \tag{8.103}$$

$$\text{state 1:} \quad 20\rho_0 + 4\rho_2 = 22\rho_1 \rightarrow \rho_2 = 50\rho_0 \tag{8.104}$$

$$\text{state 3:} \quad 10\rho_2 = 6\rho_3 \rightarrow \rho_3 = \frac{250}{3}\rho_0 \tag{8.105}$$

The solution to these equations is

$$\rho_0 = \frac{3}{433}$$

$$\rho_1 = \frac{30}{433}$$

$$\rho_2 = \frac{150}{433}$$

$$\rho_3 = \frac{250}{433}$$

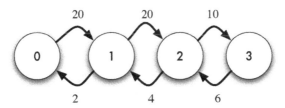

Figure 8.18 Completed rate diagram example.

Note that we can use the steady-state probabilities to obtain both machine and technician utilization. For example, the average number of machines functioning is

$$\text{AVG M} = 0\rho_0 + 1\rho_1 + 2\rho_2 + 3\rho_3 = 2.49 \text{ (83\% utilization)} \qquad (8.106)$$

and the average number of busy technicians is

$$\text{AVG S} = 2\rho_0 + 2\rho_1 + 1\rho_2 + 0\rho_3 = 0.50 \text{ (25\% utilization)} \qquad (8.107)$$

At this point, a reasonable question is, How many technicians should be assigned to maintain these machines, that is, should s equal 1, 2, or 3? To answer this question, first we modify the CTMC for the cases in which s equals 1 and s equals 3. Then, we compute the steady-state probabilities and utilization measures for each case. Then, we can use an economic model to determine the optimal value of s. Let c_s denote the cost per day of employing a technician, let c_d denote the cost per day of machine downtime, and let C denote the cost per day of system operation. Then

$$E(C) = c_s S + c_d(m - \text{AVG M}) \qquad (8.108)$$

For our example, suppose $c_s = \$200$ and $c_d = \$2500$. Then, $E(C) = \$1664.25$. For $s = 1$, the rate diagram is provided in Figure 8.19. The resulting steady-state probabilities are $\rho_0 = 0.0254, \rho_1 = 0.1271, \rho_2 = 0.3178$, and $\rho_3 = 0.5297$. Furthermore, $\text{AVG M} = 2.3518$ and $E(C) = \$1820.50$. For $s = 3$, the rate diagram is provided in Figure 8.20.

The resulting steady-state probabilities are $\rho_0 = 0.0046, \rho_1 = 0.0694, \rho_2 = 0.3472$, and $\rho_3 = 0.5787$. Furthermore, $\text{AVG M} = 2.4999$ and $E(C) = \$1850.25$. Thus, $s = 2$ is the optimal staffing level.

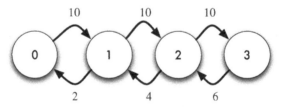

Figure 8.19   Example rate diagram, $s = 1$.

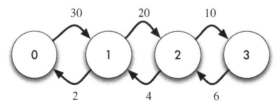

Figure 8.20   Example rate diagram, $s = 3$.

Nonidentical Machine Problems An interesting variation of the multiple-machine problem is the case in which the machines are not identical. For example, suppose a system contains two machines of different types that are repaired upon failure (no PM), and suppose two equally trained technicians maintain these machines. Let λ_i denote the failure rate for machine i, and let μ_i denote the repair rate for machine i. Modeling this problem using a CTMC requires a more complex definition of the system state.

$$X(t) = \begin{cases} 1,1 & \text{both machines are functioning} \\ 1,0 & \text{machine 1 is functioning, machine 2 is down} \\ 0,1 & \text{machine 1 is down, machine 2 is functioning} \\ 0,0 & \text{both machines are down} \end{cases} \tag{8.109}$$

The corresponding rate diagram is provided in Figure 8.21. For example, suppose $\lambda_1 = 1, \mu_1 = 8, \lambda_2 = 2$, and $\mu_2 = 10$. Construction and solution of the balance equations yields $\rho_{1,1} = 0.7407, \rho_{1,0} = 0.1481, \rho_{0,1} = 0.0926$, and $\rho_{0,0} = 0.0185$. The steady-state probabilities can then be used to compute machine availability:

$$\text{machine 1:} \quad \rho_{1,1} + \rho_{1,0} = 0.8889 \tag{8.110}$$

$$\text{machine 2:} \quad \rho_{1,1} + \rho_{0,1} = 0.8333 \tag{8.111}$$

and machine and technician utilization

$$\text{AVG M} = 2\rho_{1,1} + \rho_{1,0} + \rho_{0,1} = 1.7222 \ (86\%) \tag{8.112}$$

$$\text{AVG S} = \rho_{1,0} + \rho_{0,1} + 2\rho_{0,0} = 0.2778 \ (14\%) \tag{8.113}$$

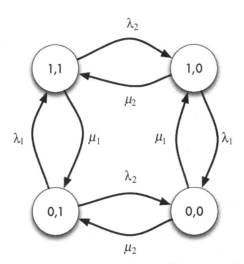

Figure 8.21 Rate diagram for two different machines.

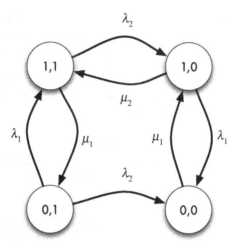

Figure 8.22   Rate diagram, multiple-machine, single technician, prioritized repair.

Another interesting variation occurs if we only have one technician available. Suppose, in addition to only having one technician, we assume that machine 1 has higher priority than machine 2. Thus, if machine 2 is being repaired and machine 1 fails, the technician will leave machine 2 to go work on machine 1. Once machine 1 is repaired, the technician will resume work on machine 2. The rate diagram for this situation is given in Figure 8.22. Notice that the rate diagram for this situation is similar to the previous example. The key difference is that when the system is in a state $(0,0)$, it can only transition to state $(1,0)$ because of the fact that machine 1 has a higher priority than machine 2 and therefore the technician must repair machine 1 as soon as it fails. Let us assume that parameters have the same values as in the previous example; therefore $\lambda_1 = 1, \mu_1 = 8, \lambda_2 = 2$, and $\mu_2 = 10$. Construction and solution of the balance equations yields $\rho_{1,1} = 0.72859, \rho_{1,0} = 0.16029, \rho_{0,1} = 0.07285$, and $\rho_{0,0} = 0.03825$. The steady-state probabilities can then be used to compute machine availability:

$$\text{machine 1:}\quad \rho_{1,1} + \rho_{1,0} = 0.8889 \tag{8.114}$$

$$\text{machine 2:}\quad \rho_{1,1} + \rho_{0,1} = 0.8015 \tag{8.115}$$

and machine and technician utilization

$$\text{AVG M} = 2\rho_{1,1} + \rho_{1,0} + \rho_{0,1} = 1.69\ (84.52\%) \tag{8.116}$$

$$\text{AVG S} = \rho_{1,0} + \rho_{0,1} + \rho_{0,0} = 0.2714\ (27\%) \tag{8.117}$$

Examining the results, we find that despite the fact that we have only one technician, we can keep machine 1's steady-state availability at the same level as if we had two technicians by giving it priority when it fails. As expected, machine 2's steady-state availability is reduced from 0.8333 to 0.8015. Another important point to take note of is that the technician utilization increases from 14% to 27%. Depending on the costs associated with an increase in downtime for machine 2, using a priority maintenance process may be more cost-effective than having two technicians available to perform maintenance.

8.7.6 Conclusions

In this section, we presented some basic techniques for modeling multistate system deterioration using Markov chains. Using Markov chains as an analysis tool for reliability and maintainability has both advantages and disadvantages, depending on the complexity of the system. A key advantage is the modeling flexibility that Markov chains give to an analyst to perform relatively quick analysis. As shown in most of our examples, Markov chains are particularly well-suited for modeling repairable systems. They are also often used to model redundancy (hot and cold standby), system dependencies, and fault tolerance systems. For most of these systems, Markov chain models are mathematically tractable and thus avoid the necessity of using simulation (see reference 21). The biggest disadvantage of Markov chain models is the "curse of dimensionality" [20]. For complex systems, the number of states required can be quite large, resulting in excessively long solution times. Fortunately, there are many commercial software tools available that help with the modeling and analysis of complex systems using Markov chains, such as Relex, ITEM, and SHARPE.

8.8 EXERCISES

8.1. Find the PDF, CDF, reliability function, and MTTF, assuming that the system has the following hazard function

$$h(t) = 0.1t^{-0.5}$$

8.2. The hazard function for a mechanical component is given by

$$h(t) = 0.0005 \left(2 + 4t + 2t^{1.5}\right)$$

(a) What is the reliability at $t = 5000$ h?

(b) What is the mean time to failure for this system?

(c) What is the expected number of failures in 1 year of operation?

8.3. Given the following failure-time data in hours:

$$50, 65, 78, 92, 99, 107, 120, 190, 200, 205$$

(a) Assuming that the system under study follows an exponential failure distribution, use the method of moments to estimate the parameter for the exponential distribution.

(b) Calculate the maximum likelihood estimator for the parameter for exponential distribution.

(c) The system under study has a 50-h warranty. What is the probability that a system in the field will fail before the warranty period?

8.4. Twelve fuel pumps are placed on test and run until failure. Their failure times are given below. Prior experience indicates that the fuel pumps follow a two-parameter Weibull failure distribution.

$$201, 1402, 351, 1078, 496, 768, 258, 480, 677, 611, 798, 802$$

(a) Use the method of moments to estimate the parameters for the Weibull distribution.

(b) Plot the PDF, reliability function, and hazard function for this system.

8.5. The failure distribution for an automotive component is modeled with a Weibull distribution with a scale parameter of 30,000 h and a shape parameter of 2.5.

(a) Find the probability that a component in the field will be operational after 2 years.

(b) Find the conditional probability that the component will fail during the third year, given that it has survived the second year of operation.

8.6. The probability that component survives a reliability test is 0.95. If 10 items are put on test, calculate the following probabilities:

(a) Exactly eight survive the test.

(b) All 10 survive the test.

(c) seven or more survive the test.

8.7. The time to failure of a component in a washing machine is known to follow an exponential distribution. The probability that the component fails during the first year is 0.975. How often should the service center expect to replace the component?

8.8. Each of the components in the system below has an exponential failure distribution. The parameters for each of the components are

$$\lambda_1 = 0.0001$$
$$\lambda_2 = \lambda_3 = 0.005$$
$$\lambda_4 = \lambda_5 = 0.0075$$
$$\lambda_6 = 0.00005$$
$$\lambda_7 = \lambda_8 = \lambda_9 = 0.025$$

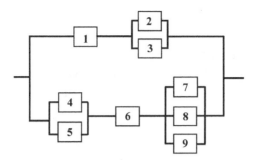

(a) Construct the system reliability function.

(b) Plot the system reliability function.

(c) What is the probability that the system will operate for one year without failure? (assume 24–7 operation)

(d) Use the component importance measures to determine which component should undergo improvement first?

8.9. The following table is intended to provide the relationships between the probability density functions, cumulative distribution function, reliability function, and hazard function. For each expression in each row of the first column, identify the appropriate relationship using the function identified across the top of the table. Please identify the appropriate relationship for the lettered block in the table.

Expressed by	$F(t)$	$f(t)$	$R(t)$	$\lambda(t)$
$F(t) =$	—	$\int_0^t f(u)du$	$1 - R(t)$	e
$f(t) =$	$\dfrac{d}{dt}F(t)$	—	c	f
$R(t) =$	$1 - F(t)$	a	—	g
$\lambda(t) =$	$\dfrac{dF(t)/dt}{1 - F(t)}$	b	d	—

8.10. Identify each of the following statements as true or false. If false correct the statement so it is true.

(a) If a component is IFR, then $R(t + T_0|T_0) < R(t)$.

(b) If a component is CFR, then $R(t + T_0|T_0) > R(t)$.

(c) If a component is DFR, then the mean residual life will be a decreasing function of T_0.

8.11. Compute each of the following quantities assuming that the time to failure of the system is exponential with a mean time to failure of 1000 h and a guaranteed life of 200 h. Be sure to include units where appropriate.

(a) the system's failure rate

(b) the system's 100-h reliability

(c) the system's 300-h reliability

(d) the system's 700-h reliability given survival up to 600 h

(e) the system's design life for a 95% reliability.

8.12. Respond to the following statements assuming that the time to failure of the system is Weibull with a shape parameter of 2.0 and a scale parameter of 100 hrs. = 2.0 and 100 h. Be sure to include units where appropriate.

(a) Classify the system's hazard function as either DFR, CFR, or IFR.

(b) Compute the system's 100-h reliability.

(c) Compute the system's MTTF.

(d) Compute the systems characteristic life.

(e) Compute the system's 150-h reliability given survival up to 100 h.

(f) Compute the systems median time to failure.

(g) Compute the probability that the system fails between 100 and 150 h.

(h) Is this system a candidate for burn-in? Why or why not?

8.13. Consider a system that consists of a collection of four components such that $R_1 = 0.95, R_2 = 0.975, R_3 = 0.99$, and $R_4 = 0.999$. Assume that the cost to improve each component by one unit of reliability is given by $C_1 = \$100, C_2 = \$200, C_3 = \$75$, and $C_4 = \$85$. Answer the following questions.

(a) Provide a lower bound on the system's reliability.

(b) Provide an upper bound on the system's reliability.

(c) Assuming that the four components are arranged in series, which component is the most important?

(d) Assuming that the four components are arranged in parallel, taking cost of improvement into consideration, which component should we improve first?

8.14. Assume that the components in the previous problem must be arranged in a series arrangement in order to function. Suppose you can afford to purchase an additional copy of each of the components.

(a) Construct a reliability block diagram assuming system level redundancy.

(b) Construct a reliability block diagram assuming component level redundancy.

(c) Which arrangement has the higher reliability?

8.15. Consider a system represented by the following reliability block diagram. Note that $R_1 = 0.9, R_2 = 0.975, R_3 = 0.999, R_4 = 0.92, R_5 = 0.95, R_6 = R_7 = 0.85$. Note the following subsystem definitions.

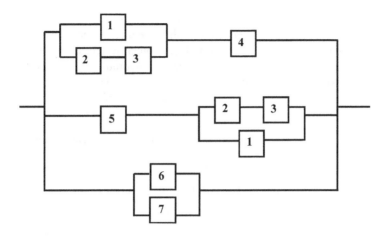

(a) Subsystem A Components 2, 3

(b) Subsystem B Component 1 and subsystem A

(c) Subsystem C Component 4 and subsystem B

(d) Subsystem D Components 5 and subsystem B

(e) Subsystem E Components 6,7

Determine the following:

(a) The reliability of subsystem B

(b) The reliability of subsystem C

(c) The reliability of subsystem D

(d) The reliability of subsystem E

(e) The reliability of the system

(f) Identify the most important component in the system.

8.16. An electronic relay has two main failure modes: premature closure (PC) and failure to close (FTC). We can model each failure mode using an exponential distribution with the following failure rates:

$$\lambda_{PC} = 5 \times 10^{-3} \text{ failures per hour}$$

$$\lambda_{FTC} = 5 \times 10^{-4} \text{ failures per hour}$$

The mean time to repair a PC failure is assumed to be 2 h, while the mean time to repair FTC failure is 48 hours. The repair times are assumed to be exponentially distributed [10].

(a) Establish a rate diagram for this system.

(b) Write the system balance equations.

(c) Calculate the steady-state availability for this system.

(d) Calculate the mean time to failure for the system.

8.17. Two identical aircraft hydraulic control systems are operated as a parallel system. During normal operation both hydraulic systems are functioning. When the first hydraulic system fails, the other system has to do the entire job alone with a higher load than when both hydraulic systems are in operation. Assume that the hydraulic systems have identical constant failure rates and that when one fails and a single system is performing the job, it has a higher failure rate.

$$\lambda_S = 5 \times 10^{-4} \text{ failures per hour (rate when systems share load)}$$

$$\lambda_{FTC} = 5 \times 10^{-4} \text{ failures per hour}$$

$$(\text{rate when single system is moving full load})$$

Assume that both systems may fail at the same time due to some external event (common cause failure). The constant failure rate with respect to common cause failures has been estimated to be $\lambda_{CC} = 9.0 \times 10^{-5}$ common cause failures per hour. This type of external stress can affect the system irrespective of how many units are functioning.

Repair is initiated as soon as one of the hydraulic systems fails. The mean downtime of a system has been estimated to be 20 h. When both systems are in a failed state, the entire process has to be shut down. In this case, the system will not be put into operation again until both pumps have been repaired. The mean downtime, when both pumps have failed, has been estimated to be 40 h (21).

(a) Establish a rate diagram for this system.

(b) Write the balance equations for the system.

(c) Calculate the steady-state probabilities.

(d) Determine the percentage of time when (i) both pumps are functioning; (ii) only one of the pumps is functioning; (iii) both pumps are in a failed state.

(e) Compute the system MTTF.

8.18. You are given a system that has five components in parallel. For the system to work, four of the five components must be working. Assume all components are identical with a failure rate of λ_1 when five components are working and a failure rate of λ_2 when only four components are operating. Assume that the repair rate for all components is μ. Assume a single repairman.

(a) Establish the reliability state diagram for the system.

(b) Write the state equations.

(c) Compute the MTTF for the system.

(d) Establish the availability state diagram

(e) Write the state equations.

(f) Calculate the steady-state availability.

8.19. Given two components in a standby configuration with components that have different failure rates. Let component 1 be the main component and component 2 the standby component. Assume component 1 has an exponential failure rate of λ_1. Let component 2 have an exponential failure rate of λ_{21} when it is in standby mode and λ_{2O} when it is operating. When component one fails a switch is used to bring component 2 online. Assume that the switch has a probability of success of p. Use a Markov modeling approach to model the system.

(a) Draw the state transition diagram

(b) Derive the system state equations

(c) Compute the reliability of the system.

(d) What is the mean time to failure for the system?

8.20. The following poem by Oliver Wendell Holmes appears in or is referenced by many reliability text books.

(a) Why do you think this poem is found in many reliability textbooks?

(b) Identify in the poem the various components of the systems decision process.

The Deacon's Masterpiece, or the Wonderful "One-Hoss Shay": A Logical Story
—by Oliver Wendell Holmes (1809–1894)

Have you heard of the wonderful one-hoss
shay,
That was built in such a logical way
It ran a hundred years to a day,
And then, of a sudden, it ah, but stay,
I'll tell you what happened without
delay,
Scaring the parson into fits,
Frightening people out of their wits,
Have you ever heard of that, I say?

Seventeen hundred and fifty-five.
Georgius Secundus was then alive,
Snuffy old drone from the German hive.
That was the year when Lisbon-town
Saw the earth open and gulp her down,
And Braddock's army was done so
brown,
Left without a scalp to its crown.
It was on the terrible Earthquake-day

That the Deacon finished the
one-hoss shay.

Now in building of chaises,
I tell you what,
There is always somewhere
a weakest spot,
In hub, tire, felloe, in spring or thill,
In panel, or crossbar, or floor, or sill,
In screw, bolt, thoroughbrace, lurking still,
Find it somewhere you must and will,
Above or below, or within or without,
And that's the reason, beyond a doubt,
A chaise breaks down, but doesn't
wear out.

But the Deacon swore (as Deacons do,
With an "I dew vum," or an "I tell yeou")
He would build one shay to beat the taown
'N' the keounty 'n' all the kentry raoun';

It should be so built that it couldn' break
daown:
"Fur," said the Deacon, "'tis mighty plain
Thut the weakes' place mus' stan' the
strain;
'N' the way t' fix it, uz I maintain,
Is only jest
T' make that place uz strong uz the rest."

So the Deacon inquired of the village folk
Where he could find the strongest oak,
That couldn't be split nor bent nor broke,
That was for spokes and floor and sills;
He sent for lancewood to make the thills;
The crossbars were ash, from the
straightest trees,
The panels of white-wood, that cuts like
cheese,
But lasts like iron for things like these;
The hubs of logs from the "Settler's
ellum,"
Last of its timber, they couldn't sell 'em,
Never an axe had seen their chips,
And the wedges flew from between their
lips,
Their blunt ends frizzled like celery-tips;
Step and prop-iron, bolt and screw,
Spring, tire, axle, and linchpin too,
Steel of the finest, bright and blue;
Thoroughbrace bison-skin, thick and wide;
Boot, top, dasher, from tough old hide
Found in the pit when the tanner died.
That was the way he "put her through."
"There!" said the Deacon, "naow she'll
dew!"

Do! I tell you, I rather guess
She was a wonder, and nothing less!
Colts grew horses, beards turned gray,
Deacon and deaconess dropped away,
Children and grandchildren where were
they?
But there stood the stout old one-hoss shay
As fresh as on Lisbon-earthquake-day!

EIGHTEEN HUNDRED; it came and
found
The Deacon's masterpiece strong and
sound.
Eighteen hundred increased by ten;

"Hahnsum kerridge" they called it then.
Eighteen hundred and twenty came;
Running as usual; much the same.
Thirty and forty at last arrive,
And then come fifty, and FIFTY-FIVE.

Little of all we value here
Wakes on the morn of its hundreth year
Without both feeling and looking queer.
In fact, there's nothing that keeps its youth,
So far as I know, but a tree and truth.
(This is a moral that runs at large;
Take it. You're welcome. No extra charge.)
FIRST OF NOVEMBER, the
Earthquake-day,
There are traces of age in the one-hoss
shay,
A general flavor of mild decay,
But nothing local, as one may say.
There couldn't be, for the Deacon's art
Had made it so like in every part
That there wasn't a chance for one to start.
For the wheels were just as strong as the
thills,
And the floor was just as strong as the
sills,
And the panels just as strong as the floor,
And the whipple-tree neither less nor
more,
And the back crossbar as strong as the
fore,
And spring and axle and hub encore.
And yet, as a whole, it is past a doubt
In another hour it will be worn out!

First of November, 'Fifty-five!
This morning the parson takes a drive.
Now, small boys, get out of the way!
Here comes the wonderful one-horse shay,
Drawn by a rat-tailed, ewe-necked bay.
"Huddup!" said the parson. Off went they.
The parson was working his Sunday's text,
Had got to fifthly, and stopped perplexed
At what the Moses was coming next.
All at once the horse stood still,
Close by the meet'n'-house on the hill.
First a shiver, and then a thrill,
Then something decidedly like a spill,
And the parson was sitting upon a rock,

At half past nine by the meet'n-house
clock,
Just the hour of the Earthquake shock!
What do you think the parson found,
When he got up and stared around?
The poor old chaise in a heap or mound,
As if it had been to the mill and ground!

You see, of course, if you're not a dunce,
How it went to pieces all at once,
All at once, and nothing first,
Just as bubbles do when they burst.

End of the wonderful one-hoss shay.
Logic is logic. That's all I say.

REFERENCES

1. Checkland, P. *Systems Thinking, Systems Practice*. West Sussex, England: John Wiley & Sons, 1981.
2. Merriam-Webster Online. Available at www.m-w.com/dictionary/effective.
3. Soban, D, Marvis, D. Formulation of a methodology for the probabilistic assessment of system effectiveness. Research Report AIAA-MSC-2000-DS, Georgia Institute of Technology, Atlanta, GA, 2000.
4. Blanchard, B. *Logistics Engineering and Management*, 5th ed. Upper Saddle River, NJ: Prentice Hall, 1998.
5. Dietrich, D. Class Notes, SIE 608, *Large Scale Systems*, University of Arizona, 1993.
6. Fuqua, N. *Reliability Engineering for Electronic Design*. New York: Marcel Dekker, 1987.
7. Barlow, RE, Proscan, F. *Mathematical Theory of Reliability*. New York: John Wiley & Sons, 1965.
8. Elsayed, E. *Reliability Engineering*. Reading, MA: Addison Wesley, 1996.
9. Ramakumar, R. *Reliability Engineering*. Englewood Cliffs, NJ: Prentice-Hall, 1993.
10. Hoyland, A, Rausand, M. *System Reliability Theory*. New York: John Wiley & Sons, 1994.
11. Kapus, K, Lamberson, L. *Reliability in Engineering Design*. New York: John Wiley & Sons, 1977.
12. Leemis, L. *Reliability: Probabilistic Models and Statistical Methods*. Englewood Cliffs, NJ: Prentice-Hall, 1995.
13. Kececioglu, D. *Reliability & Life Testing Handbook*. Englewood Cliffs, NJ: Prentice-Hall, 1993.
14. Nachlas, J. *Reliability Engineering: Probabilistic Models and Maintenance Methods*. Boca Raton, FL: Taylor & Francis, 2005.
15. Maillart, L, Pohl, E. Markov chain modeling and analysis. Tutorial Notes of the Annual Reliability and Maintainability Symposium, 2006.
16. Ascher, H, Feingold, H. *Repairable Systems Reliability*. New York: Marcel Dekker, 1984.
17. Kao, E. *Introduction to Stochastic Processes*. Belmont, CA: Duxbury Press, 1997.
18. Ross, SM. *Introduction to Probability Models*, 7th ed. San Diego, CA: Harcourt Academic Press; 1989.
19. Ross, SM. *Stochastic Processes*, 2nd ed. New York: John Wiley & Sons, 1996.
20. Minh, DL. *Applied Probability Models*. Pacific Grove, CA: Duxbury Press, 2001.

21. Fuqua, N. The applicability of Markov analysis methods to reliability, maintainability, and safety. *Select Topics in Assurance Related Technologies*, 2003;10(2):1–8.

22. Pohl, EA, Mykytka, EF. Simulation modeling for reliability analysis. Tutorial Notes of the Annual Reliability and Maintainability Symposium, 2000.

Part **III**

Systems Decision Making

Chapter **9**

Systems Decision Process Overview

GREGORY S. PARNELL, Ph.D.
PAUL D. WEST, Ph.D.

Decisions are easy, it's only the rationale that is difficult.

—Anonymous

9.1 INTRODUCTION

As we noted in Chapter 1, systems are becoming more complex, dynamic, inter-connected, and automated and face significant security challenges. As a result of these trends and the increasing involvement of stakeholders, private and public leaders are faced with difficult systems decisions. Our definition of a decision is an irrevocable allocation of resources [1]. Decision makers can change their mind, but there is a resource penalty associated with the change. For example, consider the purchase of a new car. The decision occurs when we sign the contract. If we change our mind in one month, we now have a used car. The value of the used car may be significantly lower than that of the new car. Systems decisions are very similar. We make the acquisition decision when we award a contract. If we have to cancel the contract in 1 month, there are resource penalties.

> A decision is an irrevocable allocation of resources.

Decision Making in Systems Engineering and Management, Second Edition
Edited by Gregory S. Parnell, Patrick J. Driscoll, Dale L. Henderson
Copyright © 2011 John Wiley & Sons, Inc.

In this chapter, we introduce the systems decision process (SDP) that has been developed by faculty members of the Systems Engineering Department at the United States Military Academy. The process has been applied to many military problems in all stages of the system life cycle. It is the systems engineering decision process that our seniors are expected to use for their capstone research projects. While we follow the process, we should tailor the process to the system, the stage of the system life cycle, and the problem under consideration. Shortly after graduation, our graduates will apply the process to operations and logistics problems they encounter in their military careers. Later on in their careers, they may have the opportunity to apply the process to the early stages of the system life cycle for complex systems design and development.

> The systems decision process is a general problem solving process. We need to tailor the process to the system, the decision, and the stage of the system life cycle.

We believe the process is broadly applicable in many systems engineering and engineering management domains. The systems decision process and the techniques used in the process have been applied in many private and public problem domains.

9.2 VALUE-FOCUSED VERSUS ALTERNATIVE-FOCUSED THINKING

The lead systems engineer guides the team in how it approaches the systems decision process. Two main philosophies dominate the approach strategies: *alternative-focused thinking* (AFT) and *value-focused thinking* (VFT), although hybrid strategies have been proposed. The systems decision process uses a VFT approach.

"Values are what we care about," Keeney notes in *Value-Focused Thinking* [2]. "As such, values should be the driving force for our decision-making." Values, he notes, are principles used for the evaluation of actual or potential consequences of action and inaction, of proposed alternatives, and of decisions. The VFT process differs from traditional alternative-focused thinking (AFT) in that a clear understanding of values drives the creation of alternatives, rather than the traditional approach in which alternatives are identified first. This process can be applied not only to externally generated events, or decision *problems*, but also to internally generated events, or decision *opportunities*.

The sequence of actions in VFT and AFT is shown in Figure 9.1 [2]. Columns 1 and 2 show that the major difference in reactive approaches under VFT and AFT is in the timing of when values are specified. By understanding the decision maker's values before identifying alternatives, alternatives tailored for the decision context can be generated. Keeney contrasts this with the alternative-focused approach that first identifies alternatives and then uses values to choose from what is available.

The sequences shown to create decision opportunities enable more aggressive control of a situation. Column 3 is similar to Column 2 except that rather than

React to decision problem		Identify or create decision opportunity	
Alternative-focused	**Value-focused**		
1. Recognize a decision problem	1. Recognize a decision problem	1. Identify a decision opportunity	1. Specify values
2. Identify alternatives	2. Specify values	2. Specify values	2. Create a decision opportunity
3. Specify values	3. Create alternatives	3. Create alternatives	3. Create alternatives
4. Evaluate alternatives	4. Evaluate alternatives	4. Evaluate alternatives	4. Evaluate alternatives
5. Select an alternative	5. Select an alternative	5. Select an alternative	5. Select an alternative
Column 1	**Column 2**	**Column 3**	**Column 4**

Figure 9.1 Value-focused versus alternative-focused thinking sequences. (Adapted from [2].)

reacting to a problem, the decision maker is alert to opportunities. Keeney cites a situation in which a clothing manufacturer who observes fabric scraps identifies a decision opportunity and, by working through Steps 2 through 4, decides to open a new product line built from scrap material [2]. Decision opportunities can be created once strategic objectives and values are specified.

A conceptual model of the multicriteria decision problem structure from both perspectives is shown in Figure 9.2. This figure, adapted from Buchanan [3], shows the opposing approaches. Alternatives are defined here as courses of action that can be pursued and will have outcomes measured in terms of the criteria. The criteria reflect the values of the decision maker. Attributes are defined as the objectively measurable features of an alternative. This model separates the subjective from the

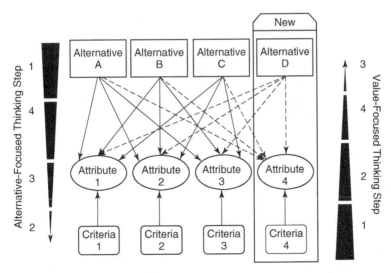

Figure 9.2 Flow of value-focused and alternative-focused thinking. (Adapted from [3].)

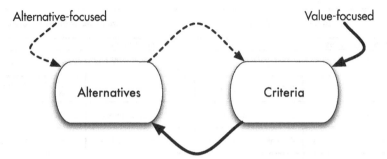

Figure 9.3   Dynamic approach to problem structuring [6].

objective components. The figure shows that the major difference of VFT is that is may result in additional criteria and new alternatives.

The greatest pitfall of a pure AFT approach is the danger of locking onto the list of alternatives and seeing only criteria that fit the alternatives while ignoring all others. Contrarily, a criticism of the VFT approach is that stakeholder values may not be sufficiently well formed in the early stages of the decision making process [4]. VFT is significantly different than AFT. First, the order is different. Second, VFT results in new criteria, new attributes, and new alternatives. Third, VFT changes the attribute ranges from local ranges to global ranges define by the existing alternatives defined by our values. Fourth, VFT takes additional effort to develop the value model, find new alternatives, and analyze the alternatives.

Finally, there is the chicken-or-egg dilemma that states that values are formed from experience with alternatives [5]. This leads some to argue that the starting point is not the issue. In their proposal for dynamic decision problem structuring, Corner, Buchanan, and Henig say, "What is important is that the decision maker learns about one (criteria or alternatives) from working with the other" [6]. Their concept is shown in Figure 9.3, where criteria and alternatives are shown as a causal loop, with entry through either an AFT (dashed line) or a VFT (solid line) approach. This suggests that by thinking about alternatives the solution designer helps identify criteria, and vice versa. This approach also recognizes the iterative nature of thinking about problems, so that both alternatives (or opportunities) and criteria can be refined during the problem definition and solution design phases.

9.3   DECISION QUALITY

Engineering managers and systems engineers need to make many decisions during the system life cycle. Since they face many complexities and significant uncertainties about the system and its environment, they should be concerned about the quality of their decisions. What makes a quality decision? Matheson and Matheson [7] used a chain illustration to identify the six elements of a quality decision for an R&D organization. Each link in the decision quality chain is important. Our systems decision process includes each of the six elements. In Figure 9.4, we identify

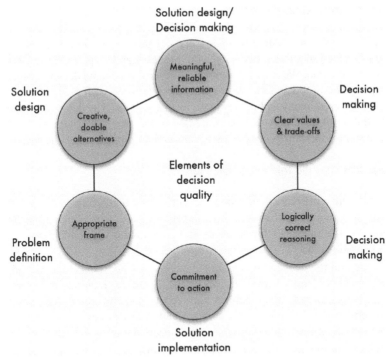

Figure 9.4 Elements of decision quality (the corresponding SDP phases are annotated in the diagram).

the systems decision process phases that address each element of decision quality. The following is a brief description of each link in the chain.

- *Appropriate Frame.* Our cognitive view of the problem must be appropriate to consider the full scope of the problem and all the needs of the stakeholders.
- *Creative, Doable Alternatives.* We want creative, feasible solutions that create value for the stakeholders and decision makers.
- *Meaningful, Realistic Data.* The data we use to generate and score candidate solutions must be understandable and credible.
- *Clear Values and Tradeoffs.* The values we use to generate and evaluate solutions must be defined and tradeoff analysis must be performed between system requirements, value measures, and resources.
- *Logically Correct Reasoning.* The mathematical techniques we use to evaluate alternatives must use a sound operations research technique. As stated in Chapter 1, the SDP uses multiple objective decision analysis (MODA) as its foundation for value tradeoffs.
- *Commitment to Action.* Finally, the decision maker(s) and stakeholders must be committed to solution implementation. Implementation barriers and risks must be identified and resolved.

9.4   SYSTEMS DECISION PROCESS

We first introduced our systems decision making process in Figure 1.7 of Chapter 1. We emphasized that the systems decision process can be applied in any stage of the system life cycle. We begin to describe further the systems decision process in this chapter. The SDP has four phases with three tasks in each phase. Chapters 10 through 13 describe many techniques to perform each task. The detailed explanation and illustration of each task will be described in the next four chapters.

The systems decision process can be tailored to any stage of the system life cycle.

The process has the following characteristics:

- Starts with efforts to understand the current system using systems thinking tools introduced in Chapter 2. The current system, or baseline, is the foundation for assessment of future needs and comparison with candidate solutions to meet those needs.
- Focuses on the decision maker and stakeholder value. Stakeholders and decision makers identify important functions, objectives, requirements (screening criteria that all potential solutions must meet), and constraints. The key stakeholders are the consumers of system products and services, the system owners, and the client responsible for the system acquisition.
- Focuses on creating value for decision makers and stakeholders and defines the desired end state that we are trying to achieve. The value modeling task of the problem definition phase plays an important role in defining the ideal solution for comparison with candidate solutions. The solution improvement task improves the alternative design solutions. Finally, we use value-focused thinking to improve the nondominated solutions.
- Has four phases (problem definition, solution design, decision making, and solution implementation) and is highly iterative based on information and feedback from stakeholders and decision makers.
- Explicitly considers the environment (historical, legal, social, cultural, technological, security, ecological, health & safety, and economic) that systems will operate within and the political, organizational, moral/ethical, and emotional issues that arise with stakeholders and decision makers in the environment.

Next, we discuss the symbolism of the SDP diagram. Although Figure 9.5 is not in color in the book here, we use colors to characterize each phase of the SDP. The colors selected have a symbolic meaning for system engineers and engineering managers.[1]

> *Problem Definition.* Stop! Don't Just Do Something—Stand There! The most important task in any systems decision process is to identify and understand

---

[1]The meanings of the colors were initially developed by one of our West Point colleagues, Daniel McCarthy.

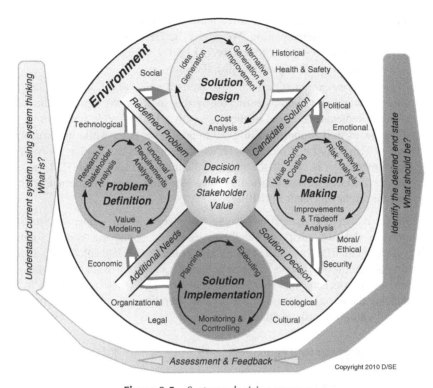

Figure 9.5 Systems decision process.

the problem which is informed by understanding the concerns, objectives, and constraints of the decision makers and stakeholders. If we fail to identify and fully understand the right problem, we may end up expending a lot of time and energy creating a great system solution that solves the wrong problem. For this very reason, in the graphical depiction of the SDP we represent the Problem Definition phase with a **RED** circle to remind ourselves to **STOP!** and make sure we have a clear understanding of the problem before we try to design a solution.

Solution Design. Proceed with Caution! Having developed a clear understanding of the problem during the Problem Definition phase of the SDP, we can now go about the business of finding a system solution to the problem. However, we must avoid the temptation to select a solution too quickly. Perhaps we have "seen this problem before" and might "know the answer." Depicting the Solution Design phase with a **YELLOW** circle reminds us to **Proceed with Caution!** We should treat each problem with a fresh perspective and give due diligence by developing new ideas and generating and improving alternative solutions to the problem.

Decision Making. Green Light—Go! During the Decision-Making phase of the SDP, we take the information we have gathered in the previous phases of the SDP to prepare and make a recommendation to our clients seeking their approval of a system solution. We depict the Decision-Making phase with a **GREEN** circle to represent the **Green Light** we are hoping to get from our client to proceed with implementing the solution.

Solution Implementation. Once we have secured a decision from our client, we turn our attention to implementing the system solution. We represent the Solution Implementation phase of the SDP with a **BLUE** circle representing the **Blue Skies and Smooth Sailing** we hope to encounter while implementing our system solution. Unfortunately, as we will learn in Chapter 13, solution implementation is one of the most challenging phases of the SDP. In fact, since we can expect smooth sailing **once in a blue moon**, an alternative color choice might be **Black and Blue**!

Decision maker and stakeholder value are placed in the center of the SDP to remind us that we must continually focus on the value the system solution will provide them. The four spokes (defined problem, candidate solutions, solution decision, and additional needs) represent major outputs of each phase that are approved by the decision makers and stakeholders. The cradle and arrows in the SDP indicate the iterative nature of the process. Iteration takes place both between phases and within the tasks defining each phase.

9.5   ROLE OF STAKEHOLDERS

Stakeholder involvement is critical to the success of the systems decision process. Stakeholders ensure that we have the appropriate frame for our decision, and they help us obtain reliable and credible information; their early involvement is essential for their commitment to action necessary to implement the decision. Without stakeholder involvement, decisions will not be sustainable and stakeholders may force costly decision changes. Stakeholders have important roles in each phase of the systems decision process.

> Decision-maker and stakeholder involvement is critical to the success of the systems decision process.

Problem Definition.   Conducting research is a key activity for identifying potential system stakeholders. Stakeholders provide information about the environment that helps frame the problem and accurately specify the problem statement. They help identify the system objectives and the requirements. They provide insights to help identify the constraints and functions. Finally, they validate the value model and the final problem statement.

Solution Design.   Stakeholders participate in the solution design, assessment, and testing. They attend design reviews and provide comments on design solutions. They continue to identify constraints and requirements. They participate in spiral development processes by evaluating each prototype solution. Finally, they participate in test and evaluation programs. Stakeholders help identify the cost elements used in system cost analysis to ensure that the alternatives will be affordable.

Decision Making. Stakeholders are involved in the value scoring and costing of candidate solutions. They provide the operational data, test data, models, simulation, and experts to obtain the scoring data. They participate in interim reviews and final decision presentations.

Solution Implementation. Stakeholders are critical to solution implementation. They help identify the solution implementation plan, tasks, and controls. All tasks in this phase require stakeholder commitment to action.

9.6 ROLE OF DECISION MAKERS

The systems decision process exists to support systems decision makers. For technology based systems, the decision maker may be the project/program manager or the engineering manager of the organization. For complex systems decisions the decision maker may be an operational manager, a functional manager, the chief technology officer, the chief information officer, or even the chief executive officer. Interaction with the decision maker(s) occurs in each phase of the systems decision process. Decision makers approve the resources to perform the systems decision process. Decision makers should "inspect what they expect" [8]. The best practice is periodic interim progress reviews during each phase in the systems decision process.

Systems engineers must communicate with decision makers and stakeholders. Decision makers should inspect what they expect.

Problem Definition. The SDP usually begins with a planned decision milestone in the system life cycle or a decision maker identifying a problem or opportunity requiring a decision. Not making a decision is also a decision. The decision maker should approve the list of stakeholders for stakeholder analysis, the functional and requirements analyses, the value model, and the final problem statement.

Solution Design. The decision maker should review the alternatives being considered and approve the candidate solutions that will be evaluated in the decision making phase. The decision maker should ensure that the alternative improvements have been considered and that cost analysis has been performed in the design process.

Decision Making. This is a major phase for decision-maker interaction. They participate in interim reviews and final decision presentations. Decision makers select the solution to implement. If required, they participate in decision reviews in their own or other organizations. Decision makers review the candidate solution value, cost, sensitivity analysis, and risk analysis. They also decide if more improvements are needed and make the difficult value versus cost tradeoff decision among the efficient solutions.

Solution Implementation. Decision makers must ensure stakeholder participation in the implementation phase, provide the implementation resources, and approve the implementation plan. They may also monitor the execution of the plan and take control actions as required to meet the performance, schedule, and cost goals.

9.7 ENVIRONMENT

During the systems decision process we must consider many factors that may affect the systems decision process and the environment that the system will operate in during its life cycle. We call all these factors the SDP environment. The factors we show on the SDP diagram and describe below are meant to be illustrative and not exhaustive. These factors are identified using the system thinking techniques described in Chapter 2. We briefly discuss some of the major environment issues that should be considered in the systems decision process.

Technological. System elements use technologies to perform functions for consumers and users. Some technologies are developed and available. New technologies may involve technical, cost, and schedule risks for the system. In addition, the consequences of technologies—for example, the health consequences of asbestos or the environmental impact of gasoline—are not always understood. A major system failure can delay a system for many years, as witnessed in the Challenger spacecraft failure.

Economic. Economic factors are almost always a major systems decision issue. Most program managers have a budget to manage. Stakeholders are concerned about the economic impact of the new system on their budgets. For example, design changes to the airline security system have dramatically impacted many government and commercial organizations.

Political. Political factors come into play for many systems decisions. Many stakeholder groups (e.g., lobby groups) exist to impact systems decisions by private or public organizations. Many public decisions require approval by U.S. government agencies and/or Congress. Press coverage can make any system a major political issue—for example, the space shuttle after the Challenger disaster.

Legal. Systems must comply with federal, state, and community legal requirements. For example, automobiles must meet federal safety and emissions standards and also state regulations.

Health & Safety. The impact of the system on the health and safety of system stakeholders are important considerations in all stages of the system life cycle. Health and safety issues can result in injury and death to system users, consumers, and affected stakeholders. Systems thinking and risk analysis are important tools to identify and mitigate potential health and safety issues.

Social. Systems can have social implications. For example, information technology (IT) systems have significantly changed how we work and how we interact with our family, friends, and associates.

Security. Systems must be secure. System owners, users, and consumers want to be sure that their system and their products and services are secure against potential threats. There are several security dimensions: Physical security and information security are very important issues for system designers.

Ecological. Systems can have significant consequences on all elements of the ecology. For example, the nuclear weapons and nuclear power industries have generated a significant amount of radioactive waste that must be properly processed and safeguarded to minimize the ecological impact.

Cultural. Many systems and products are designed for national cultural groups and international customers. System designers must consider cultural factors in their design and marketing, especially if they develop products and services for international markets with diverse customers. Cultural considerations also arise when an organization is faced with adapting to meet new challenges and desires to retain a set of cultural characteristics that define who they are or how they operate.

Historical. Some systems impact historical issues. Most states have historical preservation societies that are interested in changes that impact historical landmarks and facilities. These organizations can impact system designs and can delay solution implementation.

Moral/ethical. Many times moral or ethical issues arise in systems decisions. For example, there are privacy issues associated with information technology solutions. Also, the use of certain weapons systems (e.g., chemical, biological, or nuclear) is a moral issue to many stakeholders.

Organizational. Decisions are made within organizations. The key formal and informal organizational leaders can be important stakeholders in the decision process. Stakeholder analysis is the key to identifying and resolving organizational issues.

Emotional. Sometimes decision makers or key stakeholders have personal preferences or emotional issues about some systems or potential system solutions. For example, nuclear power is an emotional issue for some stakeholders. Systems engineers must identify and deal with these issues.

Our list of considerations in the system environment has several uses. Once we have defined the system boundary, the above list can be used as a system thinking tool to identify metasystems. It also provides a useful checklist when beginning the process of identifying research areas and stakeholders during the Problem Definition phase of the SDP.

9.8 COMPARISON WITH OTHER PROCESSES

There are other decision processes. Table 9.1 compares our systems decision process with two other problem solving processes. Both of these problem solving processes are similar to the SDP. Athey's systematic systems approach is more general. The

TABLE 9.1 Comparison of Problem-Solving Processes

Systems Decision Process	Athey's Systematic Systems Approach	Military Decision-Making Process
1. Problem definition	1. Formulate the problem	1. Receipt of mission
(a) Research & stakeholder analysis	2. Gather and evaluate information	2. Mission analysis
(b) Functional & requirements analyses	3. Develop potential solutions	3. Course of action (COA) development
(c) Value modeling	4. Evaluate workable solutions	4. COA analysis
2. Design solution	5. Decide the best solution	5. COA comparison
(a) Idea generation	6. Communicate system solution	6. COA approval
(b) Alternative generation & improvement	7. Implement solution	7. Orders production
(c) Cost analysis	8. Establish performance standards	8. Rehearsal
3. Decision making		9. Execution and assessment
(a) Value scoring & costing		
(b) Sensitivity & risk analyses		
(c) Improvements & tradeoff analysis		
4. Solution implementation		
(a) Planning		
(b) Executing		
(c) Monitoring and controlling		

SDP provides more detail about the types of steps. The military decisions process focuses on a course of action instead of a system.

9.9 WHEN TO USE THE SYSTEMS DECISION PROCESS

The systems decision process we define in this book has broad applicability for systems engineers to many systems in all system life cycle stages. In addition, since the process is very general and grounded in sound principles, it may be useful for problem solving in domains when no one has been given the title "systems engineer." In some decision applications, only one or more of the tasks may be required. For example, the stakeholder analysis techniques can be used to help define the problem in any public or private decision setting. In this section, we provide some guidance on when to use the systems decision process and when it may not be appropriate.

When considering whether or not to use the systems decision process, the engineering manager and lead systems engineer should consider three criteria: need, resources, and consequences.

TABLE 9.2 Criteria for When to Use Each Task in the Systems Decision Process

Phase	Task	When Needed	When Not Needed
Problem definition	Research and stakeholder analysis	Many research areas and multiple decision makers and stakeholders with conflicting views about the system and system life cycle.	Single decision maker and known stakeholder views.
	Functional and requirements analyses	System functions and requirements not defined and are critical to system success.	System functions and requirements defined or not applicable.
	Value modeling	System objectives and value measures not defined.	Clear objectives and value measures are known.
Solution design	Idea generation	Concern about the quantity or quality of alternatives.	Several high value alternatives are known or an established model or process is available to develop high value alternatives.
	Alternative generation and improvement	Alternatives must be designed to ensure they perform system functions.	Solution design is complete or simple to complete.
		Alternatives need improvements to achieve the desired value.	Solution optimization techniques are already available.
	Cost analysis	Resources and costs are significant drivers for the decision	Costs and resources are not significant decision drivers.
Decision making	Solution scoring and costing	Solution performance and costs are unknown.	Solution performance and costs are known or not applicable.
	Sensitivity and risk analyses	Important to understand the most sensitive assumptions or parameters and to assess the key risks.	Impact of assumptions and risks are already known or not important.
	Improvements and tradeoff analysis	Opportunity to improve the solution. Tradeoffs must be made between efficient solutions.	Ideal solution is known or no opportunity to improve the solution.

(*continued*)

TABLE 9.2 (*Continued*)

Phase	Task	When Needed	When Not Needed
Solution implementation	Planning	Many complex tasks, significant resources, and many participants.	Simple tasks, few resources, and few participants.
	Executing	Critical to success of the decision.	Not required or not critical to decision success.
	Monitoring and controlling	Schedule and resource usage are important.	No management oversight is required.

TABLE 9.3   Systems Decision Process Example

Phase	Task	Army BRAC [9]
Problem definition	Research and stakeholder analysis	Many senior leader interviews and document reviews.
	Functional and requirements analyses	Requirements analysis used to develop installation portfolio constraints.
	Value modeling	Used for the ranking of installations and the evaluation of alternatives
Solution design	Idea generation	Performed by subject matter experts informed by modeling.
	Alternative generation and improvement	Developed by mission and functional experts
	Cost analysis	Used many cost models in the analysis process.
Decision making	Solution scoring and costing	Scoring data submitted and certified by each Army installation.
	Sensitivity and risk analyses	Significant sensitivity analysis at all parts of the process.
	Improvements and tradeoff analysis	The value was to transform Army installations and save operations and support funds. Solutions were improved after analysis.
Solution implementation	Planning	Performed by BRAC division on the Army staff.
	Executing	Will be performed by major commands and the Installation Management Command.
	Monitoring and controlling	Critical since execution is monitored by the U.S. Government Accountability Office and U.S. Congress.

9.9.1 Need

The systems decision process is needed if several of the tasks meet the criteria listed in Table 9.3 for each of the four phases.

9.9.2 Resources

If the systems decision process meets the criteria stated in Table 9.3, we next consider resources. The time any systems decision process takes to analyze systems engineering data and provide reports and presentations to senior decision makers can be significant for large, complex systems. The full process we describe in this book takes time and resources. The amount of time spent on systems decision making should be proportional to the type of system and the potential consequences of the decision to the decision makers and stakeholders. The decision to select the airplane design for the next Boeing commercial airliner should require more systems decision process resources than the design of the next Sunbeam coffee maker. The consequences of the next commercial airliner design may be much more significant to the future of Boeing (including its decision makers and stakeholders) and our national economy than the next coffee maker is to the future of Sunbeam. Our process requires access to senior leaders, decision makers, stakeholders, and subject matter experts. There may be costs for developing tools to enable the process—for example, functional analysis, systems models, simulations, and life cycle cost models. In addition, depending on the experience level of the systems engineering team, there may be costs for education and training.

9.9.3 Decision Maker and Stakeholder Support

Given that the systems decision process is needed and the potential resources are reasonable for the consequences of the systems decision, the final question is the support of decision makers and stakeholders. If the senior decision makers do not support the process, the resources (people and funds) will not be available to implement the process. If the decision makers and stakeholders do not participate in the process, the recommendations will not be accepted. The use of the SDP requires the support and participation of decision makers and stakeholders.

9.10 TAILORING THE SYSTEMS DECISION PROCESS

The systems decision process can be applied in any stage in the system life cycle for a system program or project.

> The systems decision process can be tailored to any stage of the system life cycle.

The process we have introduced in this chapter will be defined and described in detail in the next four chapters. For each phase, one or more techniques will be described to perform each task. For example, in Chapter 10, we describe three stakeholder analysis techniques: interviews, focus groups, and surveys. For large, complex systems engineering projects all phases and at least one technique for all tasks may be required. However, for many systems engineering projects, the process must be tailored to the systems engineering project and the stage of the system life cycle. Some phases or tasks may not be required.

As a first task in the tailoring process, we begin by considering when to use each task in each phase of the systems decision process. Table 9.2 lists criteria for the lead systems engineer to use to decide if a task is needed or not needed. Consider the first task, research and stakeholder analysis. The key criteria are the number of decision makers and our knowledge about the stakeholders' views. If there is a single decision maker and known stakeholder views, then the stakeholder analysis task may not be needed. However, if there are many decision makers and stakeholders with unknown or conflicting views of the system or the system life cycle, stakeholder analysis will be critical to the success of the systems engineering project. The engineering manager and lead systems engineer will clearly need to think about which techniques (i.e., interviews, focus groups, and surveys) are most appropriate for stakeholder analysis and may use multiple techniques. For intermediate situations when there are several decision makers and some stakeholders with conflicting views, the lead systems engineer will need to determine the appropriate stakeholder analysis technique. Chapters 10 through 13 provide additional information about how to tailor the process to the system life cycle including how to tailor techniques to the systems engineering project.

9.11   EXAMPLE USE OF THE SYSTEMS DECISION PROCESS

Table 9.3 lists an example of a systems engineering project that used a systems decision process similar to the process described in this book. For each task, we describe how the task was performed in the project. In some cases, the phases or tasks were not done or done after the systems engineering project by another group of people. For example, the Army Base Realignment and Closure (BRAC) study was done by the Army Basing Study team in 2002–2005. The solution implementation phase is being performed in 2005–2015 by the Army's Installation Management Command.

9.12   ILLUSTRATIVE EXAMPLE: SYSTEMS ENGINEERING CURRICULUM MANAGEMENT SYSTEM (CMS)—SUMMARY AND INTRODUCTION

This chapter introduced the four-phase systems decision process that we will develop in the next chapters of the book. Each phase includes a set of tasks to accomplish to successfully complete the phase, and several alternative techniques with which to perform each task.

- Chapter 10 describes the problem definition phase.

- Chapter 11 describes the solution design phase.
- Chapter 12 describes the decision-making phase.
- Chapter 13 describes the solution implementation phase.

Throughout this chapter, we have emphasized several themes. First, we have compared value-focused with alternative-focused alternative generation. Second, we have defined the characteristics of a good systems decision. Third, we have described the critical role of decision makers and stakeholders and the system environment. Fourth, we have discussed the criteria that can be used to assess if the systems decision process is appropriate for a systems decision. Fifth, we have emphasized that the systems decision process can be used for any stage of the system life cycle. Finally, we have emphasized that the process must be tailored to the systems engineering project and the system life cycle stage.

As we explore the systems decision process, we will use a rocket design problem to provide examples of the techniques. In addition, we have an illustrative example, the curriculum management system, which we will introduce next and then develop in the final section of each of the next four chapters. We hope that this illustrative example helps to explain the tasks in each phase and shows the relationships between each of the phases. We will usually illustrate the use of at least one technique for each task.

ILLUSTRATIVE EXAMPLE: PART I, THE OPPORTUNITY

Robert Kewley, Ph.D. U.S. Military Academy

Teaching undergraduate systems engineering disciplines to cadets is the core function of the United States Military Academy's Department of Systems Engineering. Teaching includes building and managing the curriculum, delivery of that curriculum to the cadets, and continuously assessing and adjusting that curriculum to maintain excellence. The department has processes in place for curriculum management, but they can be labor-intensive, slow, and disjointed. Given the proliferation of information technology into education, there may be an opportunity for the department to leverage information technology to improve its capability to manage the curriculum and to teach cadets.

The department has a complex, evolving curriculum that must be synchronized across four academic programs in accordance with Academy education goals and accreditation requirements. Advances in systems engineering, engineering management, systems management, and operations research require continuous curriculum changes at all levels. These include program goal updates, course additions or deletions, textbook changes, and lesson changes. Any one of these changes has the potential to disrupt the synchronization an academic program in a number of ways. They can jeopardize achievement of

Academy or program goals, force cadets to use multiple texts for the same material, present duplicate material, or present material out of order from the students' perspective.

Currently, the department and the Academy prevent some of these problems through centralization and committee work. However, these processes are slow and laborious, and they do not support frequent lesson-level changes that should be synchronized across one or two courses. These low-level changes often occur outside of the purview of academic committees, but, over time, they have just as much potential to disrupt synchronization. The department's curriculum development process stands to improve flexibility and synchronization through the application of information management and decision support.

Teaching, by its nature, is a collaborative process. Instructors must develop and deliver content by a combination of textbook reading and problem assignments, in-class instruction, electronic files, and graded events. Ideally, these are well organized and stored in a structured manner that supports easy access by cadets and instructors and consistency across courses.

In reality, the department employs a variety of methods to manage the curriculum. These include:

- Network drive storage for cadet access
- Network drive storage for faculty access
- Course Web sites maintained by faculty
- Academy-wide collaboration system to support teaching.

The Problem

Each faculty member selects the method or combination of methods to use based on his or her preferences. While this is not a problem within the context of a single course, it creates difficulties for cadets with multiple courses or for instructors who must collaborate across courses. People often cannot find needed information without directly contacting its originator. As a result, collaborative opportunities are lost, and the curriculum becomes disjointed. This also complicates archiving course information at the end of the semester.

Finally, the department faculty assesses and updates the curriculum at regular intervals. These assessments ensure responsiveness to a dynamic world, maintain accreditation, support educational goals, and improve teaching methods. Because they are so data-intensive, the assessments require many man-hours to compile the necessary information. The input data is scattered across many systems:

- Cadet surveys

- Instructor assessments
- Course administrative documents
- External assessment reports
- Meeting notes
- Various internal and external Web sites

The department has an opportunity to use a structured information system and well-synchronized processes to streamline and improve the assessment process. This will ensure the curriculum remains current and responsive to its stakeholders.

The Opportunity

In order to better build and manage the curriculum, deliver material to cadets, and continuously assess against dynamic demands, the Department of Systems Engineering proposes to build an integrated Curriculum Management System (CMS). This system will seek to synchronize advanced information technology with internal management processes in order to provide leap-ahead improvements.

Because this system is a new system, the decision we will illustrate in this book is the system concept decision from the second stage in the system life cycle (see Figure 3.1). That decision requires commitment of resources. The department will choose to buy (or not buy) system development software and hire (or not hire) people to do system design and development. Illustrative sections at the end of each of the succeeding four chapters will illustrate how the department employed techniques in each phase of the systems decision process to arrive at a system concept that best supports the department's stakeholders and their values.

9.13 EXERCISES

9.1. What is the definition of a decision? Using this definition, explain how the purchase of an automobile fits this definition.

9.2. What are the six links in the decision quality chain? Which is the most important link from a systems engineering perspective? Why is the analogy of a chain used?

9.3. Describe the four phases of the SDP. What are the tasks performed in each phase? What is the role of the systems engineer in each phase?

9.4. What is the relationship between the SDP and the system life cycle?

9.5. Describe the roles of stakeholders during each phase of the SDP. Are these roles the same for active and passive stakeholders? Explain.

9.6. Describe the roles of decision makers in each phase of the SDP. Give an example of a system in which the decision maker is simultaneously the owner, a user, and consumer.

9.7. Describe the impact of each SDP environmental factor on each of the following systems decision problems.

(a) A decision by an automobile company to build a car an electric car. Would the the life cycle cost profile of an electric car be the same as a gasoline powered car? Explain.

(b) A coastal U.S. state deciding to allow off-shore oil drilling. Would the impact you propose change if the supply of oil in the United States reached a critically low level? Explain.

(c) The decision to adopt a law requiring all U.S. citizens to possess private health insurance. Define the system boundary you are working with. What type of stakeholder would you classify a lobbyist as? What is their relationship to the system?

(d) The decision by a school board to remove carbonated soft drinks with sugar from the cafeteria menu. Draw a concept map that illustrates the relationship between the following objects: school board, students, soft drink companies, local schools, parents of students, local stores.

(e) The decision by a young person to engage in a dangerous lifestyle (pick one). Describe how this person can be defined as a system. Identify an appropriate metasystem and at least one lateral system for this decision problem.

9.8. Compare the SDP to one of the other decision making processes in Table 9.3. What are the similarities and differences?

9.9. What types of systems decision problems require the full SDP? Explain and give three examples.

ACKNOWLEDGMENT

The SDP is a refinement of the Systems Engineering and Management Process developed by Dr. Daniel McCarthy of the United States Military Academy faculty.

REFERENCES

1. Howard, RA. An assessment of decision analysis. *Operations Research*, 1980;28:4–27.

2. Keeney, RL. *Value-Focused Thinking: A Path to Creative Decisionmaking*. Cambridge, MA: Harvard University Press, 1992.

3. Buchanan, JT, Henig, EJ, Henig, ML. Objectivity and subjectivity in the decision making process. *Annals of Operations Research*, 1998;80:333–345.

4. Wright, G, Goodwin, P. Value elicitation for personal consequential decisions. *Journal of Multi-Criteria Decision Analysis*, 1999;8:3–10.

5. March, JG. The technology of foolishness. In: Leavitt, HJ, et al., editor. *Organizations of the Future*. New York: Praeger, 1974.

6. Corner, J, Buchanan, J, Henig, M. Dynamic decision problem structuring. *Journal of Multi-Criteria Decision Analysis*, 2001;10:129–141.

7. Matheson, D, Matheson, J. *The Smart Organization*: *Creating Value Through Strategic R&D*. Cambridge, MA: Harvard University Press, 1998.

8. Gude, C. Personal conversation with former vice president of technology for IBM. Phrase was coined by Ginni Rometty, one of IBM's chief executives in Global Services.

9. Ewing, P, Tarantino, W, Parnell, G. Use of decision analysis in the army base realignment and closure (BRAC) 2005 military value analysis. *Decision Analysis Journal*, 2006;13(1):33–49.

Chapter **10**

Problem Definition

TIMOTHY TRAINOR, Ph.D.
GREGORY S. PARNELL, Ph.D.

A great solution to the wrong problem is ... wrong.

—Anonymous

10.1 INTRODUCTION

History is rife with examples of solutions developed to a problem incorrectly or incompletely defined by hard-working, well-intentioned individuals. For example:

- General Robert E. Lee (United States Military Academy Class of 1829) of the Army of Northern Virginia believed that the Confederacy's problem was to decisively defeat the Union Army in the Civil War to achieve the goals of the Confederate states. Lee's strategy to solve this problem was to invade the Union States to draw their Army into a decisive engagement. Ultimately this led to a disaster at the Battle of Gettysburg when Lee attacked the larger and better-equipped Union Army, which held ground favorable to the defense. Others in the Confederacy, including Lieutenant General James "Pete" Longstreet (United States Military Academy Class of 1842), believed the problem was to destroy the Union's will to continue the conflict by threatening the seat of government in Washington, DC. Through superior leadership and maneuvering, some felt the Confederate Army could capture Washington, DC and

Decision Making in Systems Engineering and Management, Second Edition
Edited by Gregory S. Parnell, Patrick J. Driscoll, Dale L. Henderson
Copyright © 2011 John Wiley & Sons, Inc.

sue for a favorable peace to achieve their goals [1]. We will never know who correctly defined the Confederacy's problem, but it is evident from history that destruction of the Union Army was not necessarily the complete problem facing the Confederacy in achieving their goals.

- The late 1970s and 1980s saw Japanese automakers gain a significant share of the U.S. auto market. U.S. auto manufacturers believed the problem was how to compete on cost against the smaller, more fuel-efficient Japanese cars that were cheaper to purchase and operate. While true, this problem definition proved to be incomplete. Japanese auto manufacturers were successful then due to both the lower costs and higher quality of their cars. U.S. auto manufacturers were slow to recognize that they could realize a positive return on investment in improved quality control measures in producing cars [2]. As history has shown, Japanese automakers gained significant market share of the U.S. auto market during this period.

- In 1983, IBM introduced the IBM PC Junior to compete in the home computer market against the then-dominant Commodore 64 and Apple II [3]. Apparently IBM's plan to meet consumer needs in the home computer market was to build a scaled-down version of its popular IBM PC, which was then very successful for business consumers. Unfortunately the changes to the business IBM PC did not make the Junior easy to use at home and did not make it affordable. The IBM PC Junior suffered from an incomplete definition of the problem for IBM to successfully compete against Apple and Commodore in the home computer market.

These examples from history show that a thorough, complete definition of a problem is crucial in forming effective solutions. This chapter provides a process for defining a problem and articulating what decision maker(s) value from a solution to his/her problem. Included in this process are the other stakeholder values that the decision maker(s) need to consider as well.

10.1.1   The Problem Definition Phase

In Chapter 9 we saw that defining the problem was the first phase in our systems decision process. We gather and process information in this phase. As with any good problem solving process, this phase requires thorough research. Performing a literature review of appropriate laws, organizational policies, applicable studies previously performed, and pertinent discipline-specific principles is necessary to effectively define a problem.

The concept diagram in Figure 10.1 shows the tasks involved in the Problem Definition phase and the relationships between them. Systems engineers define a system and determine its objectives. They also define a problem statement for a system or decision problem by analyzing the stakeholders involved, analyzing the functions of the system, and modeling what decision makers value in an effective solution. The key tasks in this phase are *research* and *stakeholder analysis*, *functional* and *requirements analyses*, and *value modeling*. While research into a systems decision problem continues throughout the SDP, it is particularly helpful early on to help the systems team gain a more comprehensive understanding of the challenge and to identify specific disciplines related to the problem. Moreover, research helps to identify those stakeholders who have a vested interest in any resulting solution.

Stakeholder analysis enables systems engineers to identify the objectives, functions, and constraints of a system or decision problem and the values of decision makers. Systems interact in an environment that affects the stakeholders of this system. As we saw in Chapter 2, systems engineers use systems thinking to understand the environmental factors affecting a system in order to identify the relevant stakeholders. These stakeholders include consumers of the products and services provided by the system. An analysis of the stakeholders and their needs helps us identify the correct and complete problem requiring a solution. This analysis also leads us to new areas to research in order to understand the problem domain. A problem is not correctly defined unless we have received input from all stakeholders. We will discuss several techniques for completing a thorough stakeholder analysis.

A system is developed to perform certain functions. These functions should be designed to meet the objectives of the system. Systems must also meet certain requirements in order to be feasible and effective. Systems engineers need to understand the functions a system is intended to perform and the requirements it must meet in order to develop effective solutions. We will describe the techniques involved in performing both functional and requirements analyses of a system.

What determines an effective solution to a problem? An effective solution is one that meets the values articulated by the key stakeholders. To determine if a potential solution meets these values, systems engineers follow a process that qualitatively and quantitatively models the stakeholder values by identifying objectives for each function and value measures for each objective. This chapter describes this process of value modeling.

10.1.2 Comparison with Other Systems Engineering Processes

Beginning our systems decision process with a deliberate, focused process for the problem definition is not unique. No matter what form of systems engineering process or life cycle is packaged, it will involve a *formulation of the problem* step in which an initial problem statement is assessed and the needs of decision makers are determined [4].

Several systems engineering and problem solving processes naturally start with some form of problem definition [5]. Wymore defines a system life cycle in seven phases, the first of which is the development of requirements. During this phase, systems engineers work on tasks such as understanding the problem situation and the customer needs [6]. The SIMILAR process for the systems design starts with a "state the problem" function [7]. The International Council on Systems Engineering includes Plowman's model of the systems engineering process in their systems engineering body of knowledge. This process includes a function for understanding what a customer wants and how a system must perform [8]. This identifies just a few of the systems engineering processes and/or life cycles that incorporate a deliberate process for defining the problem.

10.1.3   Purpose of the Problem Definition Phase

The Problem Definition phase can and should be done to support systems decision making in each stage in the life cycle. This phase provides a process for helping stakeholders define their problem before attempting to develop solutions. The initial problem defined is never the real problem that needs to be addressed. In the auto market example described at the opening of this chapter, the U.S. auto manufacturers spent considerable resources and time in attempting to get trade legislation passed that would increase the cost of Japanese cars to American consumers. While this attempted to address the cost part of the problem, it ignored the quality issues underlying the problem. The tasks and techniques described in this chapter can hopefully help prevent wasting resources by chasing potential solutions to the incorrect problem.

> The initial problem is never the real problem.

The goal for the end of this phase is to have a clearly defined problem statement that meets the approval of the key stakeholders, a set of systems requirements or constraints that alternative solutions must meet before alternatives are fully designed, modeled, and analyzed, and an initial quantitative methodology for evaluating how well alternatives meet the values of stakeholders in solving the correct problem.

10.1.4   Chapter Example

To illustrate the key points in this chapter, we will use the following systems decision problem: An aeronautics manufacturer wants to develop a new small, mobile rocket that can be used for multiple applications—for example, military or research uses. We will use this initial problem statement to demonstrate some of the tasks and techniques useful in the problem definition phase of our systems decision process.

10.2   RESEARCH AND STAKEHOLDER ANALYSIS

Given the multi- and interdisciplinary nature of systems engineering, a systems engineer will need to be an aggressive and inquisitive researcher in order to learn about a variety of topics. The research and stakeholder analysis task is an important and iterative process. When introduced to a new problem, systems engineers conduct research in order to understand the nature and domain of disciplines involved. This research leads to identifying the stakeholders impacted by the problem. Stakeholders identify additional research sources for the systems engineering team to investigate. Thorough research is critical in performing a complete definition of the problem.

Stakeholders comprise the set of individuals and organizations that have a vested interest in the problem and its solution [9]. Besides decision makers, stakeholders

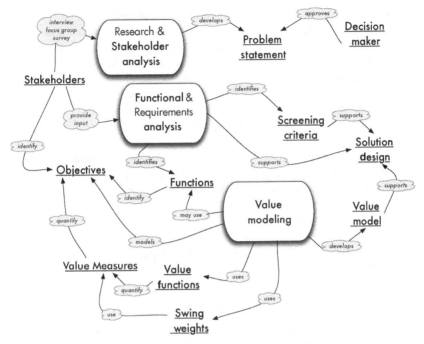

Figure 10.1   Concept diagram for problem definition.

can include the customers of the products and services output by the system, system operators, system maintainers, bill payers, owners, regulatory agencies, sponsors, manufacturers and marketers, among others. Understanding who is affected by a system or solution to a decision problem provides the foundation for developing a complete definition of the problem.

Systems engineers use research and stakeholder analysis to understand the problem they face, identify the people and organizations relevant to the problem at hand, and determine their needs, wants, and desires with respect to the problem. Through this task, stakeholders identify the functions, objectives, value measures, and constraints of the systems decision problem. It is also important in this task to understand the decision space for the problem. Systems engineers need to clearly understand up front what decisions have to be made by stakeholders in addressing the problem. Stakeholder analysis also includes a thorough research into the environmental factors impacting the systems decision problem. By considering such relevant issues as the political, economic, social, ethical, and technological factors affecting the problem, we can identify all the active and passive stakeholders within the scope of the problem. Systems engineers need to identify the requirements and constraints that all candidate solutions must meet. This task also helps identify the values that stakeholders deem important in any solution to the systems decision problem. These values help the systems engineer evaluate the quality of potential solutions.

The initial problem statement is seldom the full statement of the problem from the perspective of all stakeholders. The primary purpose of stakeholder analysis is to obtain diverse stakeholder perspectives on the problem that will provide a broader definition of the problem that captures the stakeholder perspectives. In addition, stakeholder analysis provides important insights for the systems decision process. Performing an analysis of these stakeholders will provide insight useful for developing and evaluating potential solutions for the problem.

For our rocket systems decision problem example, the list of relevant stakeholders can be very large. Stakeholders include the potential users of the rocket—for example, research organizations (like NASA) and military commanders. Senior and mid-level leaders of the manufacturer are also stakeholders, including the financial decision makers and production operations personnel. Stakeholders can also include subcontractors that may contribute to the design and production of the rocket system. Performing an analysis of these stakeholders will provide insights useful for developing and evaluating candidate solutions for the problem.

Several familiar techniques exist for soliciting input from diverse stakeholders. The three techniques we will describe are interviews, surveys, and focus groups.

10.2.1 Techniques for Stakeholder Analysis

Three general techniques are commonly used for stakeholder analysis: interviews, focus group meetings, and surveys. Each technique has different characteristics. We consider five characteristics: the time commitment of the participants, ideal stakeholder group, the preparation activities, the execution activities, and the analysis activities. Table 10.1 provides a description and a comparison of the three techniques for each of the five characteristics. The number of interviews, the number and size of focus groups, and the number of surveys required depend on the problem and the diversity of the stakeholders. Statistical tests can be used to determine the required sample sizes.

Interviews Interviews are one of the best techniques for stakeholder analysis if we want to obtain information from each individual separately. Interviews are especially appropriate for senior leaders who do not have the time to attend a longer focus group session or the interest in completing a survey. However, interviews are time-consuming for the interviewer due to the preparation, execution, and analysis time.

Since interviews take time, it is important to get the best information possible. The following are best practices for each phase of the interview process: planning, scheduling, conducting, documenting, and analyzing interviews.

Before the Interview For interviews with senior leaders and key stakeholder representatives, it is important to prepare a questionnaire to guide the interview discussion. The following are the best practices for interview preparation:

- Unless the team has significant problem domain experience, research is essential to understand the problem domain and the key terminology.

TABLE 10.1 Stakeholder Analysis Techniques

	Time Commitment of Participants	Ideal Stakeholder Group	Preparation	Execution	Analysis
Interviews	30–60 minutes	Senior leaders and key stakeholder representatives	Develop interview questionnaire(s) and schedule or reschedule interviews.	Interviewer has conversation with senior leader using questionnaire as a guide. Separate note taker.	Note taker types interview notes. Interviewer reviews typed notes. Team analyzes notes to determine findings, conclusions, and recommendations.
Focus groups	Shortest: 60 minutes Typical: 4–8 hours	Mid-level to senior stakeholder representatives	Develop meeting plan, obtain facility, plan for recording inputs. May use Group Systems software to record.	At least one facilitator and one recorder. Larger groups may require breakout groups and multiple facilitators.	Observations must be documented. Analysis determines findings, conclusions, and recommendations.
Surveys	5–20 minutes	Junior to mid-level stakeholder representatives	Develop survey questions, identify survey software, and develop analysis plan. Online surveys are useful.	Complete survey questionnaire, solicit surveys, and monitor completion status.	Depends on number of questions and capability of statistical analysis package. Conclusions must be developed from the data.

- Develop as broad a list of interviewees as possible. Identify one or more interviewees for each stakeholder group. Review the interview list with the project client to ensure that all key stakeholders are on the list of potential interviewees.
- Begin the questionnaire with a short explanatory statement that describes the reason for the interview, the preliminary statement of the problem, and the stakeholders being interviewed.
- It is usually useful to begin the interview with an unfreezing question that encourages the interviewee to think about the future and how that will impact the problem that is the subject of the interview.
- Tailor the questionnaire to help you define the problem and obtain information that will be needed in the future.
- Tailor the questionnaire to each category of interviewee. Make the questions as simple as possible.
- Do not use leading questions that imply you know the answer and want the interviewee to agree with your answer.
- Do not ask a senior leader a detailed question the answer to which can be looked up on the Internet or obtained by research.
- End the questionnaire with a closing question, for example, "Is there any other question we should have asked you?"
- Arrange to have an experienced interviewer and a recorder for each interview.
- Decide if the interviews will be for attribution or not for attribution. Usually, the information obtained during interviews is not attributed to individuals.

Schedule/Reschedule the Interview Interviews with senior leaders require scheduling and, frequently, rescheduling. The following are best practices for interview scheduling:

- It is usually best to conduct interviews individually to obtain each interviewee's thoughts and ideas on the problem and the potential solutions. Additional attendees change the interview dynamics. The senior leader may be reluctant to express ideas in front of a large audience or may defer to staffers to let them participate.
- Provide the brief problem statement to the interviewees when the interview is scheduled.
- Provide the interviewee with a short read-ahead document that clearly communicates the purpose of the interview and enables the interviewee to adequately prepare for the session. A typical read-ahead applicable for all meetings with stakeholders includes:
  1. A short summary of the background of the problem/issue being discussed. This could take the form of an abstract or executive summary. The individual(s) or organization(s) sponsoring the study that motivated the interview should be clearly identified.

2. A statement of the purpose of the interview (e.g., "... to identify all system linkages to outside agencies...", "... to identify key stakeholders related to the system design, their vested interests, and point-of-contact to schedule future interviews in support of the project ...", etc.).

3. The proposed duration of the interview.

4. A short list of desired outcomes if not clear from the purpose statement.

5. Attachments that contain pertinent, *focused* materials supporting the interview.

6. If slides are going to be forwarded as part of the read-ahead, then:

   a. Use no more than two slides per page; less is inefficient, more challenges the eyesight of the interviewee.

   b. Use the Storyline format (see Chapter 12) for slides to limit the interviewee's likelihood of misinterpreting the intended message on each slide. Remember: these slides will be present at the interviewee's location before you get a chance to conduct the interview in-person.

7. A list of the key project personnel and their contact information, identifying a primary point-of-contact for the team among those listed. Once interviewees become connected with a project via an interview, the systems team should anticipate additional information and follow-on questions to ensue. Designating a single team member to have responsibility as the centralized contact keeps this information flow (both in-and-out of the team) controlled and organized while simultaneously creating a "familiar face" on the team for the interviewees.

- Many times it is best to have the stakeholder representatives assigned to your team schedule the interview since they may have better access.

- Depending on the importance of the problem and the difficulty of scheduling, we usually request 30–60 minutes for the interview.

- The interviews can be done in person or over the phone. In-person interviews are the most effective since interaction is easier, but sometimes they are not possible and the only practical choice is a phone interview.

- The more senior the leader, the more likely scheduling will be a challenge.

One cautionary note is in order regarding advanced materials sent to the interviewee. It is usually *not* a best practice to provide detailed questions ahead of the interview for two reasons. First, remember that the purpose of the interview is to obtain information directly from the stakeholder and not an intermediary representative. If the interview questions are provided in a read-ahead packet, there is a likelihood that the stakeholder's staff will prepare responses to the questions, thereby defeating the purpose of the interview. Second, the systems team conducting the interview should have the flexibility to add questions and/or follow-up on valuable information leads should they arise. If the stakeholder is preconditioned into knowing the questions (and responses) in advance, the team might not be able to "stray from the script" without providing written questions to the stakeholder or staff.

During the Interview The interview teams' execution of the interview creates an important first impression with the senior leader about the team that will develop a solution to the problem. The goal of the interview is to obtain the stakeholder insights in a way that is interesting to the interviewee. Some thoughts for conducting interviews are as follows:

- The best number of people to conduct the interview is one interviewer and one notetaker. An alternative to the notetaker is a recorder. Some interviewees may be reluctant to be recorded. If you wish to use a tape recorder, request permission first.
- Conduct the interview as a conversation with the interviewee. Use the interview questionnaire as a guideline. Take the questions in the order the interviewee wants to discuss them.
- Make the interview interesting to the interviewee.
- Use an unfreezing question for the first question. An unfreezing question helps the interviewer focus on the problem in the future.
- Be flexible, following up on an interesting observation even if it was not on your questionnaire. Many times an interviewee will make an interesting observation you did not anticipate in your questionnaire. It is critical to make sure you understand the observation and the implications.
- Ask simple open-ended questions that require the interviewee to think and respond. Avoid complex, convoluted questions that confuse the interviewee.
- Respect the interviewee's time. Stay within the interview time limit unless the interviewee wants to extend the interview period.
- When the interviewee's body language signals that they have finished the interview (e.g., fold up paper, look at their watch), go quickly to your closing question, and end the interview.

After the Interview Documentation of the interview is the key to providing the results of the interview to the problem definition team. The best practice for documenting the interviews is the following:

- As soon as possible after the interview, the recorder should type the interview notes.
- The questions and the answers should be aligned to provide proper context for the answers.
- It is best to record direct quotes as much as possible.
- The interviewer should review the recorder's typed notes and make revisions as required.
- Once the interview notes are complete, they should be provided to the interview team.
- The documentation should be consistent with the decision to use the notes with or without attribution.

Analysis of the Interview Notes The interview notes are a great source of data for the entire systems engineering team. The key to interview analysis is binning (i.e., categorizing) the comments, summarizing observations, and identifying unique "nuggets" of information that only one or two interviewees provide. The best practice for analysis of interview notes is the following:

- The most common analysis approach is to bin the interviewee responses by the questions.
- The most challenging task is to identify unique "nuggets" of information that only one or two interviewees provide.
- The best way to summarize interviews is by findings, conclusions, and recommendations. Findings are facts stated by the stakeholders. Conclusions are a summary of several findings. Recommendations are what we recommend we do about the conclusion.
- It is important to integrate research findings with the interview findings. Many times an interviewee will identify an issue that we must research to complete our data collection.
- Identifying the findings for a large number of interviews is challenging. One approach is the preliminary findings approach. Here is one way to do the approach:
  - Read several of the interview notes.
  - Form preliminary findings.
  - Bin quotes for the interviews that relate to the preliminary findings.
  - Add research information to the quotes.
  - Revise the preliminary findings to findings that are fully supported by the interview and research data.

Table 10.2 is an illustrative example of a set of findings and conclusions that lead to a recommendation.

As the findings are being identified, it is important not to get distracted by focusing on potential findings that are interesting but unrelated to the purpose of the stakeholder analysis. If appropriate, these findings should be presented separately to the decision makers.

Follow up with Interviewees Many times the interviewee will request follow-up information. The following are examples of appropriate follow-up:

- Thank you note or e-mail to the interviewee and/or the stakeholder representative that scheduled the meeting.
- A revised statement of the problem after the problem definition is complete.
- A copy of the findings, conclusions, and recommendations from the interviews.
- A briefing or copy of the report at the end of the project.

TABLE 10.2   Illustrative Example of Findings, Conclusions, and Recommendations

Findings	Conclusions	Recommendations
20 of the 30 interviewees were concerned about the cost of the solution.	Many of the stakeholders believe that cost will be an important criteria in our study.	The system analysis team should consider cost and should use the System Life Cycle Cost Model for the cost analysis.
Policy directive 10.123 recommends the consideration of life cycle costs for all acquisition decisions [research].	Policy directives recommend the consideration of life cycle costs.	
The cost expert stated that the most appropriate cost model for solution evaluation is the System Life Cycle Cost Model		

Examples of Studies Using Interviews The following are a selection of studies conducted by the authors that have used interviews as a key technique in the stakeholder analysis task:

- Ewing, P, Tarantino, W, Parnell, G. Use of decision analysis in the army base realignment and closure (BRAC) 2005 military value analysis. *Decision Analysis Journal*, 2006;(1)l3:33–49 [about 40 interviews with senior Army leaders].

- Powell, R, Parnell, G, Driscoll, PJ, Evans, D, Boylan, G, Underwood, T, Moten, M. Residential communities initiative (RCI) portfolio and asset management program (PAM) assessment study, Presentation to Assistant Secretary of the Army for Installations and Environment, 15 December 2005 [72 interviews with Army senior leaders, installation leaders, and RCI personnel].

- Parnell, G, Burk, R, Schulman, A, Westphal, D, Kwan, L, Blackhurst, J, Verret, P, Karasopoulos, H. Air Force Research Laboratory space technology value model: Creating capabilities for future customers, *Military Operations Research*, 2004, 9(1):5–17 [about 50 interviews with senior Air Force leaders].

- Parnell, G, Engelbrecht, J, Szafranski, R, Bennett, E. Improving customer support resource allocation within the National Reconnaissance Office. *Interfaces*, 2002;32(3):77–90, [about 25 interviews with senior leaders and key stakeholders].

- Trainor, T, Parnell, G, Kwinn, B, Brence, J, Tollefson, E, Downes, P. Decision analysis aids regional organization design. *Interfaces*, 2007;37(3):253–264 [about 50 interviews with senior Army leaders and key stakeholders].

Focus Groups Focus groups are another technique for stakeholder analysis. Focus groups are often used for product market research; however, they can also be useful for determining relatively quickly how groups of stakeholders feel about a specific systems decision problem. While interviews typically generate a one-way flow of information, focus groups create information through a discussion between the group members who typically have a common background related to the problem being studied. For example, if the problem involved designing an efficient production plant, the team would form separate focus groups for plant management and for plant laborers. As a general rule, focus groups should comprise 6–12 individuals. Too few may lead to too narrow a perspective, while too many will lead to some individuals not able to provide meaningful input. As with interviews, the focus group facilitation team needs to devote time to the preparation of, execution of, and analyzing data from focus groups [10].

Preparing for the Focus Group Session As with any stakeholder analysis technique, developing the goals and objectives of the focus group session is critical to success. A few best practices for preparing for a focus group session [11] include the following:

- Develop a clear statement of the purpose of the focus group and what you hope to achieve from the session. This should be coordinated with the project client and provided to the focus group participants.
- Develop a profile of the type of participant that should be part of the session and communicate that to the project client.
- Select a participant pool with the project client.
- Select and prepare moderators that can facilitate a discussion without imposing their own biases on the group. If resources permit, hire a professional moderator.
- Schedule a time and location during which this group can provide 60–90 minutes of uninterrupted discussion.
- Develop a set of questions that are open-ended and will generate discussion. Do not use "Yes/No" questions that will yield little discussion. The most important information may come out of discussion about an issue ancillary to a question posed to the group.

Conducting the Focus Group Session The most important components of conducting the session are the moderator and the recording plan. Here are some thoughts for the execution of a focus group session [12]:

- The moderator should review the session goals and objectives, provide an agenda, and discuss the plan for recording the session.
- Ask a question and allow participants a few minutes to discuss their ideas. The moderator should ensure even participation from the group to prevent a few individuals from dominating the group.

- A good technology solution for facilitating focus groups is the GroupSystems software [13]. This technology facilitates groups in brainstorming activities and generating ideas. It helps mitigate the impacts from individuals who tend to dominate discussions because participants type their ideas on a computer in response to questions generated by the moderator. It also significantly helps the team in recording the information from the session and sets them up for analysis of the data.
- Do a video and audio recording of the session if possible. If not, use multiple notetakers.
- The moderator may steer the discussion to follow a particular issue brought up that impacts the problem being studied.
- On closing, tell the participants they will receive a record of the session to verify their statements and ideas.
- Follow up the session with an individual thank you note for each participant.

Analyzing the Information Focus groups can provide a great source of qualitative data for the systems analysis team to analyze and create useful information. The recorders should first verify the raw data that was generated during the session. These data should then be processed into findings, conclusions, and recommendations using the methods discussed in the interview section of this chapter. If you run more than one focus group, realize that you cannot necessarily correlate the data between the groups since they represent different subgroups of the stakeholders.

Surveys Surveys are a good technique for collecting information from large groups of stakeholders particularly when they are geographically dispersed. Surveys are appropriate for junior to mid-level stakeholders. If the problem warrants, surveys can be used to gather quantitative data that can be analyzed statistically in order to support conclusions and recommendations. A great deal of research exists on techniques and best practices for designing effective surveys. Systems engineers can distribute and collect survey data via mail, electronic mail, or the World Wide Web for many of the problems they face. This section provides an overview of survey design and methods for conducting surveys. As with any stakeholder analysis technique, surveys require detailed planning to accomplish the team's goals. These steps can be followed to plan, execute, and analyze surveys [14]:

- Establish the goals of the survey.
- Determine who and how many people you will ask to complete the survey, that is, determine the sample of stakeholders you will target with the survey.
- Determine how you will distribute the survey and collect the survey data.
- Develop the survey questions.
- Test the survey.
- Distribute the survey to the stakeholders and collect data from them.
- Analyze the survey data.

Preparing an Effective Survey Determine your goals, survey respondents, and means of distributing and collecting survey data. The stakeholder analysis team needs to clearly articulate the goals of the survey and the target sample of stakeholders whom they want to answer the survey. Often surveys for systems engineering decision problems will be used to collect textual answers to a standard set of questions. However, if the team plans to collect and analyze data from questions with standard answer scales (e.g., "Yes/No" or multiple choice answer scales), it is important to determine the appropriate sample size needed to draw valid statistical conclusions from the survey data. Sample size calculations are described in basic statistics books, and online tools are available to do these calculations [15]. The team needs to work with the project client in determining the appropriate stakeholders to survey. The method for implementing a survey needs to be selected before the survey is designed. Popular methods for systems engineers are mail, electronic mail, and web surveys. Table 10.3 provides a listing of some of the advantages and disadvantages of these survey methods [14].

The ability to collect survey responses in a database when using a web survey instrument can be extremely beneficial to the stakeholder analysis process. Several online programs now exist to help teams design web surveys, collect responses, and analyze the results. Some popular programs include surveymonkey.com [16], InsitefulSurveys.com [17], and the SurveySystem.com [18].

Executing a Survey Instrument Developing the survey questions, testing, and distributing the survey. Surveys should be designed to obtain the information that will help the stakeholder analysis team meet the goals of the survey. To maximize response, the survey should be short with clearly worded questions that are not ambiguous from the respondent's perspective. Start the survey with an overview of the purpose of the survey and the goals that the team hopes to achieve from the information provided by the respondents. Here are some general principles that can be followed in developing effective survey questions [19]:

- Ask survey respondents about their first-hand experiences, that is, ask about what they have done and their current environment so that they can provide informed answers. Respondents should not be asked hypothetical questions, nor should they be asked to comment on things outside their working environment.
- Ask only one question at a time.
- In wording questions, make sure that respondents answer the same question. If the question includes terms that could be interpreted differently by respondents, provide a list of definitions to clarify any possible ambiguities. This list of definitions should precede the questions.
- Articulate to respondents the kind of acceptable answers to a question. For objective questions, the answer scales can be set up as multiple choice answers from a rating scale or level-of-agreement scale. For certain questions and stakeholders, it may be appropriate to provide benchmark examples for the

answer scales. For example, the responses to a question regarding the respondent's level of effort on a project may include a benchmark statement like "full time effort equates to 40 hours of work per week." For open-ended text response questions, the question should be worded so that respondents provide information germane to the question. Close the survey with a statement allowing respondents to provide any additional information they believe is pertinent to the goals of the survey.

- Format the survey so that it is easy for respondents to read the questions, follow instructions, and provide their answers. For example, answer scales should follow a similar pattern in terms of the order in which they are presented (e.g., the least desirable answer is the first choice ascending to the most desirable answer).

- Orient the respondents to the survey in a consistent way. This can be accomplished with a set of instructions that describe the goals of the survey, the

TABLE 10.3   Advantages and Disadvantages of Popular Survey Methods

Survey Method	Advantages	Disadvantages
Mail	• Can include extensive supporting graphics • Respondents have flexibility in completing the survey	• Takes a great deal of time • Hard to check compliance and conduct follow-up with respondents • Response data will have to be transformed by the analysis team into a format for analysis
Electronic mail	• Fast to distribute and get responses • Low cost • Easy to check compliance and do follow-up	• Need to obtain e-mail addresses for the survey sample • Cannot program automatic logic into the survey (e.g., "skip over the next set of questions if your answer is No to this question") • Respondent e-mail programs may limit the type of information that can be sent in the survey • Response data will have to be transformed by the analysis team into a format for analysis
Internet Web survey	• Extremely fast • Can include special graphics and formatting • Can collect responses in a database to facilitate analysis	• May be hard to control who responds to the survey due to worldwide Internet access • Respondents can easily provide only a partial response to the survey

method for completing their responses, and the means for submitting the completed survey.

Once the survey questions are written, test the survey instrument with a few individuals outside the team. Ask them to complete the survey using the same medium that respondents will use (e.g., by e-mail, mail, or on the web). Ask for input from the test sample regarding the instructions and wording of the questions and answer scales. If a web survey is used, test the method for collecting responses, for example, in a database. Use the input from the test sample to improve the survey. Once improvements are made, distribute the survey to respondents using the method chosen. Develop a plan for monitoring the response rate and establish when reminders will be sent to respondents who have not completed the survey. The team should also have a standard way to thank respondents for their time and efforts, for example, a thank you note or e-mail.

Analyzing Survey Data A key part of the analysis effort will be in formatting the survey data that is received. If a web survey is used, the team can program the survey instrument to put responses directly into a database file. This will allow the team to perform statistical analysis on objective-type questions relatively quickly. For text answer questions, a database file provides a means to bin the responses quickly. The goals of the analysis are the same as for interviews and focus group sessions. Similar to the process discussed earlier in this section, the team should bin the responses by survey question and analyze these responses to develop findings. These findings will lead to forming conclusions, which then will lead the team to form recommendations.

10.2.2 Stakeholder Analysis for the Rocket System Decision Problem

The systems engineering team primarily relied on the interview technique for the stakeholder analysis for the rocket system. They also conducted focus groups with potential military and research stakeholders. Table 10.4 provides a partial (notional) example of the analysis of the information obtained from stakeholders using the binning process described in Section 10.2. Through research and stakeholder analysis we also discover any "must meet" requirements affecting the problem. For our rocket example, the stakeholders identified that the rocket system must meet the following constraints:

- The rocket must have a minimum effective range of 5 km.
- The rocket must be launched from a four-wheeled drive vehicle.
- The rocket must be able to move through a variety of terrain conditions, from open fields to forested areas.

These requirements become screening criteria that candidate solutions must meet. These are discussed in greater detail in Chapter 11.

TABLE 10.4 Partial Stakeholder Analysis Results for the Rocket System Decision Problem

Findings	Conclusions	Recommendations
15 of the 20 interviewees said the current need of potential users was for a smaller mobile rocket system.	Stakeholders believe that mobility of the rocket system will be a key design criterion.	The systems engineering team should consider mobility and the ability to operate at short-range key design criteria for the new rocket system.
The military has a call for proposals out to manufacturers to build a rocket system that can be easily moved around the battlefield [from research and interviews].		
Current rocket systems are only useful at ranges of greater than 10 kilometers [research].	Stakeholders think the rocket system should be designed to operate at short ranges.	
Most interviewees said that a market exists for rockets that can deploy an object (e.g., a warhead or sensor) between 5 and 10 km.		

10.2.3 At Completion

When research and stakeholder analyses are completed, the systems engineering team should have a thorough understanding of the stakeholder objectives and decisions needed in arriving at a solution to the systems decision problem. The team will also understand the facts and assumptions that are needed in further analysis of the problem. The team should also understand what stakeholders value from solutions to the problem. This is discussed further in Section 10.4. Stakeholder analysis should help the team identify the key life-cycle cost and risk factors that should be considered in any solution to the problem along with any screening criteria.

While stakeholder analysis helps identify initial requirements, objectives, and values for developing candidate solutions to the systems decision problem, we use functional analysis and requirements analysis to identify the key functions and requirements that the solution must be designed to perform.

10.3 FUNCTIONAL AND REQUIREMENTS ANALYSES

In Chapter 2 we introduced functional hierarchies as a technique for describing the functions a system will be designed to perform. In Section 6.5 we concluded that defining the system functions and requirements was one of the three fundamental tasks of systems engineers. In Section 10.2 we described the systems engineering activities involving systems requirements analysis and functional analysis. In this section we describe the central role of functional and requirements analyses in the SDP. We focus first on functional analysis in some depth and then provide an overview of requirements analysis.

10.3.1 Terminology

We begin by introducing some key terminology used in functional analysis.

Function. "A characteristic task, action, or activity that must be performed to achieve a desired outcome. For a product it is the desired system behavior. A function may be accomplished by one or more system elements comprised of equipment (hardware), software, firmware, facilities, personnel, and procedural data" [19].

Functional Analysis. A systematic process to identify the system functions and interfaces required to achieve the system objectives.

Functional Hierarchy. A hierarchical display of the functions and subfunctions that are necessary and sufficient to achieve the system objectives.

Functional Flow Diagram. A flow diagram that depicts the interrelationships of the functions.

IDEF0. IDEF0 stands for Integrated Definition for Function Modeling and is a modeling language (semantics and syntax) with associated rules and techniques for developing structured graphical representations of a system or enterprise [20].

Functional Architecture. "The hierarchical arrangement of functions, their internal and external functional interfaces and external physical interfaces, their respective functional and performance requirements, and the design constraints" [19].

Models and Simulations. See Chapter 4 for definitions.

Requirements Analysis. "The determination of system specific characteristics based on analysis of customer needs, requirements and objectives; missions; projected utilization environments for people, products and processes; and measures of effectiveness" [21].

10.3.2 Importance of Functional Analysis

Systems accomplish functions with system elements. Functional analysis is the process to identify the system functions and interfaces required to meet the system performance objectives. Functional analysis of the system design is performed in the first several stages of the system life cycle. If we do not identify all the system functions and interfaces, the system will not be designed to perform all the functions and will not work with its environment. The functional architecture is the allocation of the functions and interfaces to systems elements. If we do not allocate the functions to system elements, the functions will not be performed. If we do not identify the external and internal system interfaces, the system elements will not be able to perform the functions. If we do validate the system design using models and simulations, we may have to make costly changes to meet system performance objectives. Functional analysis can also be performed later in the system life cycle.

Since functional analysis plays such a critical systems engineering role in the system life cycle, we have included functional analysis in the SDP. We believe that systems engineers should use the most appropriate functional analysis information for the stage of the system life cycle that the SDP is being used.

10.3.3 Functional Analysis Techniques

There are many functional analysis techniques. The INCOSE Systems Engineering Handbook lists 13 functional analysis techniques [19]. In this section we introduce, in the order of increasing detail, four of the most useful techniques. Table 10.5 summarizes the purposes, uses, and limitations of the four techniques. Each of these will be discussed in the following sections.

Functional Hierarchy Functional hierarchies can be developed using affinity diagrams. An affinity diagram is a collection of ideas binned into logical groupings for a specific purpose (Figure 10.2) [22]. Affinity diagramming is a simple creative technique that has many valuable uses. Affinity diagramming is usually a group process used to generate ideas and provide new groupings of the ideas for a specific purpose. The affinity diagram is similar to the KJ Method originally developed by Kawakita Jiro [23]. The affinity diagram was made popular as a quality management technique. It is one of the most widely used Japanese management and planning tools.

Affinity diagramming can be done on any vertical or horizontal surface with Post-It™ notes (hence the nickname "the yellow-stickee" drill) or it can be done with specialized collaborative decision support software—for example, GroupSystems [13].

Affinity diagramming can be used for many systems engineering activities that require working with a group to generate new ideas and grouping the ideas into logical categories. Affinity diagramming can be combined with other techniques to determine priorities or actions for each group of ideas. Our interest in using affinity diagramming in functional analysis is to develop a functional hierarchy.

Steps for Affinity Diagramming for Functional Analysis We describe each of the steps for developing a functional hierarchy in what follows.

1. *Invite Required Stakeholders or Their Representatives to Attend.* For the affinity diagramming exercise to be successful, stakeholders must participate in the process either directly or through a representative who can clearly articulate the viewpoint of the stakeholder. All key stakeholder groups should

TABLE 10.5 Selected Functional Analysis Techniques

Functional analysis techniques	Purpose	Uses	Limitations
Functional hierarchy	Identify the system functions and subfunctions.	Provides functional hierarchy to guide concept development, design, and help identify performance measures.	Does not define functional relationships and interfaces, nor does it validate the system design.
Functional flowdiagram	Identify and show the relationships of system functions and subfunctions.	Defines the relationship of functions and subfunctions to guide concept development, design and help identify performance measures.	Does not define or validate system interfaces, nor does it validate the system design.
IDEF0	Model the decisions, actions, and activities of an organization or system.	Provides detailed information on the functions including inputs, outputs, mechanism, and controls to support the system design and development. Helps refine performance measures.	Does not validate system design.
Models and simulations	Model the system and/or its operation	Understand and support system design and evaluation. Helps refine performance measures.	Will not validate all aspects of system design.

be represented. If key stakeholders are not represented, important functions may not be identified.

2. *Define the System.* The scope of the system is critical for obtaining the appropriate functions. The following systems will have very different functions: the U.S. transportation system, an urban transportation system, an automobile, or a car engine. We recommend using a system boundary diagram (Figure 2.3) or an input–output diagram.

3. *Generate System Functions.* Each function should be specified with a verb and an object. The verb defines the activity of the function. The object provides the context. Both are required! The function should be specified without specifying the system element that will perform the function. For a vehicle, some appropriate functions might be transport passengers, store luggage, avoid collision with an object, and so on. An inappropriate function would

Figure 10.2   Affinity diagramming in action.

be "step on the brakes" since this function assumes that brakes will be an element and that the foot will be used to activate the brakes. The brainstorming
can be done by individuals in the groups on the basis of their knowledge and
experience. In some settings, it is appropriate (if not essential) to have the
individual use organizational documents that provide required capabilities,
functions, or requirements for the system. It is usually a good idea to get
10–20 functions from each individual.

4. *Rapidly Group (Affinitize) Similar Functions.* Once the functions have been
   recorded they should be displayed. Next, a few individuals should bin the
   functions into logical groups. Usually the verbs are the most helpful for binning the functions. For example, the following functions might be binned
   together: relocate passengers, take family to the store, seat passengers, and
   transport kids to soccer practice. As the affinitizing process is being performed, participants should continue to add new functions as ideas come to
   their mind or the discussion keys an idea. These function groups will become
   the lowest tier in the functional hierarchy. At this point each function group
   should be named. The name should be the most general name that captures
   the activity of all the functions in the group. The functional group should be
   named with a verb and an object. You can use one of the function names
   or develop a new name. In our example, "transport passengers" might be a
   good function group name.

5. *Develop Preliminary Functional Hierarchy.* The next step is to affinitize the
   function groups into the next higher level. Again, similar function groups will
   be binned together. The higher level function groups will need to be named
   with a verb and object that capture the meaning of the function groups below

it in the hierarchy. For example, transport passengers and move material, might binned into transport people and objects. For some systems, two levels may be sufficient. For complex systems, many levels may be required. This step would be repeated until the first tier of the hierarchy has about three to five functions. Three to five is a useful guideline since it is relatively easy to remember this number of functions. At each level of the hierarchy, the functions should be presented in the most logical order. For example, time sequencing may be appropriate. At this point the group activity is complete.

6. *Refine the Functional Hierarchy.* The lead systems engineer will need to refine the hierarchy and vet it with stakeholders who could not attend the affinity diagramming workshop and with system decision makers. During the process the function names on each tier may change as reviewers provide insights on a clearer or more acceptable way to name the functions.

Uses of Functional Hierarchy Once the functional hierarchy is complete, it has several uses. First, the functional hierarchy can (should!) be used to provide a clear understanding of the functions the system is being designed to perform. This makes it useful for presentations to participants in the systems engineering process, including decision makers and stakeholders. Second, the functional hierarchy is an important first step for more detailed functional analysis, which is required to identify and define interfaces and requirements. These systems engineering activities are described in Chapter 7. Third, the functional hierarchy can serve as the foundation for the assessment of the candidate solution designs (see Section 10.4). Fourth, the functional hierarchy can be used to help develop models and simulations. The functional hierarchy should be updated throughout the system life cycle and especially at every application of the systems decision process. Fifth, the functional hierarchy can be used to support system architecting and system design. Both of these processes involve the allocation of system elements to system functions.

USING THE AFFINITY DIAGRAMMING TECHNIQUE TO DEVELOP A FUNCTIONAL HIERARCHY FOR THE ROCKET SYSTEM EXAMPLE

To continue our notional example, the systems engineering team assembled a group of stakeholders representing users of the rocket (military, commercial, and research organizations) and manufacturers. After providing the system description, the group used the affinity diagramming process to develop a list of functions. The team grouped the resulting list into these top level functions:

Function 1.0: Launch rocket.

Function 2.0: Transport payloads.

Function 3.0: Achieve desired effect.

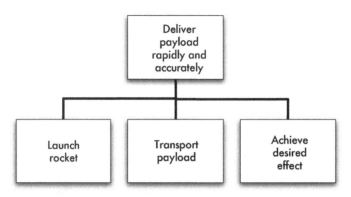

Figure 10.3   Functional hierarchy for the rocket example.

While this is a short list, it is useful to illustrate the affinity diagramming process. At the most basic level, the functions of a rocket are to launch, carry an object in flight, and put that object when and where needed to achieve some desired effect. Figure 10.3 provides a basic functional hierarchy for this example. We will continue to develop the example further.

Limitations of Functional Hierarchy   Hierarchy is a first step to identify and structure the system functions; however, additional information is required to identify the interrelationships of functions and interfaces. The functional hierarchy can also identify the functions that may need to be modeled and/or simulated, but additional information will be required to develop the models and simulations for the development and evaluation of the system design.

Functional Flow Diagram   Once the top-level functions have been identified their relationships can be depicted in a functional flow diagram. As the functions are decomposed their interfaces become more specific and more complex. The functional decomposition can be continued until discrete tasks have been defined that can be allocated to system elements. Trade studies are performed to allocate tasks to system elements. Figure 10.4 illustrates the functional decomposition for the top level of the National Aeronautical and Space Agency's Space Transportation System (STS) Flight Mission [19]. The flight mission is composed of eight tasks.

Each of the top-level functions in Figure 10.4 can be decomposed into the functions that are required to perform the top level functions. Figure 10.5 [19] shows the functional decomposition for the Perform Mission Operations 4.0 function into 10 functions and provides more information about their interfaces. The functional decomposition of each of the functions could continue to the next level.

Functional flow diagrams are the most useful in the system concept, design, and development stages of the systems life cycle. The functional flow diagram defines the relationships of the functions but does not define the interfaces needed to complete the design. The next technique provides additional functional analysis data.

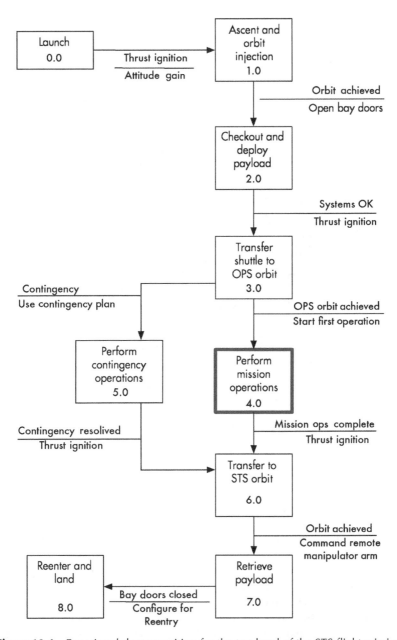

Figure 10.4 Functional decomposition for the top level of the STS flight mission.

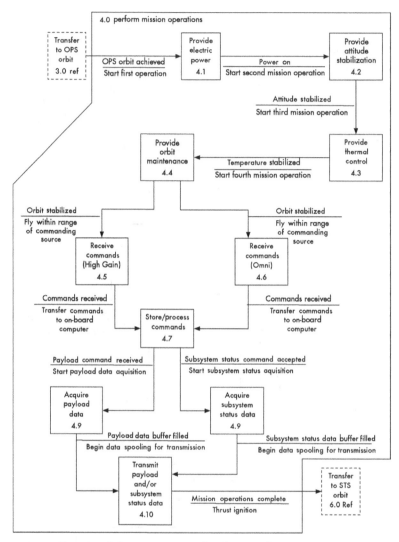

Figure 10.5 Functional decomposition—level 4.0 perform mission operations [19].

Uses of the Functional Flow Diagram Once the functional flow diagram is complete it has several uses. First, the diagram provides a better understanding of the relationships of the functions and tools to support the identification of the requirements. The top-level functions are useful for presentations to participants in the systems engineering process, including decision makers and stakeholders. Second, the functions in the diagram can serve as a foundation for the assessment of the alternative solution designs (see Section 10.4). Third, the functional flow diagram

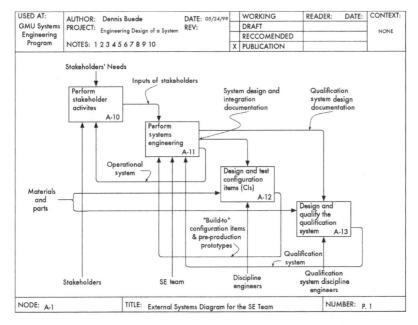

USED AT:	AUTHOR: Dennis Buede		DATE: 05/24/99	WORKING	READER:	DATE:	CONTEXT:
GMU Systems Engineering Program	PROJECT: Engineering Design of a System		REV:	DRAFT			NONE
				RECCOMENDED			
	NOTES: 1 2 3 4 5 6 7 8 9 10			X PUBLICATION			

Figure 10.6 IDEF0 Level 1 functional analysis example.

can be used to support system architecting and system design. Fourth, the functional flow diagram can be used to help develop models and simulations to design and evaluate systems.

Limitations of the Functional Flow Diagrams The diagrams identify the functions and the interrelationships of the functions. This information is useful in identifying the models required. However, the diagrams do not identify all the inputs, outputs, controls, and mechanisms that are required for system design or the detailed development of models and simulations.

IDEF0 Functional Modeling Method IDEF0 models were discussed in detail in Chapter 2. These models can help the systems engineer communicate with key stakeholders to identify the functions, inputs, outputs, mechanisms, and controls of the system. IDEF0 is a useful method for functional analysis. The simple graphical approach helps involve subject matter experts early in the system development process. IDEF0 can be implemented with widely available tools (e.g., Microsoft® Visio [24]) or can be implemented with specialized software. Figure 10.6 provides an example from Buede's book of a high-level IDEF0 functional analysis for performing an engineering design of a system [25].

Uses of IDEF0 The model has several uses. First, the model can support the system design. IDEF0 helps identify and structure the system functions. IDEF0 models can be used to define 'As-Is' and 'To-Be' architectures. As-Is models are usually developed bottom-up, whereas To-Be models are developed top-down. For each function the inputs, outputs, controls, and mechanisms (ICOMs) are specified using a standard modeling language. The model is developed as a hierarchy of functions. The modeler can stop when functions are specified at the level of detail required to allocate components to functions. Second, the functions in the diagram can serve as a foundation for the assessment of the alternative solution designs (see Section 10.4). Third, the models can be used to help develop models and simulations to design and evaluate systems.

Limitations of IDEF0 These models provide detailed information to support system design and qualitative evaluation of candidate solutions. However, the models do not validate that the design will work in the environment or that the design has the capacity to provide the inputs and outputs of each function to meet the quantitative system performance parameters. Models, simulations, development tests, and operational tests are needed to overcome these limitations.

Models and Simulations   Models and simulations have been described in Chapter 4. Models and simulations play an important role in system design and evaluation. We believe developing models will save time and money in system design and evaluation. If you cannot develop a model, you probably do not have enough information to design the system. In Chapter 4, we also described the uses and limitations of models and simulations. While simulation can replace some of the testing and evaluation, we believe that in almost all cases[1] some live testing is required to ensure that the system design meets the system performance objectives.

10.3.4   Requirements Analysis

Requirements analysis involves determining the specific characteristics of a system and is critical in designing an effective and cost efficient solution to a problem. This topic can be the subject of whole courses in universities however we will provide only a cursory introduction to the subject. For a more complete discussion see Martin [21], Wasson [26], or Kossiakoff [27].

Requirements, which are also called specifications or specification requirements, describe the technical capabilities and levels of performance desired in a system. These requirements can be binned into several types such as stakeholder, operational, capability, interface, verification, and validation requirements, to name a few. Requirements are not established randomly and the systems engineering team should maintain auditable records that trace requirements to objective information defining and describing the requirement. Similar to functional hierarchies, requirements can be derived from top down so the lowest level of requirements can be mapped back to some higher level operational requirement [26]. Software products,

---

[1]Nuclear weapons systems are an obvious example of systems that cannot be fully tested in an operational environment.

such as CORE® exist to create and trace requirements hierarchies [28]. A good way to categorize requirements is to describe them as constraints or capabilities. A constraint is a requirement that must be met while a capability is a desired feature, trait or performance characteristic [21]. Constraints will be used in a screening matrix in the Solution Design phase in order to weed out alternative solutions that do not meet requirements categorized as constraints. A key objective of requirements analysis is to transform operational requirements, or required outcomes of a system, into performance requirements that can be defined as engineering characteristics of the system [27].

Wasson [26] provides a helpful list of questions that systems engineers can think about while performing requirements analysis:

1. Do the list of requirements appear to be generated as a feature-based "wish list" or do they reflect a structured analysis?
2. Do the requirements appear to have been written by a seasoned subject matter expert?
3. Do the requirements adequately capture user operational needs? Are they necessary and sufficient?
4. Do the requirements unnecessarily constrain the range of viable alternative solutions?
5. Are all the system interface requirements identified?
6. Are there any critical operational or technical issues that require resolution or clarification?

Systems engineers need to follow a process to elicit, understand, document and trace the requirements of a system throughout the system life cycle. At this point in the Problem Definition phase, critical stakeholder and operational requirements should be defined and constraints identified so that alternative solutions can be screened for feasibility.

10.3.5 At Completion

A key output from functional and requirements analyses is a thorough understanding of the functions required from the solution to the systems decision problem. This task also provides an understanding of the interrelationships between the system functions. The system requirements are also defined and classified into constraints (must be met) and desired capabilities.

At this point, research, stakeholder, functional, and requirements analyses have enabled us to identify the objectives, functions, and constraints of the system being designed. The requirements identified as constraints become the screening criteria, which all candidate solutions must satisfy. However, how do we evaluate the goodness of candidate solutions to our problem? We create evaluation criteria based on what stakeholders value from the solution. These evaluation criteria become the value model that we use to evaluate candidate solutions. We use a process called value modeling to develop the candidate solution evaluation criteria.

10.4   VALUE MODELING

Value modeling provides the systems engineering team an initial methodology for evaluating candidate solutions. This task employs the concepts of value-focused thinking discussed in Chapter 9. This methodology will continue to be refined in the solution design and decision-making phases of the systems decision process. We use the information developed from the research and stakeholder analysis and the functional and requirements analysis tasks to create the evaluation methodology. These tasks enable us to develop a qualitative value model that captures the most important functions and objectives for the system. We use the qualitative value model to build a quantitative value model, which provides a measurable method to evaluate how well candidate solutions meet the fundamental objective of our systems decision problem. In essence, we are evaluating the future value of the implemented solution to our problem. We use the concepts of multiple objective decision analysis to build the mathematical framework for the value model.

The key question is, whose values are in the value model? Three important stakeholder groups must be considered: consumers, users, and customers. The consumers use the systems products and services. The users operate the system to provide products and services. The customers acquire the systems and provide them to users. For individual investing financial Web sites, the consumer is the investor, the user is the Web site manager, and the customer is the IT department that manages the design and development of the Web site. One individual or organization can perform multiple roles. For example, for an automobile system, the user (driver) is one of the consumers. Passengers are also consumers.

10.4.1   Definitions Used In Value Modeling

Constructing a value model requires a good deal of attention to detail in research and sensitivity to what stakeholder groups are expressing during stakeholder analysis. It is important to clearly define some of the key terminology used in value modeling [29].

Fundamental Objective. The most basic high level objective the stakeholders are trying to achieve.

Value Measure. Scale to assess how well we attain an objective. Alternate terms include the following: evaluation measures, measures of effectiveness, performance measures, and metrics.

Qualitative Value Model. The complete description of the stakeholder qualitative values, including the fundamental objective, functions (if used), objectives, and value measures.

Value Hierarchy/Value Tree. Pictorial representation of the qualitative value model.

Tier. Levels in the value hierarchy.

Weights. The weight assigned to a value measure reflects the measure's importance and the range of its measurement scale. We assess swing weights by assessing the impact of "swinging" the value measure from its worst to its best level.

Score. A number in the range of the value measure that reflects the estimated future performance of a candidate solution.

Value Function. A function that assigns value to a value measure's score. A value function measures returns to scale over the range of the value measure.

Quantitative Value Model. The value functions, weights, and mathematical equation used to evaluate candidate solutions.

Global (Measure) Weights. The measure weights for each value measure. Global (measure) weights sum to 1.

Local Weights. The weights at each node in the value hierarchy. Local weights below each node sum to 1.

10.4.2   Qualitative Value Modeling

In value modeling, we create both qualitative and quantitative value models. Of these the qualitative value model is the most important because it reflects the key stakeholder values regarding the systems decision problem. If the qualitative model does not accurately capture these values, the quantitative value model will not be useful in evaluating candidate solutions. In this section we describe how to develop a qualitative value model and then demonstrate it using our rocket example.

Criteria for a Value Model   Kirkwood describes the criteria for a good value hierarchy as "completeness, nonredundancy, decomposability, operability, and small size" [30]. The concept of completeness basically means that the value model, represented by all its objectives and value measures, must be sufficient in scope to evaluate the fundamental objective in the systems decision problem. Nonredundancy means that functions or value measures on the same tier in the hierarchy should not overlap. The criteria of completeness and nonredundancy are often referred to as the value model being "collectively exhaustive and mutually exclusive." Value independence means that value of the scores on one value measure to not depend on the scores on any of the other value measures. Operability means that the value hierarchy is easily understood by all who use it. Finally, a value hierarchy should contain as few measures as possible while still meeting the requirement to be mutually exclusive and collectively exhaustive. This will help the team focus their analysis on the most important value measures.

Developing a Qualitative Value Model   We use the information developed in the research and stakeholder analysis and functional and requirements tasks to determine the functions, objectives, and value measures that comprise the value model. These are the basic steps in developing the qualitative value model [29]:

1. *Identify the Fundamental Objective.* This is a clear, concise statement of the primary reason we are undertaking the decision problem. For example, the authors worked on a decision problem for the U.S. Army concerning the organizational structure of a key subordinate element of the army. The fundamental objective of this decision problem was "develop the most effective and efficient organizational structure to support the Army's mission" [32].

2. *Identify Functions that Provide Value.* We did this in developing the functional hierarchy in Section 10.3. For many systems, the functional hierarchy provides a basis for the value hierarchy.

3. *Identify the Objectives that Define Value.* An objective provides a statement of preference, for example, we may want to "maximize efficiency" or "minimize time."

4. *Identify the Value Measures.* Chapter 4 provides a detailed discussion on measures of effectiveness (MOE), including the characteristics of a good MOE. Value measures are the same as MOEs. Value measures tell us how well a candidate solution attains an objective. Value measures are developed based on their alignment with the objective and their scale of measurement. Kirkwood [30] describes value measures as either direct (can directly measure attainment of an objective) or proxy (measures attainment of an associated objective). For example, profit would be a direct measure of an objective for maximizing profit. For "maximize safety," a reasonable proxy measure might be the estimated number of fatalities. Value measure scales can be either natural or constructed scales. A natural scale for "maximize profit" would be dollars. A constructed scale for "maximize profit" would be thousands, millions, tens of millions, hundred of millions, and billions of dollars. Parnell [29] provides a good synopsis of preferences in developing value measures. Table 10.6 shows that value measures with natural scales that directly measure attainment of an objective are the most preferred while proxy measures with constructed scales that measure attainment of an associated objective are least preferred, but may be necessary given the problem. At this point, we can build a value hierarchy using the structure depicted in Figure 10.7.

5. *Discuss the Value Model with the Key Stakeholders.* At this point it is very important to get approval of the value model from the key stakeholders. This will ensure that future system development efforts are on track.

TABLE 10.6   Preferences for Types of Value Measures

Type of Measure	Direct	Proxy
Natural	1	3
Constructed	2	4

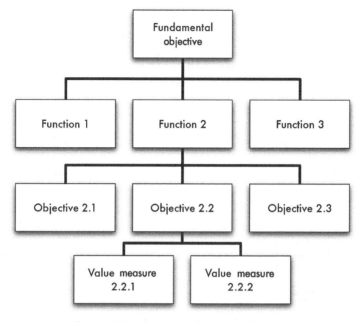

Figure 10.7 Structure of a value hierarchy.

QUALITATIVE VALUE MODEL FOR THE ROCKET EXAMPLE

Recall that our initial systems decision problem was "to develop a new small, mobile rocket that can be used for multiple applications." Our stakeholder and functional analysis yielded the screening criteria listed in Section 10.2.2, and the functions that provide value depicted in the functional hierarchy in Figure 10.3. Throughout stakeholder analysis we also sought to identify stakeholder values in this decision problem. Here is a list of the values identified by the systems engineering team:

1. Mobility of the launch platform through a variety of terrain is important.
2. The support requirements for launch platform should be as small as possible.
3. The rocket should be able to carry heavy payloads.
4. The rocket needs to be flexible enough to carry different types of payloads.
5. The rocket needs to be as accurate as possible.
6. The rocket should be able to carry payloads as far as possible.

These values can be translated into objectives, the attainment of which can be evaluated using the value measures depicted in our value

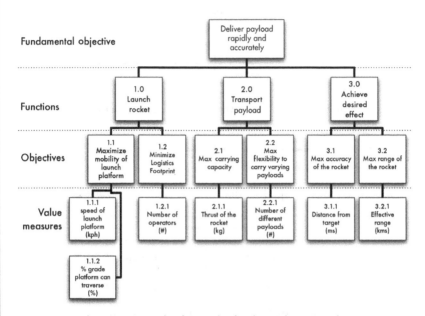

Figure 10.8   Value hierarchy for the rocket example.

hierarchy in Figure 10.8. Our fundamental objective in this problem is to develop a small, mobile rocket that delivers payloads rapidly and accurately to the intended destination. The functions are the functions from the functional hierarchy. The objectives provide a statement of preference regarding the values articulated by the stakeholders. The seven value measures provide a quantitative means to evaluate how well a candidate rocket system attains the stated objectives. The objectives and associated value measures are further defined as follows:

Objective 1.1 Maximize mobility of the launch platform. Mobility is difficult to directly measure, so it is broken down into two components here: speed and the ability to move up/down hills.

> **Measure 1.1.1** Speed of the launch platform in kilometers per hour (kph). Speed of candidate solution platforms in kph is a natural measure for attainment of an associated objective (proxy) to maximize the mobility of the launch platform.

> **Measure 1.1.2** Percent grade of hill the launch platform can traverse (% grade). This is a proxy measure. Again since attainment of the mobility objective is not directly measured, this measure is classified as a "natural-proxy" measure. Modeling, simulation, and testing can provide the percent of hill grade (natural measure) candidate solution platforms can traverse.

Objective 1.2 Minimize the logistical footprint of the launch platform. The size of the total support requirement for the system will be difficult to directly measure, so it will be measured in terms of the number of people needed to put the rocket into operation.

> **Measure 1.2.1** Number of people required to operate the rocket system. Since attainment of the logistics footprint objective is not directly measured, this measure is also classified as a "natural-proxy" measure.

To illustrate a different type of value measure, here is the classification for objective 3.2:

Objective 3.2 Maximize range of the rocket. The range of the candidate solutions can be directly evaluated through testing.

> **Measure 3.2.1** Effective range of the rocket (kilometers). Kilometers that candidate rockets travel in testing is a natural measure that can be used to directly evaluate attainment of this objective; so this is an example of a "natural-direct" measure.

The classifications of the remaining value measures are left as chapter exercises.

10.4.3 Quantitative Value Model

Quantitative value modeling allows us to determine how well candidate solutions to our systems decision problem attain the stakeholder values. We introduce here the basics of forming a mathematical model to assess the values of candidate solutions. The emphasis for our quantitative value model is at the bottom tier of the value hierarchy. We build functions for each value measure to convert a candidate solution's score on the measure to a standard unit called "value." We also weight the value measures to reflect their importance to the overall problem and the impact of the variability in their measurement scales on the decision.

The Multiple Objective Decision Analysis Mathematical Model In our rocket example, the objective to maximize mobility of the launch platform would guide designers to make the rocket as small as possible so that it can be moved around easily. However, the objective to maximize the carrying capacity of the rocket would steer designers to build a large rocket that can generate a high level of thrust. These are conflicting objectives in our decision problem that need to be resolved.

Multiple objective decision analysis (MODA) provides a quantitative method for trading off conflicting objectives [30]. MODA has many different mathematical relationships to do this. We will focus on the most common method called the additive value model to calculate how well candidate solutions satisfy stakeholder values for the problem.

The mathematical expression for the additive value model used to compute the total value for competing solution is given by

$$v(x) = \sum_{i=1}^{n} w_i v_i(x_i) \tag{10.1}$$

where $v(x)$ is the total value of a candidate solution, $i = 1$ to n for the number of value measures, x_i is the score of the candidate solution on the ith value measure, $v_i(x_i)$ is the single-dimensional value of the candidate solution on the ith value measure, w_i is the measure weight (normalized swing weight) of the ith value measure and $\sum_{i=1}^{n} w_i = 1$.

The additive value model makes some assumptions about the structure of the problem to which it is applied. Kirkwood [30] provides an excellent discussion of the concepts of mutual preferential independence, measurable value, and utility for the reader who wishes to gain a deeper understanding of this topic. If two value measures are preferentially dependent, we can combine the measures into one measure and then use the additive value model [31]. We will continue to develop the quantitative value model assuming that the additive value model is applicable to most systems decision problems.

Developing Value Functions   Value measures have varying units in their measurement scales. In our rocket example, the effective range of the rocket is measured in kilometers while the thrust is measured in pounds force (lbf). We use value functions to convert candidate solution scores on the value measures to a standard unit. For a single-dimensional value function, the x-axis is the scale of the value measure (e.g., kilometers) while the y-axis is a standard unit of value scaled from 0 to 1. Other scales for value can be used—for example, 0–10 or 0–100 depending on what stakeholders are comfortable with; however, the scale needs to be consistent for all value measures.

Value functions measure returns to scale on the value measure. Value functions can be discrete or continuous. Continuous functions typically follow four basic shapes shown in Figure 10.9. The curves on the left of the figure are monotonically increasing from a minimum to a maximum level of the value measures. The curves on the right are monotonically decreasing. Kirkwood [30] provides several techniques to develop the shape of value curves using input from subject matter experts. One technique starts with having an expert identify the general shape of the curve. Then proceed to have the expert identify their increase (or decrease depending on the shape of the curve) in value from a specific incremental increase in the measure scale. Doing this for multiple increments up to the maximum on the measure scale will lead to the shape of the value function. This could produce a piecewise linear function.

As an example we will use value measure 1.2.1, the number of people to operate the system, from our rocket example. Since the objective is to minimize the logistic footprint, less people are better. We first set the limits of the x-axis of our function for the number of people. These limits are set based on what we expect

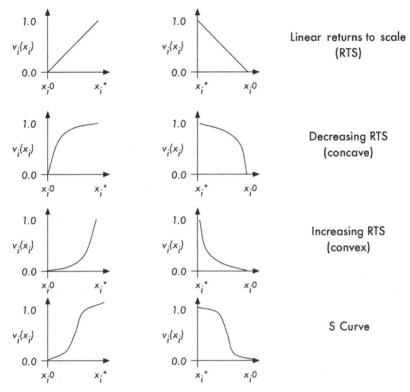

Figure 10.9 Typical shapes of value function.

an ideal solution and worst feasible solution would have for the number of people required. The best we can achieve is to both drive the platform and fire the rocket with no people (total remote control). The maximum number of people for a candidate solution would be seven since more people than that would require either an additional four-wheel drive vehicle, or require larger than a four-wheel drive vehicle. Seven is therefore a screening criterion.

Experts say the general shape of the value curve is an S-curve. With two or less people, the candidate solution has high value, but the value decreases rapidly beyond two people. Figure 10.11 demonstrates the process of building this value function. Using a scale of 0 to 100 for value, subject matter experts said that the incremental decrease in value from 0 to 1 person was only 3 units, while the incremental decrease in value from 2 to 3 people was 30 units of value. The table in Figure 10.10 shows the remaining increments of value. The resulting value function is piecewise linear, generally following the S-curve shape depicted in Figure 10.10. Once this process is complete, a candidate solution's value for a measure can be derived from the value function using the candidate solution score.

Weighting the Value Model Typically, not all value measures are equal in the view of stakeholders. The measures are weighted in the additive value model

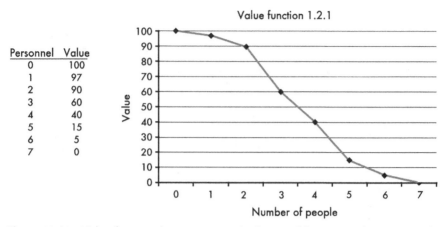

Figure 10.10 Value function for minimizing the logistical footprint value measure for the rocket example.

to arrive at an overall value for candidate solutions. These weights enable us to quantify the tradeoffs between conflicting objectives. The weights depend on both the importance of the value measure and the impact of changing the value of the score within the range on the x-axis scale of value measures [30]. Therefore, we assess measure weights using swing weights by "swinging" the value measure score from its worst to its best level.

The concept of swing weights is one of the most important concepts in MODA and, for many, the most difficult to understand. Kirkword [30] provides a mathematical proof for the definition of swing weights and Parnell and Trainor [36] provide summary of the issue and several examples of the swing weight matrix. Weights should be generated bottom up to ensure that the importance and variation are considered. Weights must sum to one at each tier of the hierarchy. Kirkwood [30] and Clemen and Reilly [33] provide multiple methods to elicit weights from stakeholders. We will describe a swing weight matrix method developed by Parnell [29] that can be used to generate weights relatively quickly from the bottom up.

To develop weights, first create a matrix in which the top scale defines the value measure importance and the left side represents the impact of changing the value measure range on the decision. A measure that is very important in arriving at a solution to the decision problem and is one in which changes across its range of values has a large impact on the decision would be placed in the upper left of the matrix. A value measure that is deemed not important and whose changes across its range have little impact on the decision would be placed in the lower right of the matrix. Figure 10.11 illustrates an example 3×3 matrix. While this is a fairly standard size used in applications, the dimensions of the matrix could expand if a higher level of discrimination between value measures (e.g., 5×5) is required. Once all the value measures are arrayed in the matrix, we could use any swing weight technique to obtain the weights. The levels of importance and variability (high, medium, low) should be defined in terms appropriate for the systems decision.

		Level of importance of the value measure		
		High	Medium	Low
Variation in measure ranges	High	measure *X* $f_1 = 100$		
	Medium			
	Low			measure *Y* $f_9 = 1$

Figure 10.11   Example of swing weighting matrix for determining measure weights.

Using the method introduced by Parnell [29], we begin by assigning measure *x* (the upper left-hand corner cell) an arbitrary non-normalized swing weight, for example, 100. Using the value increment approach [30] we assign the weight of the lowest weighted measure, measure *y* (the lower right-hand corner) the appropriate swing weight, for example, 1. This means the weight of measure *x* is 100 times more than that of measure *y*. Non-normalized swing weights are assigned to all the other value measures relative to the weight of 100 by descending through the very important factors, then through the important factors, then through the less important factors. We then normalize the resulting swing weights to obtain measure weights on a scale of 0.0–1.0 using Equation (10.2). This converts the swing weights into *measure weights*, which are also referred to as *global weights*. We use the terms interchangeably in what follows.

The mathematical expression for obtaining measure weights by normalizing swing weights in order to express the relative weighting of value measures in the quantitative value model is given by

$$\text{Measure weights for value measures}(w_i): \qquad w_i = \frac{f_i}{\sum_{i=1}^{n} f_i} \qquad (10.2)$$

where f_i is the non-normalized swing weight assigned to the *i*th value measure, $i = 1$ to n for the number of value measures, and w_i are the corresponding measure weights.

The basic concept of this method in determining weights is relatively straightforward. A measure that is very important to the decision problem should be weighted higher than a measure that is less important. A measure that differentiates between candidate solutions—that is, a measure in which the changes in candidate solution scores have a high impact on the decision—is weighted more than a measure that does not differentiate between candidate solutions.

Since the swing weights may be provided by different individuals, it is important to check for the consistency of the weights assigned. It is easy to understand that a very important measure with a high impact for changes in its range will be weighted more than a very important measure with a medium impact for changes in its range. It is harder to trade off the weights between a very important measure

		Level of importance of the value measure		
		Very important	Important	Less important
Variation in measure range	High	A	B_2	C_3
	Medium	B_1	C_2	D_2
	Low	C_1	D_1	E

Figure 10.12 Value measure placement in swing weight matrix: consistency example.

with a low impact on the decision for range changes and an important measure with a high impact on the decision for changes in its range. Weights should descend in magnitude as we move in a diagonal direction from the top left to the bottom right of the swing weight matrix. For clarity, consider the matrix in Figure 10.12. For the weights to be consistent, they need to meet the following relationships:

- A measure placed in cell A has to be weighted greater than measures in all other cells.
- A measure in cell B_1 has to be weighted greater than measures in cells C_1, C_2, D_1, D_2, and E.
- A measure in cell B_2 has to be weighted greater than measures in cells C_2, C_3, D_1, D_2, and E.
- A measure in cell C_1 has to be weighted greater than measures in cells D_1 and E.
- A measure in cell C_2 has to be weighted greater than measures in cells D_1, D_2, and E.
- A measure in cell C_3 has to be weighted greater than measures in cells D_2 and E.
- A measure in cell D_1 has to be weighted greater than a measure in cell E.
- A measure in cell D_2 has to be weighted greater than a measure in cell E.

No other strict relationships hold. The stakeholders will need to express their tradeoffs between level of importance and the impact of range changes in value measure as weights are assigned. More concisely, the strict relationships in inequalities 10.3 through 10.10 must hold:

$$A > \text{ all other cells} \tag{10.3}$$
$$B_1 > C_1, C_2, D_1, D_2, E \tag{10.4}$$
$$B_2 > C_2, C_3, D_1, D_2, E \tag{10.5}$$
$$C_1 > D_1, E \tag{10.6}$$
$$C_2 > D_1, D_2, E \tag{10.7}$$

$$C_3 > D_2, E \tag{10.8}$$

$$D_1 > E \tag{10.9}$$

$$D_2 > E \tag{10.10}$$

No other specific relationships hold.

Quantitative Value Model for the Rocket Example The quantitative value model consists of the objectives and weighted value measures with definitions, scales, and functions. This model is passed to the Solution Design phase of the systems decision process as an initial means of evaluating alternatives to identify the candidate solutions. Figure 10.13 provides the functions for the value measures, minus measure 1.2.1 that is depicted in Figure 10.10. Of course, Figures 1.2.1 and 2.2.1 could also be plotted as bar charts since their scores are discrete. The systems engineering team used the swing weight matrix method to elicit

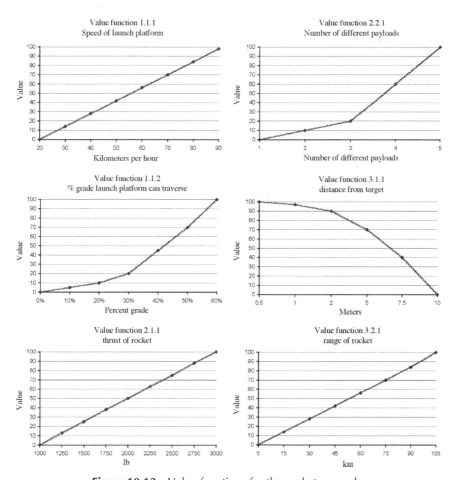

Figure 10.13 Value functions for the rocket example.

		Level of importance of the value measure		
		Mission critical	Mission enabling	Mission enhancing
Variation in measure ranges	Large capability gap	3.1.1 Accuracy 100	3.2.1 Range 50	
	Significant capability gap	1.1.1 Speed of platform 85	2.1.1 Thrust 45	1.1.2 % Grade 5
	Small capability gap	2.2.1 Number of payloads 60	1.2.1 Number of people 20	

Figure 10.14   Completed swing weight matrix for the rocket example.

global weights for the value measures from the key stakeholders. Figure 10.14 provides the completed matrix. Importance was defined as the importance of that value measure to accomplishing the mission. Variation was defined as the impact of the capability gap between the lowest value score and the ideal score for each value measure. The numbers appearing under each measure represent the swing weight assigned by the key stakeholders. The value measures are indexed by numbers assigned during the creation of the value hierarchy shown in Figure 10.8.

The measure, or global, weights for value measures are then calculated from the data in the swing weight matrix by using Equation (10.2). Figure 10.15 shows both the stakeholder assigned swing weights and their corresponding measure, or global, weights. The key information about the quantitative value model can be concisely summarized as in Table 10.7. The systems engineering team uses this value model to evaluate alternatives and candidate solutions developed during the Solution Design and Decision Making phases of the systems decision process. Candidate solutions receive a value for each of the value measures using the value functions. These values combined with their corresponding measure global are used to calculate a candidate solution's total value using the additive value model given in Equation (10.1).

Value measures	Swing weight	Measure global weight
1.1.1 Speed of platform	85	0.23
1.1.2 % Grade	5	0.01
1.2.1 Number of people	20	0.06
2.1.1 Thrust	45	0.12
2.2.1 Number of payloads	60	0.17
3.1.1 Accuracy	100	0.27
3.2.1 Range	50	0.14
Total =	365	1.00

Figure 10.15   Global weights of the value measures rocket example.

TABLE 10.7 Quantitative Value Model for the Rocket Example

Function	Objective	Value Measure	Definition of Value Measure	Measure Type	Global Weight	Shape of Value Curve
1.0 Launch rocket	1.1 Maximize mobility of launch platform	1.1.1 Speed of platform	Speed of launch platform in kph (more is better)	Proxy, natural	0.23	Linear
		1.1.2 % Grade	% Grade platform can traverse (more is better)	Proxy, natural	0.01	Convex
	1.2 Minimize logistics footprint	1.2.1 Number of people	of people to operate system (less is better)	Proxy, natural	0.06	S-curve
2.0 Transport payloads	2.1 Max carrying capacity of the rocket	2.1.1 Thrust	Thrust of the rocket in pounds force (more is better)	Proxy, natural	0.12	Linear
	2.2 Max flexibility to carry varying payloads	2.2.1 Number of payloads	Number of different payloads rocket can carry (more is better)	Direct, natural	0.17	Convex
3.0 Achieve desired effects	3.1 Max accuracy of the rocket	3.1.1 Accuracy	Average distance from target in testing in meters (less is better)	Direct, natural	0.27	Concave
	3.2 Max range of the rocket	3.2.1 Range	Effective range of the rocket in kilometers (more is better)	Direct, natural	0.14	Linear

10.4.4   At Completion of Value Modeling

The value modeling process provides the framework for evaluating how well candidate solutions attain the objectives articulated by the stakeholders for the decision problem. We want to develop this model <u>before</u> developing candidate solutions so that we can focus on developing solutions that meet stakeholder values. We create a value hierarchy using information developed during the research and stakeholder, requirements, and functional analysis tasks. The value hierarchy depicts the fundamental objective in solving the problem, the key functions, the stakeholder objectives the solution should attain, and the measures the team can use to quantify stakeholder values. We develop value functions as the means for evaluating how well candidate solutions meet these values. We weight these value measures based on their relative importance and impact of the variation in the measurement scale. This value modeling process can be refined as the systems engineering team continues their work in the Solution Design and Decision Making phases of the SDP.

10.5   OUTPUT OF THE PROBLEM DEFINITION PHASE

The Problem Definition phase of the systems decision process prepares the systems engineer to develop effective candidate solutions to a problem. The tasks of performing research and conducting stakeholder analysis, functional analysis, and requirements analysis focus the effort on defining the right problem that needs to be solved. The output from these tasks is used in value modeling to provide an initial evaluation methodology for solutions. At this point, the systems engineering team should have a refined problem statement, constraints that will serve as screening criteria, and a value model. We cannot overstate the need for the team to provide these to the key stakeholders for review before getting deeply involved in the Solution Design phase.

10.5.1   Discussion

Screening criteria are simply requirements that any potential solution to the problem must meet in order to be a feasible solution. These requirements constrain solutions by establishing characteristics, along with upper or lower bounds on value measures that must be met. We developed the value model in great detail in Section 10.4 to map stakeholder values for a solution to a methodology for evaluating how well candidate solutions attain these values. The problem statement also flows from the Problem Definition phase.

Problem statements can come in several forms. In the military, problem statements are usually expressed as a mission statement. A mission statement has as its basic elements who, what, when, where, and why. The military groups these elements into two basic components: the purpose of the mission and the key tasks to accomplish. The who, what, when, and where elements form the key task(s) component while the why element becomes the purpose. The purpose is the most important part of the mission statement. There may be several ways to complete the stated mission task(s); however, all ways must lead to accomplishing the overall purpose of the mission.

The problem statement is similar to the military mission statement. The problem statement contains functions that must be accomplished in order to achieve some purpose. INCOSE provides this discussion of a problem statement:

> The problem statement starts with a description of the top-level functions that the system must perform, the operational environment in which the functions must be performed, and the key items (physical, informational, and energy entities) that the system must output. This problem statement might be in the form of a vision, mission statement and mission requirements, a set of scenarios for how the system will be used and interact with other systems, and a set of contexts in which the system will be used. Acceptable systems must satisfy all mandatory requirements. The problem statement should be in terms of *needs* or what must be done (e.g., the key characteristics of the items to be produced by the system), not how to do it (e.g., what resources the system should have) [34].

This problem statement captures the needs of the stakeholders and the key objectives that must be attained by its solution. For our rocket example, the problem statement could be stated as follows:

> Develop a small, mobile rocket system that can accurately deliver a variety of payloads rapidly in order to meet a market niche that is currently not met.

This problem statement addresses both the broad functions of the system and the needs and values of the stakeholders. It also provides a purpose for the rocket system, which we had not discussed in previous sections. For another example of a problem statement, see the problem statement for the illustrative example at the end of Section 10.6.

10.5.2 Conclusion

Through research, stakeholder analysis, functional analysis, requirements analysis, and value modeling, we now have a refined problem statement, screening criteria that candidate solutions must meet, and a quantitative value model that can be used to evaluate any candidate solutions. In the Solution Design phase that follows, we develop candidate solutions that meet stakeholder objectives to solve this refined problem.

10.6 ILLUSTRATIVE EXAMPLE: SYSTEMS ENGINEERING CURRICULUM MANAGEMENT SYSTEM (CMS)—PROBLEM DEFINITION

In this section we use techniques for the three tasks in the Problem Definition phase of the systems decision process: research and stakeholder analysis, functional and requirements analyses, and value modeling.

RESEARCH AND STAKEHOLDER ANALYSIS

Robert Kewley, Ph.D. U.S. Military Academy

To determine the proper functions and performance measures for the system, the development team had to first identify the system's stakeholders.

- Cadets (consumers) receive the entire curriculum from the department and must apply what they learned upon graduation in a dynamic world with proliferating technology.
- Instructors (users) prepare and deliver that curriculum to cadets.
- Program directors (users) synchronize that curriculum across courses to support a major. They are also responsible for program accreditation and assessment.
- The department's accreditation officer (user) works with ABET, Inc. and various Academy committees to ensure their programs meet accreditation requirements.
- The department operations officer (user) synchronizes execution of the department's entire curriculum from semester to semester.
- The department leadership (decision authority, owner) sets and enforces standards, allocates resources to the academic programs, ensures alignment with the Academy's overall academic program goals, and ensures that programs achieve and maintain accreditation.

The department employed all three stakeholder analysis techniques to elicit needs, wants, and desires from the stakeholders listed above.

Cadet Survey

The development team issued a Web survey to cadets. This method was selected based on the considerations in Tables 9.1 and 9.3. This was a large group of very important stakeholders. They selected cadets enrolled in a junior year Computer Aided Systems Engineering course as our sample. The sample of 70 cadets represented about 35% of the total population. However, because one cannot control respondents in a web survey, only 48 cadets completed it. The primary focus of the cadet survey was to determine their needs for curriculum delivery. Questions asked them to compare and rank the various delivery methods used by the department. They were also asked for suggestions to improve content delivery. Analysis of survey data produced the findings shown in Tables 10.8 and 10.9.

One survey question described a structured web-based portal to cadets and asked them to rank it with respect to the other methods currently used.

Free text questions asked cadets to describe the reasoning for their rankings and to provide suggestions as to how the department can improve content

TABLE 10.8   Percentage of Cadets Who Prefer Different Methods of Electronic Content Delivery

Content Delivery Method	Percentage of Cadets Who Ranked This Method 1st	95% Confidence Interval
Course Web site	56%	± 12.3%
Course network folder	33%	± 11.6%
Academy wide portal	10%	± 7.4%

TABLE 10.9   Percentage of Cadets Who Would Rank a Newly Developed Portal Against the Current Methods

Ranking against Current Methods	Percentage of Cadets Who Ranked This Ranking	95% Confidence Interval
1st	58%	± 12.2%
2nd	27%	± 10.1%
No preference	15%	± 8.8%

delivery. There were a few trends in their responses. Several cadets mentioned availability as a reason for ranking criteria. The Academy-wide portal failed frequently. They also did not like having to do a separate logon to get content. One thing they liked about course folders was the ability for them to deposit files there while in class and access them later from their rooms. Finally, they wanted consistency across courses.

From these findings, we can conclude with strong confidence that most cadets prefer Web-based content delivery. However, they do not like the current academy-wide implementation of that delivery. They show a strong preference for a proposed redesigned portal. However, they are concerned about system availability, extra logon steps, ability to post content, and consistency.

These conclusions yield a recommendation to consider redesigning the department's content delivery system as a web-based portal that is reliable, easy to use, allows two-way communication, and is used consistently across the department.

Faculty Focus Groups

The primary users of the system are the Department of Systems Engineering faculty. To elicit needs, wants, and desires from this group, the design team brought them together as a focus group in the department's Systems Methodology and Design Lab. This lab is specially configured for the use of GroupSystems collaboration software. GroupSystems enables much more productive focus groups because it allows anonymous parallel input from many people. One person cannot dominate the session, so the group is less susceptible to "groupthink". Finally, the system automatically records all input and generates a report for future reference. Before the focus group meeting, the lead engineer developed an agenda:

1. Brainstorm functions in the Topic Commenter. This allows faculty to list functions in a brainstorming fashion. Other faculty can comment on the ideas.
2. Bin functions in the Categorizer. This develops related groups of functions.
3. Brainstorm values for this system in the Topic Commenter.
4. Bin values in the Categorizer.

The entire focus group took about 90 minutes and resulted in the following categories of functions and values:

• Functions	→ Lesson management
	→ Compatibility with other Academy and department IT systems
	→ Scheduling
	→ Resources
	→ Course administration
	→ Course project management
	→ Interface
	→ Accreditation
• Values	→ Easy to use interface
	→ Automatic administrative report generation
	→ Compatibility and non-redundancy with other systems
	→ Accessibility
	→ Security and stability
	→ Usability
	→ Ad hoc query capability
	→ Ease of maintenance

Note that there is some overlap between the above lists of functions and values. As is usually the case, the focus group output cannot be directly translated into a functional hierarchy and values hierarchy. However, the GroupSystems report is a valuable resource to ensure that the needs, wants, and desires of the faculty are reflected in the follow-on steps.

Stakeholder Interviews

Because of their unique positions and critical relationship with this system, the following people were interviewed to elicit their needs, wants, and desires:

- Department leadership—establishes department vision, mission, and objectives, allocates department resources. The results of this interview contributed to development of top-level values in the values hierarchy and approval of top-level functions for the functional hierarchy.
- Department operations officer—responsible for scheduling and execution of department courses. The results of this interview contributed to the development of the Integrate Department Academic Operations function.
- Department accreditation officer—responsible for coordinating accreditation for department programs. The results of this interview contributed to development of the Assess Program function.

FUNCTIONAL AND REQUIREMENTS ANALYSES

Once complete with research and stakeholder analysis, the design team was able to develop a functional hierarchy for this system. Using the general trends from the surveys, focus group, and interviews, they developed the high-level functions for the system. They then used the detailed information from these components of the stakeholder analysis to break these high-level functions into subfunctions. Many of the subfunctions were broken down as well. Once complete, the design team presented the functional hierarchy to the department leadership for approval.

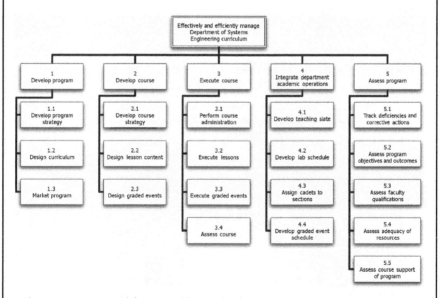

Figure 10.16 Partial functional hierarchy for curriculum management system.

Figure 10.16 shows the first two levels of the functional hierarchy for the CMS system. It is important to note that the functions depicted in this hierarchy are not simply the functions of the CMS information system. They are all of the functions that the department faculty has to perform in order to manage the curriculum. A subset of these functions will be supported by the CMS information system. This design technique ensures that we are developing an information system that helps us perform our core functions and not simply designing an information system to pass data around hoping someone will use it for something.

VALUE MODELING

Upon completion of functional analysis, the design team was able to align curriculum management functions with top-level department values solicited from the department head. Once they had this alignment, they developed supporting objectives that would enable execution of the CMS functions. Figure 10.17 illustrates this alignment. The team assigned weights to each of the objectives based on the degree to which that capability supports the critical CMS functions in the diagram. These objectives will distinguish the different alternatives for the system concept decision.

The development team used swing weighting to calculate the global weights assigned to each value measure for the supporting IT objectives. These swing weights were based on importance values determined

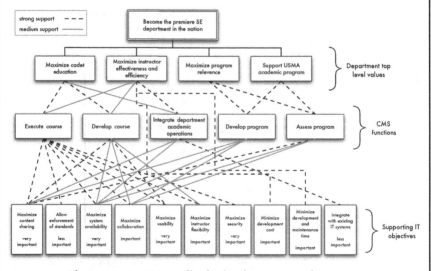

Figure 10.17 Crosswalk of value functions to objectives.

		Level of importance of the value measure		
		Very important	Important	Less important
Variation in measure ranges	High	Usability 100	Collaboration 75 Development time 75 Development cost 50	
	Medium	Content sharing 90 Availability 75	Instructor flexibility 40	Enforce standards 10 Integrate 5
	Low	Security 45		

Figure 10.18 Swing weight matrix for system concept decision.

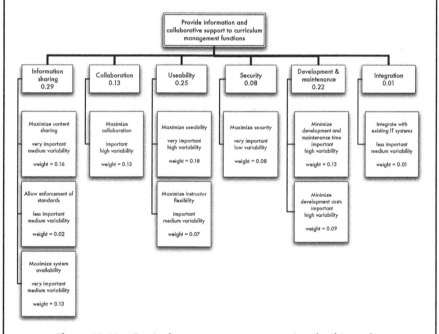

Figure 10.19 Curriculum management system's value hierarchy.

by the values to functions to objectives crosswalk and the degree to which the alternatives varied with respect to these measures. Figure 10.18 shows the swing weight matrix for this decision.

The development team organized the objectives into categories shown in the hierarchy of the values in Figure 10.19.

For each of the supporting objectives, the design team developed a value measure by which they would assess the degree to which each alternative achieved that objective. In most cases, they used a constructive scale to compare the alternative to the current system for an objective. The constructive scale had the following values:

−1 Worse than current system

 0 Same as current system

+1 Marginal improvement to current system

+2 Some improvement to current system

+3 Significant improvement to current system

For two of the objectives, development cost and development time, the design team used the available data. They had total cost estimates for each alternative. They also had estimates of the time required (in months) to achieve course-level functionality for each system.

For each value measure, the design team developed a value function to convert the individual scores (raw scores) to a value between 0 and 10. These value functions are shown in Figure 10.20. Some of the criteria, such as content sharing, enforcement, and collaboration, have increasing returns-to-scale. This reflects the exponential increase in collaborative interactions as more and more people join the collaboration network. Some criteria, such as availability and security, have decreasing returns-to-scale. This reflects the negative consequences of not achieving at least a baseline performance. Others, such as development cost and development time, have a linear return-to-scale.

The requirements analysis also exposed several constraints for the system. The total development budget, including software and development effort is $250,000. The system has to operate on the United States Military Academy IT network. To support an upcoming ABET accreditation visit, the system must achieve course-level functionality by December 2006 and program assessment functionality by June 2007.

The revised problem statement is as follows:

The United States Military Academy Department of Systems Engineering will develop an integrated curriculum management system to support the following core functions: develop programs, develop courses, execute courses, integrate department academic operations, and assess programs. In order to support these functions, the system requires capabilities for information sharing, collaboration, usability, security, efficient development and maintenance, and integration with other IT systems. In order to meet accreditation time lines, the system must fully support course-level functionality by December 2006 and assessment functionality by June 2007. The department Chief Information Officer will lead development using a combination of internal IT support, capstone students, and contractor support as required. The total cost of system development shall not exceed $250,000.

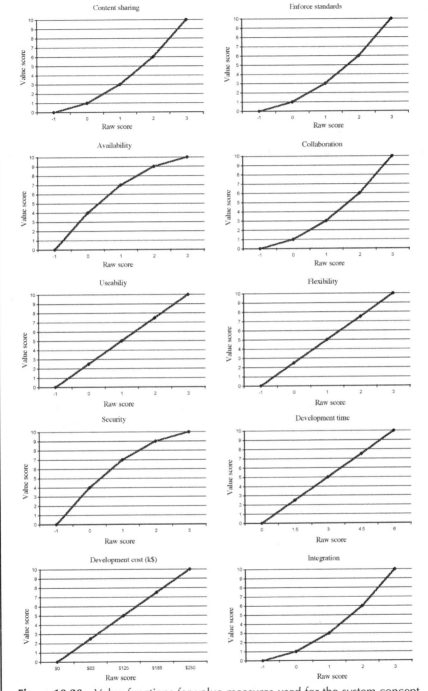

Figure 10.20 Value functions for value measures used for the system concept decision.

10.7   EXERCISES

10.1. Define the three main tasks of the problem definition phase of the systems decision process.

10.2. Compare and contrast the three main techniques for stakeholder analysis. Which technique would you use primarily in designing a new cell phone?

10.3. In Chapter 7 you developed a list of functions for a clock. Build a functional hierarchy for this clock.

10.4. Compare and contrast the four main techniques presented for functional analysis.

10.5. Use Microsoft® Visio to draw an input–output diagram using the IDEF0 framework from Figure 10.6 for the function "launch rocket" from our rocket example.

10.6. Classify value measures 2.1.1, 2.2.1, and 3.1.1 from our rocket example (Figure 10.8). Discuss your rationale for each classification.

10.7. Develop one additional value measure for the rocket example. Define the measure, classify it, and develop a value function for this new measure. Create your function and curve in Microsoft® Excel.

10.8. Create a new swing weight matrix for the rocket example incorporating your additional value measure created for Exercise 9.7. Develop new weights for all the value measures using this matrix.

10.9. Generate a creative candidate solution to the rocket example problem that meets the screening criteria listed in Section 10.2.2. Create quantitative parameters of this candidate solution that can be evaluated with the seven value measures developed in Section 10.4.

10.10. Generate a total value (Equation 10.1) for your candidate solution from Exercise 10.9 using the quantitative value model developed in Section 10.3.

10.11. Write a problem statement that describes your decision problem of whether to pursue graduate education or not in the future. Define the stakeholders for your decision.

10.12. Perform all three tasks of the Problem Definition phase for the purchase decision for your next car.

REFERENCES

1. Shaara, M. *The Killer Angels*. New York: Ballantine Books, 1974.
2. Reid, RD, Sanders NR. *Operations Management: An Integrated Approach*, 2nd ed. New York: John Wiley & Sons, 2005.
3. Old Computers.com. IBM PC Junior. Available at http://www.old-computers.com.
4. Sage, AP, Rouse, WB. *Handbook of Systems Engineering and Management*. New York: John Wiley & Sons, 1999.

5. Information & Design. What is Affinity Diagramming? Available at http://www.info design.com.au/usabilityresources/general/affinitydiagramming.asp. Accessed June 16, 2006.

6. Wymore, AW. *Model-Based Systems Engineering*. New York: CRC Press, 1993.

7. Bahill, AT, Gissing, B. Re-evaluating systems engineering concepts using systems thinking. *IEEE Transactions on Systems, Man, and Cybernetics—Part C: Applications and Reviews*, 1998; 28(4): 516–527.

8. International Council on Systems Engineering (INCOSE). Plowman's Model of the Systems Engineering Process. INCOSE Guide to the Systems Engineering Body of Knowledge, G2SEBoK. Available at http://g2sebok.incose.org/.

9. Sage, AP, Armstrong, JE, Jr. *Introduction to Systems Engineering*. New York: John Wiley & Sons, 2000.

10. Proctor, C. What are Focus Groups? Section on Survey Research Methods, American Statistical Association, 1998. Available at http://www.surveyguy.com/docs/surveyfocus.pdf.

11. Greenbaum, TL. Focus Groups: A help or a waste of time ? *Product Management Today*, 1997; 8(7): 00–00 Available at http://www.groupsplus.com/pages/pmt0797.htm.

12. McNamara, C. Basics of Conducting Focus Groups. Free Management Library, 1999. Available at http://www.managementhelp.org/evaluatn/focusgrp.htm.

13. Group Systems Corporation. Collaborative Thinking and Virtual Meetings: The New Way to Work! Available at http://www.groupsystems.com/page.php?pname=home.

14. Creative Research Systems. The Survey System—Survey Design, 2005. Available at http://www.Surveysystem.com/sdesign.htm.

15. Creative Research Systems. The Survey System—Sample Size Calculator, 2005. Available at http://www.Surveysystem.com.

16. SurveyMonkey.com. Available at http://surveymonkey.com/home.asp?Rnd=0.5588495.

17. Insiteful Surveys. Available at http://insitefulsurveys.com/.

18. Creative Research Systems. The Survey System. Available at http://www.Surveysystem.com.

19. INCOSE-TP-2003-016-02. 2004. *INCOSE SE Handbook*, Version 2a.

20. Draft Federal Information Processing Standards Publication 183, December, 21, 1993, Standard for Integration Definition For Function Modeling (IDEF0). Available at http://www.itl.nist.gov/fipspubs/idef02.doc. Accessed June 2, 2006.

21. Martin, JN. *Systems Engineering Guidebook: A Process for Developing Systems and Products*. Boca Raton, FL: CRC Press, 1997.

22. Affinity Diagrams, Basic Tools for Process Improvement. Available at http://www.saferpak.com/affinity_articles/howto_affinity.pdf. Accessed June 1, 2006.

23. SkyMark Corporation. Available at http://www.skymark.com/resources/tools/affinity_diagram.asp. Accessed June 1, 2006.

24. Lempke, J. *Microsoft Visio 2003, Step-by-step*. Redmond, Washington: Microsoft Press, 2003. Available at http://office.microsoft.com/visio. Accessed June 2, 2006.

25. Buede, DM. *The Engineering Design of Systems: Models and Methods*. Wiley Series in Systems Engineering. New York: Wiley-Interscience, 2000.

26. Wasson, CS. *System Analysis, Design, and Development: Concepts, Principles, and Practices*. Hoboken, NJ: John Wiley & Sons, 2006.

27. Kossiakoff, A, Sweet, NS. *Systems Engineering: Principles and Practice*. Hoboken, NJ: John Wiley & Sons, 2003.

28. CORE Software, Vitech Inc. http:\\www.vitechcorp.com Information about the CORE® Software from Vitech, Incorporated is available online.

29. Parnell, GS. Value-focused thinking. In: Loerch, AG, and Rainey, LB, editors. *Methods for Conducting Military Operational Analysis* Military Operations Research Society, 2007. pp. 619–656.

30. Kirkwood, CW. *Strategic Decision Making: Multiple Objective Decision Analysis with Spreadsheets*. Pacific Grove, CA: Duxbury Press, 1997.

31. Ewing, P, Tarantino, W, Parnell, G. *Use of decision analysis in the army base realignment and closure (BRAC) 2005 military value analysis. Decision Analysis Journal* 2006; (1)13: 33–49.

32. Trainor, T, Parnell, G, Kwinn, B, Brence, J, and Tollefson, E, Downes, P. The US Army uses decision analysis in designing its installation regions. *Interfaces*, 2007; 37(3): 253–264.

33. Clemen, RT, Reilly, T. *Making Hard Decisions with Decision Tools Suite*. Pacific Grove, CA: Duxbury Press, 2004.

34. International Council on Systems Engineering (INCOSE), State the problem. INCOSE Guide to the Systems Engineering Body of Knowledge, G2SEBoK. Available at http://g2sebok.incose.org/. Accessed June 16, 2006.

35. Fowler, FJ, Jr. *Improving Survey Questions—Design and Evaluation*. Applied Social Research Methods Series, Volume 38. Thousand Oaks, CA: Sage Publications, 1995.

36. Parnell, G. and Trainor, T., *"Using the Swing Weight Matrix to Weight Multiple Objectives."*. *Proceedings of the INCOSE International Symposium*, Singapore, July 19–23, 2009

Chapter 11

Solution Design

PAUL D. WEST, Ph.D.

We can't solve problems by using the same kind of thinking we used to create them.
—Albert Einstein

You can observe a lot just by watching.

—Yogi Berra

11.1 INTRODUCTION TO SOLUTION DESIGN

The solution to any well-formed problem exists, needing only to be found. Once the decision maker's needs, wants, and desires have been identified, the solution design process develops a pool of candidates in which the "best" solution can be expected to be found. The pool is refined as it grows, checked constantly against the problem definition and measured against stakeholder objectives until the best solution emerges. The process is fluid and may iterate often across the spectrum of define, design, and decide loops, as shown in Figure 11.1.

The resulting *solution* is a process for solving a problem [1]. Given the stakeholders' needs, it is a process of determining candidate *solutions* to present to the decision maker. These present the decision maker with a choice limited to one of two or more possibilities [2]. Each carries with it some degree of uncertainty and risk, so the solution must include techniques to reduce these to make the decision maker's job easier.

Decision Making in Systems Engineering and Management, Second Edition
Edited by Gregory S. Parnell, Patrick J. Driscoll, Dale L. Henderson
Copyright © 2011 John Wiley & Sons, Inc.

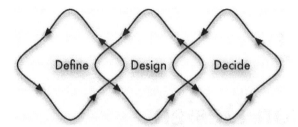

Figure 11.1   Define, design, decide loops.

Solution design is a deliberate process for composing a set of feasible alternatives for consideration by a decision maker.

In this chapter we explore techniques for casting a sufficiently broad net to confidently capture the best solution in a raw form, reduce the catch to most likely candidates, and present a compelling argument for how each candidate solution meets the needs of the decision maker.

Identifying *who* contributes to the solution design is a critical task of the systems engineer, as was seen in Chapter 7. Figure 11.2 shows a part of the Solution Design Concept Map that illustrates the role of the solution designer. The lead systems engineer must identify early on the best mix of skills needed to accomplish the tasks shown here (see Section 7.4).

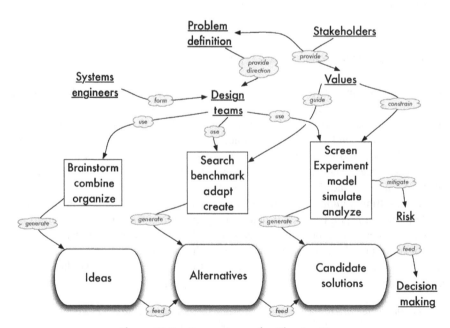

Figure 11.2   Concept map for Chapter 11.

11.2   SURVEY OF IDEA GENERATION TECHNIQUES

Albert Einstein offers solution designers good advice in the opening quote of this chapter. Innovative thinking is essential for successful solution design, but should not be confused with haphazard design, which more often leads to observations like Thomas Edison's that "I have not failed. I've just found 10,000 ways that won't work."

There is no single right way to generate alternatives within a solution design. Much of the art of solution design is in knowing which tools to draw from the toolkit for a specific problem. However, all methods follow the basic model shown in Figure 11.3. The circular arrows indicate feedback and iteration loops, as necessary.

11.2.1   Brainstorming

Brainstorming capitalizes on the idea that a panel of creative thinkers can create a pool of ideas that will include the nucleus of a solution. It adopts the adage that two heads are better than one, or in what early operations researchers proclaimed as the *n-heads rule* [3]: *n heads are better than n heads* -1. The sole product of a brainstorming session is a list of ideas. No attempt at analysis is made except to reduce the total list of ideas to general categories by using such techniques as affinity diagramming, described later in this chapter.

The brainstorming concept was developed by advertising executive Alex Osborn in 1941, who observed that conventional business meetings were conducted in a way that stifled the free and creative voicing of ideas. The term *brainstorming* evolved from his original term to *think up* for his new process [4]. In a series of later writings, he described brainstorming as "a conference technique by which a group attempts to find a solution for a specific problem by amassing all the ideas spontaneously by its members." Osborn became a prolific writer on the Creative Problem Solving process in works including *Your Creative Power: How to Use Imagination* [5], in which he introduced the brainstorming term,*Wake Up Your Mind: 101 Ways to Develop Creativeness* [6], and *Applied Imagination: Principles and Procedures of Creative Thinking* [7].

Osborn developed the following four basic rules for guiding brainstorming sessions:

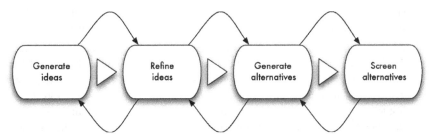

Figure 11.3   The ideation process.

No Criticism of Ideas. All judgment on ideas is deferred and participants feel free to voice all ideas without fear of judgment of the idea or themselves.

Encourage Wild and Exaggerated Ideas. Wild and crazy ideas often contain nuggets of unique insight that may be built upon. Wild ideas will be refined and filtered during later concept evaluation.

Seek Large Quantities of Ideas. Brainstorming sessions are fast-paced and spontaneous. Ideas may not only be useful in their own right, but also serve as catalysts for other ideas.

Build on Each Other's Ideas. Ideas offered from one person's perspective often trigger wholly new ideas from others.

Brainstorming sessions generally take one or more of three forms: *structured, unstructured* (free form), or *silent*. The main difference between the first two forms involves the level of control by the facilitator, who may combine any of the three to meet problem needs.

A typical brainstorming session has six steps and lasts for about an hour. Care is needed in selecting the right mix of participants (see Section 10.2). Groups generally consist of 5–10 participants known for their creativity and initiative and represent a broad range of experience with the problem. A structured session may take the following form [8]:

1. *Problem Definition.* The problem statement should be simple, often a single sentence, and identified in advance. The problem should be more specific than general, and complex problems should be split into functional subproblems that can be addressed in multiple sessions. Osborn recommends that a one-page background memo should accompany the invitation to participants, along with some sample ideas.

2. *Organization.* Seating participants around a round table is ideal for promoting interaction and for enabling the fast pace of the actual session. Distracters such as food and books should be avoided.

3. *Introduction.* Facilitators introduce the problem statement and context. The process is explained and any questions on the problem or process are answered.

4. *Warm-up.* A five-minute warm-up with sample problems may help jump-start the session to a high-tempo level. Sample questions may include "What if restaurants had no food?" and "What if pens wrote by themselves?" [9].

5. *Brainstorming.* Restate the problem and begin. Limit the session to 20–30 min. Pass a My Turn object around the table, with a 10-second time limit for the holder to offer an idea before passing it to the next person. The idea is recorded on a 3-by-5 index card or Post-It® -like note by the holder, a recorder, or facilitator. Ideas should not exceed about 10 words. If the stated idea spawns one from another member of the group, the turn goes

out of sequence to the second person; if the holder cannot think of an idea, the My Turn object moves to the next person in sequence. Trends toward an exhaustion of ideas may be broken by radical ideas interjected by the facilitator. The session ends when time expires or no more ideas are offered. When the session is complete, idea cards are grouped into categories using affinity diagramming and central ideas are identified.

6. *Wrap-up.* Conduct a brief after-action review to identify strengths and weaknesses from the session. Use these comments to describe the environment in which ideas were made during the solution design process and how to improve future sessions.

Unstructured sessions use a more free-form approach, abandoning the My Turn object. Participants contribute ideas as they come to mind.

Brainstorming critics soon emerged. By the 1950s, detractors declared that the effectiveness of brainstorming had been overstated and the tool simply did not deliver what it promised [10]. Much of the criticism, however, was based on poorly planned or conducted sessions, or focused not on brainstorming itself, but on the use of individuals or groups in brainstorming, as was the case of the report that became known as the Yale Study done in 1958 [11].

Critics generally agree on seven main disadvantages of Osborn's process, summarized by Sage and Armstrong [12] below. Several of these factors contribute to what Yale social psychologist Irving Janis called *groupthink*, when group members' striving for unanimity overrides their motivation to realistically appraise alternative courses of action [13].

Misinformation. There is no guarantee that incorrect information will not influence the process.

Social Pressure. A group may pressure individuals to agree with the majority, whether they hold the majority view or not.

Vocal Majority. A strong, loud, vocal majority may overwhelm the group with the number of comments.

Agreement Bias. A goal of achieving consensus may outweigh one for reaching a well thought-out conclusion.

Dominant Individual. Active or loud participation, a persuasive personality, or extreme persistence may result in one person dominating the session.

Hidden Agendas. Personal interests may lead participants to use the session to gain support for their cause rather than an objective result.

Premature Solution Focus. The entire group may possess a common bias for a particular solution.

The ensuing decades witnessed numerous variations on Osborn's technique, each seeking to overcome one or more of the identified flaws. Some of these are described below.

11.2.2   Brainwriting

Brainwriting is a form of silent brainstorming that attempts to eliminate the influence of dominant individuals and vocal majorities. The two most popular variations are written forms of the structured and unstructured brainstorming techniques.

- In structured brainwriting, ideas related to the given problem are written on a paper that is passed from member to member in much the same way as the My Turn object. New ideas may be added or built from others.
- Unstructured brainwriting does not follow a sequential path. Ideas are written on note cards and collected in a central location.

11.2.3   Affinity Diagramming

This technique was introduced in Chapter 10 and categorizes the ideas into groups that can then be rated or prioritized. This is also a silent process, with participants creating logical groups of ideas with natural affinity. Duplicates are first removed from the collection, and notes are stuck to a wall or large board so that all participants can see them. Without speaking, participants move notes into unnamed groups using logical associations until everyone is satisfied with the organization. This often results in notes being moved to different clusters many times until the groups reach equilibrium. The groups are then labeled with descriptive headers, and often form the basis of alternatives, with the header as the alternative name and the member ideas as attributes.

11.2.4   Delphi

Delphi methods seek to minimize the biasing effects of dominant individuals, irrelevant or misleading information, and group pressure to conform. Delphi introduces three unique variations on the brainstorming concept [14].

1. Anonymous response—where opinions are gathered through formal questionnaires.
2. Iteration and controlled feedback—processing information between rounds.
3. Statistical group response—with group opinion defined as an aggregate of individual opinions on the final round.

Delphi methods also differ from traditional brainstorming techniques in their stricter reliance on subject matter experts. Development of the tool began in the mid-1940s shortly after General Henry H. "Hap" Arnold, Commanding General of the Army Air Forces, pushed for the creation of a Research and Development (RAND) project within the Douglas Aircraft Company of Santa Monica, California. Arnold, a 1907 graduate of West Point, who was taught to fly in 1911 by Orville Wright and rose to lead the Army Air Forces during World War II, wanted expert forecasts of technologies reaching decades into the future.

The Air Force's Project RAND, which became the RAND Corporation in 1948, developed a statistical treatment of individual opinions [15], and researchers evolved the process that became known as Delphi. By 1953, Dalkey and Helmer had introduced the idea of controlled feedback [14]. Their project was designed to estimate, from a Soviet strategic planner's view, the number of atomic bombs necessary to reduce United States' munition output by a given amount.

Sessions typically follow 10 basic steps, described by Fowles [16]:

1. Form the team to undertake and monitor the session.
2. Select one or more panels of experts to participate.
3. Develop the first round questionnaire.
4. Test the questionnaire for proper wording.
5. Submit the first questionnaire to the panelists.
6. Analyze the first round responses.
7. Prepare the second round questionnaire.
8. Submit the second questionnaire.
9. Analyze the second round responses. Repeat steps 7 through 9 as necessary.
10. Prepare the report on the findings.

The 10 steps highlight another benefit of the approach in that panelists most likely are not assembled in a central location. However, this also suggests a danger inherent in this process. A Delphi session in which there is insufficient buy-in by participants or is poorly conceived or executed may suffer from participant frustration or lack of focus. Other critics of this technique note that the aggregate opinion of experts may be construed as fact rather than opinion, facilitators may influence responses by their selection and filtering of questions, and future forecasts may not account for interactions of other, possibly yet unknown, factors.

Selecting the right questions is key for Delphi success. The sidebar, *In Case of Nuclear War...*, extracted from Dalkey and Helmer's reflections on applications of the Delphi method [3], shows questions from the first two of five rounds of their 1953 study. Panelists included four economists, a physical vulnerability expert, a systems analyst (precursor of the systems engineer), and an electronics engineer. The paper was published 10 years after the event for security reasons, and even then parts remained classified and were omitted.

IN CASE OF NUCLEAR WAR—A DELPHI STUDY

Questionnaire 1: This is part of a continuing study to arrive at improved methods of making use of the opinions of experts regarding uncertain events. The particular problem to be studied in this experiment is concerned with the effects of strategic bombing on industrial parts in the United States.

Please do not discuss this study with others while this experiment is in progress, especially not with the other subject experts. You are at liberty, though, to consult whatever data you feel might help you in forming an opinion.

The problem with which we will be concerned is the following:

Let us assume that a war between the United States and the Soviet Union breaks out on 1 July 1953. Assume also that the rate of our total military production (defined as munitions output plus investment) at that time is 100 billion dollars and that, on the assumption of no damage to our industry, under mobilization it would rise to 150 billion dollars by 1 July 1954 and to 200 billion dollars by 1 July 1955, resulting in a cumulative production over that two-year period to 300 billion dollars. Now assume further that the enemy, during the first month of the war (and only during that period), carries out a strategic A-bombing campaign against U.S. industrial targets, employing 20-KT bombs. Within each industry selected by the enemy for bombardment, assume that the bombs delivered *on target* succeed in hitting always the most important targets in that industry. What is the least number of bombs that will have to be delivered on target for which you would estimate the chances to be even that the cumulative munitions output (exclusive of investment) during the two-year period under consideration would be held to no more than one quarter of what it otherwise would have been?

Questionnaire 2: As the result of the first round of interviews, it appears that the problem for which we are trying with your help to arrive at an estimated answer breaks down in the following manner.

There seem to be four major items to be taken into consideration, namely:

a. The vulnerability of various potential target systems,
b. The recuperability of various industries and combinations of industries,
c. The expected initial stockpiles and inventories, and
d. Complementarities among industries.

Taking all these into account, we have to:

1. Determine the optimal target system for reducing munitions output to one fourth, and
2. Estimate for this target system the minimum number of bombs on target required to create 50 percent confidence of accomplishing that aim.

We would like to establish the background material consisting of A, B, C, D more firmly. With regard to A and B, the interviews have suggested the following tentative breakdown of possibly relevant factors: two lists of factors were given, related to vulnerability and recuperability, respectively.

11.2.5 Groupware

This is increasingly popular for conducting brainstorming and other collaboration using networked computers. Applications such as GroupSystems [37] mirror the brainstorming process and emphasize anonymity. Other groupware tools include e-mail, newsgroups and mailing lists, workflow systems, group calendars, server-based shared documents, shared whiteboards, teleconferencing and video teleconferencing, chat systems, and multiplayer games.

11.2.6 Lateral and Parallel Thinking and Six Thinking Hats

These are unique creativity techniques for thinking about problems differently. "You cannot dig a hole in a different place by digging the same hole deeper," says concept developer Edward DeBono [17]. He insists that basic assumptions be questioned in lateral thinking. For example, he notes that chess is a game played with a given set of pieces, but challenges the rationale for those pieces. Lateral thinking, he says, is concerned not with playing with the existing pieces, but with seeking to change those very pieces. Parallel thinking [18], he says, focuses on laying out arguments along parallel lines instead of adversarial. Techniques for accomplishing this are described in his Six Thinking Hats approach. In these sessions, participants display or wear one of six colored hats that determine the person's role in the session.

- *White hat*: neutral, focused on ensuring that the right information is available.
- *Red hat*: intuitive, applies feelings, hunches, emotion, and intuition.
- *Black hat*: cautious, provides critical judgment and says why things cannot be done.
- *Yellow hat*: optimist, finds ways that things can be done, looks for benefits.
- *Green hat*: creative, encourages creative effort, new ideas, and alternative solutions.
- *Blue hat*: strategic, organizes and controls the session.

11.2.7 Morphology

Morphology explores combinations of components through the study of forms and structure. This approach, called morphological analysis, was developed by astrophysicist and aerospace scientist Fritz Zwicky in the mid-1960s for studying multidimensional, nonquantifiable complex problems [19]. Zwicky, a scientist at the California Institute of Technology (CalTech), wanted to find a way to investigate the total set of possible relationships between system components, then to reduce the total set to a feasible solution space. He summarized the process in five steps:

1. Concisely formulate the problem to be solved.
2. Localize all parameters that might be important for the solution.
3. Construct a multidimensional matrix containing all possible solutions.

4. Assess all solutions against the purposes to be achieved.

5. Select suitable solutions for application or iterative morphological study.

Zwicky proposed finding "complete, systematic field coverage" by constructing a matrix with all possible system attributes, then producing a feasible set by eliminating combined attributes that are inconsistent with the stated purposes of the system. These inconsistencies could be:

- Logically inconsistent, such as "achieve Earth orbit in an underwater environment."
- Empirically inconsistent, based on improbability or implausibility, such as "build an aircraft carrier using personal savings."
- Normatively inconsistent, based on moral, ethical, or political grounds.

Solution designers should focus initially on the first two inconsistencies, Zwicky says, but they cannot ignore the third, which includes legal and regulatory constraints. The U.S. Department of Defense (DOD), for example, publishes DOD Directive 5000.1 and DOD Instruction 5000.2, specifying processes for the Defense Acquisition System.

Although Zwicky's model is actually an n-dimensional matrix, his original illustration, shown in Figure 11.4, makes clear why it has become known as Zwicky's Morphological Box or simply Zwicky's Box. This figure shows a $5 \times 5 \times 3$ matrix containing 75 cells, with each cell containing one possible system configuration. Continuing the rocket design problem from Chapter 10, solution designers might develop the 4×4 Zwicky Box shown in Figure 11.4 to conduct a morphological analysis. Columns are labeled with parameter names and rows with variable names. The matrix is filled in with possible values.

Figure 11.5 shows an alternative generation table for the rocket problem. The systems engineers have decided four system design decisions: the number of fins,

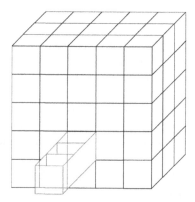

Figure 11.4 Zwicky's morphological box [19].

Solution design parameters			
Fins	**Thrust**	**Seeker**	**Guidance**
2	1000	Forward looking infared	Inertial
3	1667	Laser	Global positioning system
4	2334	Audio	Wire
5	3000	None	Optical

	Solution design parameters			
Strategy	**Fins**	**Thrust**	**Seeker**	**Guidance**
Global Lightening	5	2334	Laser	Global positioning system
Hot Wired	4	3000	None	Wire
Sight and Sound	2	3000	Audio	Optical
Slow poke	3	1000	Forward looking infared	Inertial
Star Cluster	3	1667	Forward looking infared	Global positioning system

morphological box   generated alternatives

Figure 11.5   From morphological box to alternatives.

the thrust, the seeker, and the type of guidance. On the basis of the number of options in each column (5, 4, 4, and 4), 320 alternatives can be developed. However, all of these may not be feasible. Systems engineers select system alternatives that span the design space. Five designs are shown in Figure 11.5.

11.2.8   Ends–Means Chains

Thinking about how means support goals challenges solution designers to think creatively by presenting a higher-level objective and seeking different means for achieving it. This is repeated for means—seeking means to achieve previously stated means—to generate lower-level ideas. In this process, described by Keeney as a means–ends objectives network [20], lower-level objectives answer the question of how higher level objectives can be achieved. This differs from a fundamental objectives hierarchy (used in Chapter 10), he says, in which lower level objectives identify aspects of higher level objectives that are important. The process of finding new ways to achieve successive levels of objectives assists solution designers in structuring alternatives. Such a chain for the rocket design is shown in Figure 11.6. Arrows in this figure indicate the direction of influence.

11.2.9   Existing or New Options

Not all ideas must be completely new. Choices for using new or existing options are summarized by Athey [21] in Table 11.1, focusing on a problem to buy or upgrade a residence.

11.2.10   Other Ideation Techniques

Other techniques are cited in MacCrimmon and Wagner's presentation on Supporting Problem Formulation and Alternative Generation in Managerial Decision Making [22]. These include the following:

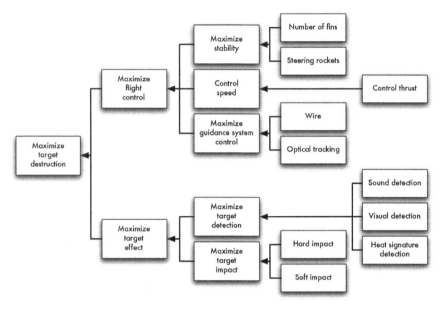

Figure 11.6   Rocket design objectives structure.

TABLE 11.1   Housing Upgrade Example

Method	Housing Example
Existing system	Present house—three bedrooms, new rug, new dishwasher.
Modified existing system	Remodel present house. Add-on a fourth bedroom. Repaint outside.
Prepackaged design	
• Off-the-shelf	• New housing tract.
• Learn from others	• Buy neighbor's four-bedroom house.
• Influential people suggest	• House mother-in-law suggests.
New system design	
• Idealized	• Design custom house.
• Parallel situation	• Build beehive house.
• Morphological	• Consider house, apartments, condo, mobile home.
• Cascading	• Select home by bedroom first, then family room, then backyard, etc.

Metaphoric connections to link disparate contexts by associating the problem with fragments of modern poems.

Loto modifications that alter existing ideas using modifiers such as "make it bigger, make it smaller."

Relational combinations and juxtaposition seek to create new ideas in ways similar to Zwicky's. The first technique applies sentences such as "[Process] by means of [Object 1] *relational word* [Object 2]." The latter technique uses up to three different problem elements at the same time.

11.3 TURNING IDEAS INTO ALTERNATIVES

Converting freewheeling ideas into reasonable alternatives occurs in the conceptual design phase of the systems life cycle. Here, ideas are compared with requirements and constraints and a feasible subset emerges for further analysis. Sage notes that "potential options are identified and then subjected to at least a preliminary evaluation to eliminate clearly unacceptable alternatives" [23]. Surviving alternatives, he adds, are then subjected to more detailed design efforts, and more complete architectures or specifications are obtained.

11.3.1 Alternative Generation Approaches

There is no clear line marking where idea generation ends and alternative screening begins. As seen earlier, this is built into many ideation techniques and the entire process normally cycles through several iterations. While the previous section focused on generating original ideas, practice shows that this represents only a fraction of the tactics used by organizations to uncover alternatives. A study of 376 strategic decisions showed that only about a quarter of the alternatives considered were developed using an original design tactic for generating alternatives [24]. In "A Taxonomy of Strategic Decisions and Tactics for Uncovering Alternatives," Nutt describes a study of 128 private, 83 public, and 165 private nonprofit organizations and the tactics they used to generate alternatives. Organizations ranged from NASA to Toyota and from Hertz–Penske Rental to AT&T. He distilled their tactics into six categories:

1. The *existing idea* tactic draws on a store of fully developed, existing solutions and follows a solution-seeking-a-problem approach with subordinates on the lookout for ways to put their ideas and visions to use.

2. *Benchmarking* tactics draw alternatives from practices of others who are outside of the organization rather than inside, as with the original idea tactic. Nutt draws a distinction between *benchmarking*, which adapts a single practice used by another organization; and,

3. *Integrated benchmarking* uses a collection of ideas from several outside sources.

4. *Search* tactics outsource the process through requests for proposals (RFPs) from vendors, consultants, and others who seem capable of helping.

5. *Cyclical searches* use an interactive approach, while *simple searches* use a one-time RFP.

6. The *design* tactic calls for custom-made alternatives that stress innovation to achieve competitive advantage. These normally require more time and other resource commitment.

Of the decisions studied, either only 36 could not be classified using Nutt's taxonomy, or decision makers switched methods during the process. The remaining 340 decisions are categorized in Table 11.2. Examining these techniques from alternative-focused thinking (AFT) versus value-focused thinking (VFT) perspectives (see Section 9.2) shows that *design* and *search* techniques favor a VFT approach, while *benchmarking, integrated benchmarking*, and *existing idea* techniques use AFT. The *cyclical search* technique could use either or both, depending on circumstances in each iteration.

11.3.2 Feasibility Screening

Feasibility screening techniques were introduced in the discusson of Zwicky's Box in Section 11.2.7. The objective is to reduce the number of alternatives that must be considered by refining, combining, or eliminating those that do not meet critical stakeholder requirements identified during the Problem Definition phase of the SDP (see Chapter 10).

The feasibility screening process can be thought of as a series of increasingly fine screens that filter out alternatives that fail to meet the stakeholders' needs, wants, and desires, as shown in Figure 11.7.

TABLE 11.2 Remaining 340 Decisions

Tactic	Number	Percent	How Alternatives Were Generated
Cyclical search	9	3	Multiple searches in which needs are redefined according to who is available
Integrated benchmarking	21	6	Amalgamation of ideas from several outside sources
Benchmarking	64	19	Adapt a practice used by another organization
Search	69	20	A single search cycle with a decision after RFP responses received
Design	82	24	Develop a custom solution
Existing idea	95	28	Validate and demonstrate benefits of a preexisting idea known in the organization

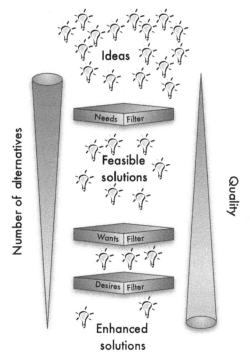

Figure 11.7 Feasibility screening.

- *Needs* are those essential criteria that must exist for the alternative to be considered. These are "must have" requirements identified by stakeholders.
- *Wants* are additional features or specifications that significantly enhance the alternative, but do not cause an alternative to be rejected if missing. These are "should have" requirements identified by stakeholders.
- *Desires* are features that provide a margin of excellence. These are "nice to have" requirements identified by stakeholders.

Solution designers may choose to initially screen at only the *needs* level, or may add screens for *wants* and *desires*.

In all cases, screening is evaluated on a go or no-go basis—an alternative either passes through the screen or it does not. Before rejecting a no-go alternative, however, designers should examine it to see if the offending feature can be modified or deleted so that the otherwise feasible alternative can make it through a repeat screening. The concept of alternative reduction also may be enhanced by combining alternatives or ideas, as described for affinity diagramming in Section 11.2.3. Revisiting the rocket design from Section 11.2.7, assume that the minimum standard for such a rocket is that it must have an effective range of 5 km, be launched from a standard four-wheel drive vehicle, and operate in terrain ranging from open fields to dense forest. A *feasibility screening matrix* with six of the 16 possible

TABLE 11.3   Partial Rocket Design Feasibility Screening Matrix

Alternative	Range Criterion (> 500 km)	Mobility Criterion (> med)	Terrain Criterion (All)	Overall Assessment
Global lightening	Go (800)	Go (high)	Go (All)	Go
Hot wired	Go (600)	Go (high)	Go (All)	Go
Sight and sound	Go (700)	Go (high)	No-Go (limited)	No-Go
Slow poke	Go (550)	Go (very high)	Go (All)	Go
Star cluster	Go (600)	Go (high)	Go (All)	Go
5-fin, 2234# thrust, none, wire	No-Go (400)	Go (high)	No-Go (limited)	No-Go

alternatives identified earlier versus criteria is shown in Table 11.3. Alternatives often take on descriptive names by this point, but in this case we list the parameters of *number of range criterion, mobility criterion, terrain criterion, and overall assessment*. This example shows that four of these alternatives met all of the minimum requirements. A single *no-go* for any criterion triggers an overall *no-go* assessment and the alternative is flagged for revision or rejection. For example, Sight and Sound fails the terrain criterion.

Feasibility screening matrices support alternative generation by quickly eliminating ideas that are clearly not feasible. Recall that two of the four guidelines for brainstorming described in Section 11.2.1 are "encourage wild and exaggerated ideas" and "seek large quantities of ideas." This will naturally lead to many alternatives that are clearly not feasible. For example, the development of an aircraft requires that it must be able to fly. Any alternative that does not meet this fundamental criterion is eliminated during the feasibility screening process.

Screening criteria should avoid targeting feasible but less desirable alternatives. Feasibility screening sifts out alternatives based on non-negotiable criteria. For all others, finding the best trade-offs will lead the preferred solution. For example, a raw value of 90 for variable A may be acceptable when variable B is 30, but not when it is 50. Setting a No-go criterion for A at 90, then, would unduly restrict the solution space.

All feasible alternatives become solution candidates. Balancing tradeoffs to find the preferred solution is accomplished by enhancing and measuring solution candidates.

11.4   ANALYZING CANDIDATE SOLUTION COSTS

Cost constraints are vital considerations at all stages of a system's life cycle. Chapter 5 introduced life cycle costing, cost estimating techniques, and the life cycle stages appropriate for the various techniques. In previous phases in this process, the systems engineer identified the components of the life cycle costs of the system under development. Now, the systems engineer must review these cost components and ensure that they completely cover the candidate solutions' costs.

The systems engineer should know a great deal more about the candidate solutions at this point than at the beginning of the process when the cost components were developed. What has changed? Are there hidden or higher costs, such as development costs, manufacturing costs, or operational costs? Does the cost model consider all components of the newly developed candidate solutions? Once the systems engineer has ensured that the cost model is indeed complete, costs are computed for each candidate solution still under consideration. Although overall costs are needed for the next step, it is important to include all of the components in detail to sufficiently document the work and to answer any questions the decision maker may have about the analysis. It is also useful to note that Monte Carlo simulation can be conducted for the cost model just as it was for the value model.

11.5 IMPROVING CANDIDATE SOLUTIONS

The solution designer, armed with a list of feasible alternatives, must choose which of them to present to the decision maker for action. This choice is based on both qualitative and quantitative measures so that only the best alternatives make the cut to become final *candidate solutions*—the best the solution design team can offer. Key tools at this stage are the models and simulations described in Chapter 4 and a deliberate experimental design strategy for analyzing alternatives.

11.5.1 Modeling Alternatives

Models have been used both directly and indirectly throughout the problem-solving process. Early in the problem definition stage, key stakeholder values were assembled into a value model that identified five key aspects of value:

1. Why is the decision being made (the fundamental objective)?
2. What has a value (functions and objectives)?
3. Where are objectives achieved?
4. When can objectives be achieved?
5. How much value is attained (the value function)?

The value modeling process identified many of the requirements used during feasibility screening. It continues to be used at this stage to assess the value of each feasible alternative as candidate solutions emerge.

Feasible alternatives must also be evaluated using mathematical, physical, or event models developed for the system and assessed using measures of effectiveness or performance, as described in Chapter 4.

11.5.2 Simulating Alternatives

Complex systems with no closed-form solution or that have many interactive variables that change during operation often require simulations to generate effectiveness or performance values. As noted in Chapter 4, a simulation is a model of a system's operation over time.

Models and simulations provide solution designers and decision makers with insights into possible futures of a system given specific conditions. It is important that these conditions are well defined to support the study and its outcomes. In its Defense Acquisition Guidebook [25], the DOD's Defense Acquisition University recommends that modeling and simulation analyses specify the following at the beginning of any study:

- *Ground rules*, including operational scenarios, threats, environment, constraints, and assumptions.
- *Alternative descriptions*, including operations and support concepts.
- *Effectiveness measures*, showing how alternatives will be evaluated.
- *Effectiveness analysis*, including methodology, models, simulations, data, and sensitivity analysis.
- *Cost analysis*, including life cycle cost methodology, models and data, cost sensitivity, and risk analysis.

11.5.3   Design of Experiments

Having identified alternatives that meet stakeholders' needs, wants, and desires, the solution design team must develop a plan for getting the most information about the operation of those alternatives with the least amount of effort. A mathematical process for accomplishing this is the *Design of Experiments* (DOE), developed in the 1920s and 1930s by Ronald A. Fisher [26], a mathematician who worked at the Rothamsted Agricultural Experimentation Station, north of London (after whom the statistical *F-test* was named), wanted to understand the effects of fertilizer, soil, irrigation, and environment on the growth of grain. His technique, published in 1936, examined the main and interaction effects of key variables (factors) in a system design. The development of Fisher's process is chronicled in *Lady Tasting Tea: How Statistics Revolutionized Science in the Twentieth Century* [27].

Concepts   DOE provides solution designers with a way to simultaneously study the individual and interactive effects of many factors, thus keeping the number of experiment iterations (replications) to a minimum. The basic question addressed by DOE is, "What is the average outcome (effect) when a factor is moved from a low level to a higher level?" Since more than one factor is at play in a complex system, efficiencies can be gained by moving combinations of factors simultaneously.

Consider an experiment for the new rocket design developed in this text. The solution design team may have initially determined that the key factors are engine thrust and number of stabilizing fins. They also determined the constraints for each factor: The engine thrust must fall between 1000 and 3000 lb at launch, and competing concepts call for 3-fin and 5-fin designs. The primary measure of effectiveness is "distance from target impact point." In this design, the two factors,

A and B, are shown with both low (−) and high (+) levels, representing the upper and lower bounds on the constraints (Figure 11.8). Four conditions are possible in this design: point a, where both A and B are at their lower levels; point b, where A is low and B is high; point c, where both A and B are high; and d, where A is high and B is low. Table 11.4 shows a design in which each of these points can be tested. Experiments designed this way are called 2^k *factorial designs*, since they are based on factors and each factor has two possible levels. The k exponent refers to the number of factors being considered—in this case, two. This way of looking at designs tells the experimenter how many possible states—or design points—exist, and it also lays the groundwork for determining the main and interactive effects of combining the factors.

The order in which the table is constructed is important for subsequent calculations. Notice that the column labeled Factor 1 alternates between − and +. This will always be the case, no matter how many design points there are. The second column, labeled in this case Factor 2, will always alternate between two minuses, followed by two pluses. Note that this is because the design moves one factor from its low to its high point while holding other factors constant.

Calculating Main and Two-Way Interaction Effects

The *main effect* of a factor is the *average change in the response* (the performance measure) that results from moving a factor from its − to its + level, while holding all other factors fixed.

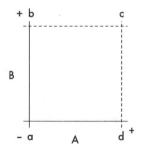

Figure 11.8 Two-factor, two-level design.

TABLE 11.4 A 2 × 2 Design Matrix

Design Point	Factor 1	Factor 2	Response
1	−	−	R_1
2	+	−	R_2
3	−	+	R_3
4	+	+	R_4

The main effect is the effect that a particular factor has, without regard for possible interactions with other factors. In a 2^2 *factorial design*, this is calculated using Equation (11.1) for Factor 1 (e_1).

$$e_1 = \frac{(R_2 - R_1) + (R_4 - R_3)}{2} \qquad (11.1)$$

Figure 11.9 shows the relationship between Table 11.4 and Equation (11.1).

Some designers prefer to use capital letters of the alphabet to label factors, while others use numbers. When letters are used, the *effect* nomenclature is the lowercase equivalent of the factor letter. So, the main effect for factor A would be written e_a.

Assume that tests were run on the experimental rocket design and yielded the results shown in Table 11.5. These results show the distance, in meters, from the desired and actual impact points for rockets built using the given constraints. Using Equation (11.1), the main effects of this rocket design can be calculated as follows:

$$e_1 = \frac{(13.0 - 13.6) + (14.8 - 17.6)}{2} = -1.7 \qquad (11.2)$$

$$e_2 = \frac{(17.6 - 13.6) + (14.8 - 13.0)}{2} = 2.9 \qquad (11.3)$$

This means that by changing from a 3-fin to a 5-fin design, holding thrust constant, there should be an average decrease in error of 1.7 m. Increasing thrust alone

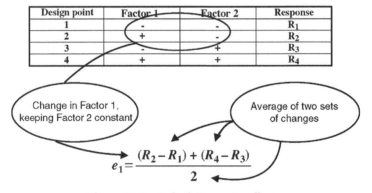

Figure 11.9   Calculating main effects.

TABLE 11.5   Rocket Example Initial Test Results

Design Point	Factor 1	Factor 2	Response
1	−	−	13.6
2	+	−	13.0
3	−	+	17.6
4	+	+	14.8

should result in an average increase in error of 2.9 m. In the equation for e_2, the response for DP 3 is subtracted from that of DP 1, and DP 4 is subtracted from DP 2 before the sum is averaged.

A shortcut that achieves the same mathematical result is to apply the signs of the factor column to the corresponding result (R), sum them, then divide by 2^{k-1}, as shown in Equation (11.4) for calculating the main effect of Factor 2 (e_2).

$$e_2 = \frac{-R_1 - R_2 + R_3 + R_4}{2} \qquad (11.4)$$

This method provides a simpler approach for achieving the same result, as shown below:

$$e_1 = \frac{-13.6 + 13.0 - 17.6 + 14.8}{2} = -1.7 \qquad (11.5)$$

$$e_2 = \frac{-13.6 - 13.0 + 17.6 + 14.8}{2} = 2.9 \qquad (11.6)$$

A strength of DOE is that it also allows the solution design team to understand the *interaction effects* between factors. These effects show the synergistic relationships between factors by measuring how the effect of one factor may depend on the level of another.

> The two-way *interaction effect* of a two-level design is half the difference between the average effects of Factor 1 when Factor 2 is at its + level and when it is at its − level. (All other factors held constant.)

A design matrix with interaction effects is created by adding an interaction column to the standard design. The "sign" values for this column are determined by multiplying the signs of the factors of interest. A matrix for the rocket example that includes the interactions of Factors 1 and 2 (labeled e_{12}) is shown in Table 11.6.

If letters are used to indicate factors, the interaction effect of AB would be written as e_{ab}. Using the definition for interaction effects, the equation for two-way interactions becomes

$$e_{12} = \frac{1}{2}[(R_4 - R_3) - (R_2 - R_1)] \qquad (11.7)$$

$$e_{12} = \frac{1}{2}[(14.8 - 17.6) - (13.0 - 13.6)] = -1.1 \qquad (11.8)$$

TABLE 11.6 DOE Matrix with Interactions

Design Point	Factor 1	Factor 2	1 × 2	Response
1	−	−	+	13.6
2	+	−	−	13.0
3	−	+	−	17.6
4	+	+	+	14.8

This tells the solution designer that by combining both *treatments*, the overall effect is that the rocket hits 1.1 meters closer to the aim point than without the higher-level alternatives. Following the shortcut logic described earlier, the same result can be found by multiplying the signs of the interaction column (1 × 2) by the response, as shown below:

$$e_{12} = \frac{R_1 - R_2 - R_3 + R_4}{2} \tag{11.9}$$

$$e_{12} = \frac{13.6 - 13.0 - 17.6 + 14.8}{2} = -1.1 \tag{11.10}$$

Designs with More Than Two Factors It is rare for complex systems to be dominated by only two factors. The solution designer must carefully balance the complexity of the design with an ability to find meaningful results. The earlier design could be represented as a two-dimensional square with four corners (design points). The addition of a third factor changes the design from a square to a cube, with 2^3, or eight design points, shown in Figure 11.10. The addition of factor C results in the 2^k factorial design with eight design points shown in Table 11.7. Notice that the repetition of minuses and pluses for Factors 1 and 2 follows a similar pattern as with a 2^2 design. If there were four factors, the pattern of Factor 3 would repeat to fill the 2^4 (16) design points, and Factor 4 would have a series of eight minuses, followed by eight pluses, and so on for however many factors that are being considered.

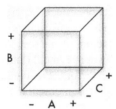

Figure 11.10 Three-factor, two-level design.

TABLE 11.7   2 × 3 Design Matrix

Design point	Factor 1	Factor 2	Factor 3	Response
v1	−	−	−	R1
2	+	−	−	R2
3	−	+	−	R3
4	+	+	−	R4
5	−	−	+	R5
6	+	−	+	R6
7	−	+	+	R7
8	+	+	+	R8

Determining the main and interaction effects of 2^3 designs follows the same logic as for the 2^2 design, although the complexity of the non-shortcut method increases significantly. The two methods for calculating the main effect of Factor 1 described earlier are expanded below to consider three factors.

$$e_1 = \frac{(R_2 - R_1) + (R_4 - R_3) + (R_6 - R_5) + (R_8 - R_7)}{4} \tag{11.11}$$

$$e_1 = \frac{-R_1 + R_2 - R_3 + R_4 - R_5 + R_6 - R_7 + R_8}{4} \tag{11.12}$$

Notice that the denominator is now four, reflecting the four comparisons in the numerator. As before, this value will always be 2^{k-1} for full factorial designs.

When more than three factors are considered, the geometry of the design gets more complex, with the three-way cube becoming part of a larger shape known as hypercube. A four-factor design, for example, can be viewed as two cubes at either end of a line having minus and plus ends that provide for the 16 design points, as shown in Figure 11.11. All 16 design points have unique locations in this geometry. For example, the lower right corner of the right cube has these signs: $A = +, B = -, C = -, D = +$. Similarly, a 2^5 hypercube design has 32 design points and would appear to have cubes on the corners of a 2^2 (square) design.

Blocking and Randomization A major concern facing solution designers as they consider their strategic approach is that of controllable and uncontrollable factors. This is more of a problem in physical experiments than in simulation-based experiments, where all factors are controllable. Physical experiments generally have an initialization period before reaching a steady operational state. Likewise, samples taken from different lots may show significant variation not found within a lot. Two methods for managing these variations are *blocking and randomization*.

A *block* is a portion of the experimental material that is expected to be more homogeneous than the aggregate [28]. *Blocking* takes known, unavoidable sources of variation out of the design picture. For example, a situation in which conditions are expected to be the same at a given time of day could be *blocked on time-of-day*,

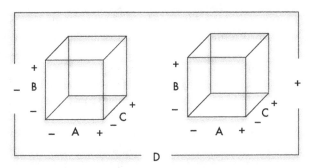

Figure 11.11 Geometry for a 2^4 factorial design.

taking that out of the analysis. Another example might be an experiment comparing hand–eye coordination between dominant and nondominant hands. An approach *blocked on people* would be to select 100 people at random and have each one throw 10 darts at a target using the right hand and 10 with the left. Then compare the difference in results for each person, thus taking individual skill, ability, and similar person-specific variables out of the mix. The result is an experiment isolated on the test issue and avoiding influence by *noise* factors. By confining treatment comparisons with such blocks—say Box, Hunter, and Hunter [28]—greater precision can often be obtained.

The sequence in which trials are run within an experiment must also be planned by the solution design team. A single trial of an experiment that contains random variables is not sufficient to base findings. Concluding that a coin flip will always result in "heads," based on a single flip (trial), is not supportable. "Block what you can and randomize what you cannot" [28] is a useful mantra for the solution designer.

11.5.4   Fractional Factorial Design

Factorial techniques explored so far are reliable ways to identify all main and interaction effects in solution designs when the number of factors is small. But even with increasing computing speed, designs quickly become unmanageable as the number of factors grows. A 2^5 design, for example, has 32 design points, while a 2^7 design has 128. It is not uncommon for designs to be affected by 15 or more factors, which would require 32,768 or more design points. Designs with random variables further increase complexity since they require multiple replications to determine central tendencies. A two-level design with 15 factors, with each design point requiring 10 replications, would require 327,680 separate iterations. Some designers use the terms *replications, runs, and design points* interchangeably. Care must be taken to differentiate between single design points, as used here, and *iterations*, which denote multiple trials to account for randomness.

Fractional factorial designs present two useful solutions to the problem of scale. First, they allow designers to achieve nearly the same results with a fraction of the effort, and second, they provide designers with a tool to further screen out factors that do not make a significant contribution to the outcome.

Fractional designs provide similar results as from a 2^k design, but with 2^{k-p} design points, where p determines what fraction of the total design will be used. We have seen that a full three-factor design requires $2^3 = 8$ design points. If p is set to 1, half as many design points are required, since $2^{3-1} = 4$. This is known as a half-fraction design. A p of 2 produces a quarter-fraction design that would require only a quarter as many design points to achieve nearly the same result.

Having fewer design points does not come without a cost, however. Precision is increasingly lost as the fraction increases. Solution designers must be aware of these tradeoffs and balance the loss of precision with savings in resources (time, material, etc.).

Consider the full factorial design with two levels and four factors shown in Table 11.8. There is one mean effect, four main effects, six two-way effects, four three-way effects, and one four-way effect, for a total of 16 design points.

A half-fraction factorial design promises to provide nearly the same outcome with half as many design points. To do this, start with a full factorial design with $2^{k-p} = 2^{4-1} = 2^3 = 8$ design points, as shown in Table 11.9. Provide for the fourth factor by multiplying the signs of 1, 2, and 3.

The half-fraction matrix directly accounts for eight of the 16 original effects: one mean effect (no contrast between design points—all values are minus), four main effects, and three two-way interaction effects. The "missing" effects—three two-way, four three-way, and one four-way—can be found by examining them separately, shown in Table 11.10.

TABLE 11.8 Full 2^4 Factorial Design

	1	2	3	4	12	13	14	23	24	34	123	124	134	234	1234
DP1	−	−	−	−	+	+	+	+	+	+	−	−	−	−	+
DP2	+	−	−	−	−	−	−	+	+	+	+	+	+	−	−
DP3	−	+	−	−	−	+	+	−	−	+	+	+	−	+	−
DP4	+	+	−	−	+	−	−	−	−	+	−	−	+	+	+
DP5	−	−	+	−	+	−	+	−	+	−	+	−	+	+	−
DP6	+	−	+	−	−	+	−	−	+	−	−	+	−	+	+
DP7	−	+	+	−	−	−	+	+	−	−	−	+	+	−	+
DP8	+	+	+	−	+	+	−	+	−	−	+	−	−	−	−
DP9	−	−	−	+	+	+	−	+	−	−	−	+	+	+	−
DP10	+	−	−	+	−	−	+	+	−	−	+	−	−	+	+
DP11	−	+	−	+	−	+	−	−	+	−	+	−	+	−	+
DP12	+	+	−	+	+	−	+	−	+	−	−	+	−	−	−
DP13	−	−	+	+	+	−	−	−	−	+	+	+	−	−	+
DP14	+	−	+	+	−	+	+	−	−	+	−	−	+	−	−
DP15	−	+	+	+	−	−	−	+	+	+	−	−	−	+	−
DP16	+	+	+	+	+	+	+	+	+	+	+	+	+	+	+

TABLE 11.9 2^{4-1} Design Matrix

	1	2	3	4(123)	12	13	23
DP1	−	−	−	−	+	+	+
DP2	+	−	−	+	−	−	+
DP3	−	+	−	+	−	+	−
DP4	+	+	−	−	+	−	−
DP5	−	−	+	+	+	−	−
DP6	+	−	+	−	−	+	−
DP7	−	+	+	−	−	−	+
DP8	+	+	+	+	+	+	+

TABLE 11.10   The Lost 2^4 Effects

	14	24	34	123	124	134	134	1234
DP1	+	+	+	−	−	−	−	+
DP2	+	−	−	+	−	−	+	+
DP3	−	+	−	+	−	+	−	+
DP4	−	−	+	−	−	+	+	+
DP5	−	−	+	+	+	−	−	+
DP6	−	+	−	−	+	−	+	+
DP7	+	−	−	−	+	+	−	+
DP8	+	+	+	+	+	+	+	+

Notice the relationships of the signs in the factor columns. The sequence for factor 14 (shorthand for 1×4) is the same as that for 23 in Table 11.9. Comparing all the columns provides the data for Figure 11.2. A striking observation in Figure 11.2 is that every effect involving Factor 4 is algebraically the same as another effect not using factor 4. These identical terms are said to be *aliases* of each other. They are also described as being *confounded* by one another, since the effect is calculated exactly the same way and therefore it is impossible to know which effect is producing the result. Notice that Figure 11.12 now accounts for all 16 main and interaction effects.

This suggests that the same result should be found using a 2^{4-1} design as with a full 2^4. This relies on the *sparsity of effects* assumption that states that if an outcome is possible from several events, the less complex event is most likely the cause. So, if faced with two identical outcomes, one resulting from a single main effect and one resulting from the interaction of several effects, this assumption claims that the single main effect is most likely the cause and the complex effect can be ignored. This assumption, though convenient, is a source of loss of precision and must be used with caution.

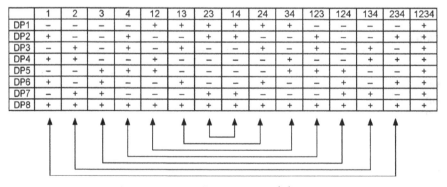

	1	2	3	4	12	13	23	14	24	34	123	124	134	234	1234
DP1	−	−	−	−	+	+	+	+	+	+	−	−	−	−	+
DP2	+	−	−	+	−	−	+	+	−	−	+	−	−	+	+
DP3	−	+	−	+	−	+	−	−	+	−	+	−	+	−	+
DP4	+	+	−	−	+	−	−	−	−	+	−	−	+	+	+
DP5	−	−	+	+	+	−	−	−	+	+	+	−	−	−	+
DP6	+	−	+	−	−	+	−	−	+	−	−	+	−	+	+
DP7	−	+	+	−	−	−	+	+	−	−	−	+	+	−	+
DP8	+	+	+	+	+	+	+	+	+	+	+	+	+	+	+

Figure 11.12   Fully expanded 2^{4-1} matrix.

Another observation from Figure 11.12 is that the interaction effect 1234 is composed of all plus signs. Comparing this table with Table 11.8, it becomes apparent that the fractional matrix is identical to the full matrix design points 1, 10, 11, 4, 13, 6, 7, and 16. The annotated hypercube in Figure 11.13 confirms that the fractional design symmetrically targets half of the possible points of a full design. The corners not selected in Figure 11.13 represent the half fraction where factor 1234 is a column of all minus signs. Either half could be used, although the one shown uses the principal fraction (all pluses) and is most common. The column with all plus signs is also known as the *identity* column, and it has a number of unique attributes. Most obvious, it is the only column that does not have an equal number of plus and minus signs. But more importantly, it reflects the fact that any sign multiplied by itself results in a positive value. In the fractional design above, Factor 4 was determined by multiplying factors 1, 2, and 3. Therefore, $4 = 123$, which produced the identity column I, which equals 1234, or $I = 1234$. This is called the *design generator*, because it can be used to generate the entire aliasing structure for the design, eliminating the need for pattern-match tables of pluses and minuses.

To find the alias for a factor, simply multiply the factor times I. Finding the alias for factor 2 in the previous design, for example, is accomplished as shown in steps below:

- Step 1: $I = 1234$, establish the design generator
- Step 2: $I = 2 \times 2$
- Step 3: $2I = (2)1234$
- Step 4: $2(2 \times 2) = (2 \times 2)134$
- Step 5: $2 = 134$, which is verified in Figure 11.12

A simplified rule of thumb for finding aliases is to drop the column in question from the *generating relation*. Therefore, in this example, $1 = 234$; $2 = 134$; $3 = 124$; $12 = 34$; and so forth. Higher-order fractions (quarter, eighth, etc.) have more than one generator, but the fundamental process is the same.

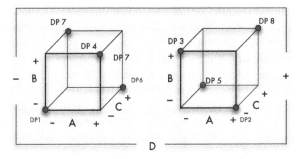

Figure 11.13 2^{4-1} design plot (principal fraction).

Calculating main and interaction effects in fractional factorial designs is accomplished in the same way as with full factorial designs, except that the denominator is 2^{k-p-1} instead of 2^{k-1}. The main effect of Factor 1 for the half-fraction design shown above is developed in Equation (11.14).

$$e_1 = \frac{-R_1 + R_2 - R_3 + R_4 - R_5 + R_6 - R_7 + R_8}{2^{4-1-1}} \tag{11.13}$$

$$= \frac{-R_1 + R_2 - R_3 + R_4 - R_5 + R_6 - R_7 + R_8}{2^4} \tag{11.14}$$

Given a choice of conducting a full factorial experiment with many design points or a fractional design with many fewer, why would a solution designer consider anything but the one with the fewest? The answer is in the balancing of resources available and the loss of precision inherent in fractional designs. The key to achieving that balance lies in the concept of design *resolution*.

> Design resolution is a measure of the degree of confounding that exists in a fractional factorial design.

Generally, and with notable exceptions, the highest resolution design the experimenter can afford to conduct is the preferred choice. Screening designs, discussed later, are the main exceptions to the rule.

The level of resolution is shown using roman numerals to distinguish it from other numbers in the design notation. The numeral indicates the level at which fractional designs are clear of confounding. Two effects are not confounded if the sum of their *ways* is less than the resolution of the design. Main effects are considered one-way effects, while interaction effects such as 12 and 34 are two-way, 134 and 234 are three-way, and so on. In a resolution IV design, no main effects are aliased with any two-way effect ($1 + 2 < 4$), but two-way effects are aliased with other two-way or higher order effects. ($2 + 2$ or more < 4).

Knowing that higher-resolution designs have less confounding and therefore less precision loss, yet will require more design points, the solution designer can balance available resources against precision in constructing an experimental design. It is also important that the designer indicate this tradeoff in discussing the design. The notation for fractional factorial designs is shown in Figure 11.14.

Tables for quickly determining which designs best meet the resolution criteria exist in much of the DOE literature, including *Statistics for Experimenters* [28] and *Simulation Modeling and Analysis* [29]. The table of designs where $k \leq 7$, shown in Table 11.11, illustrates how greater confounding occurs at lower resolutions.

Lower-resolution designs are particularly useful as *screening designs* to eliminate factors that do not contribute significantly to the system's operation. Caution and experience are necessary to prevent too low of a resolution from being selected and a resulting greater loss of precision leading to faulty conclusions.

While feasibility screening, described in Section 11.3.2, provides a *criteria-based* tool for narrowing alternatives, factorial screening provides a *merit-based*

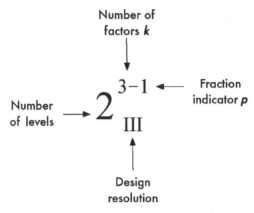

Figure 11.14 Fractional factorial design notation.

TABLE 11.11 Confounding in Fractional Factorial Design

		Resolution	
k	III	IV	V
3	$2_{III}^{3-1} \rightarrow 3 = \pm12$		
4		$2_{IV}^{4-1} \rightarrow 4 = \pm123$	
5	$2_{III}^{5-2} \rightarrow 4 = \pm12$ $5 = \pm13$		$2_{V}^{5-1} \rightarrow 5 = \pm1234$
6	$2_{III}^{6-3} \rightarrow 4 = \pm12$ $5 = \pm13$ $6 = \pm23$	$2_{IV}^{6-2} \rightarrow 5 = \pm123$ $6 = \pm234$	
7	$2_{III}^{7-4} \rightarrow 4 = \pm12$ $5 = \pm13$ $6 = \pm23$ $7 = \pm123$	$2_{IV}^{7-3} \rightarrow 5 = \pm123$ $6 = \pm234$ $7 = \pm134$	

tool to further refine the solution space. Revisiting the rocket scenario described earlier, the solution designers' feasibility screening matrix shows that five factors pass the initial test, as shown in Table 11.12.

A full factorial design is developed and tested, with the response being a score of 0–100 where 100 is a bull's-eye target hit. The results are shown in Table 11.13.

Several statistical software packages, such as Minitab®, allow solution designers to construct and analyze factorial designs. Analysis of the responses reveals that only the *seeker type, guidance system*, and *thrust* main effects and the interactions of *seeker-guidance* and *guidance-thrust* have a statistically significant effect on system performance, given an alpha value of 0.05.

Skin type and *number of fins* did not make a significant difference in rocket performance, as seen in the Minitab® chart shown in Figure 11.15. The vertical line at 3.12 on the horizontal axis indicates the threshold for significance.

TABLE 11.12 Modified Rocket Design Factors

Factor	Name	Low (−)	High (+)
1	Skin composition	Aluminum	Composite
2	Seeker	Forward-looking infrared	Laser-guided
3	Fins	3	5
4	Guidance	Inertial navigation	Global positioning system
5	Thrust	1000 lb	3000 lb

TABLE 11.13 Rocket Design Test Results

R1 = 61	R2 = 53	R3 = 63	R4 = 61
R5 = 53	R6 = 56	R7 = 54	R8 = 61
R9 = 69	R10 = 61	R11 = 94	R12 = 93
R13 = 66	R14 = 60	R15 = 95	R16 = 98
R21 = 59	R22 = 55	R23 = 67	R24 = 65
R25 = 44	R26 = 45	R27 = 78	R28 = 77
R29 = 49	R30 = 42	R31 = 81	R32 = 82

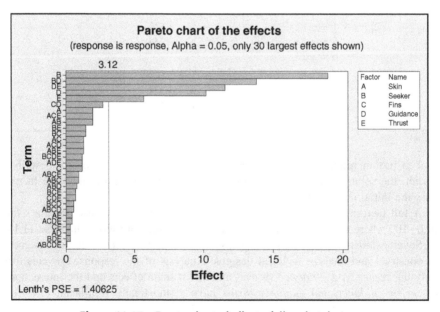

Figure 11.15   Pareto chart of effects, full rocket design.

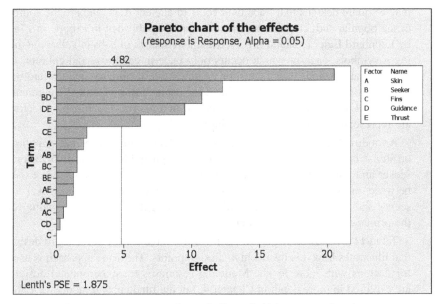

Figure 11.16 Pareto chart of effects, half-fraction rocket design.

A 2_V^{5-1} design should produce essentially the same findings, but with half the physical testing and the associated time and material costs required by the full factorial experiment. Figure 11.16 shows the Minitab® results. Once again, *seeker, guidance*, and *thrust* main effects and *seeker-guidance* and *guidance-thrust* interaction effects are shown to be significant. *Skin type* and *number of fins* could be screened out of the design based on the rocket performance.

Survey of Other Design Strategies Full and fractional factorial designs as described earlier are common, general-purpose approaches to experimental design. However, many others are available to the solution designer, each designed to meet unique situations that may be encountered.

Plackett–Burman designs are two-level, resolution III fractional factorial designs often used for screening when there is a large number of factors and it is assumed that two-way interactions are negligible. Recall that in a resolution III design, main effects are confounded with two-way interactions $(1 + 2 \geq 3)$. Introduced in 1946 [30], it is based on the number of design points being a multiple of four instead of a power of two, as seen earlier. The number of factors must be less than the number of design points, so a 20 DP design can estimate main effects for up to 19 factors. In general, these designs will estimate main effects for n factors using $n + 1$ design points, rounded up to the nearest multiple of four.

Latin Squares designs are used when there are more than two factors and there are assumed to be only negligible interaction effects. They are an offshoot of magic square designs that trace their roots to ancient Asia, and are the source of the popular sudoku puzzles. The concept was published in Europe in 1782 by Leonhard Euler [31] and involved a square matrix of n-by-n cells containing n symbols, each of which occurs only once in each row and column. In Latin Square designs, one factor, or treatment, is of primary interest and the remaining factors are considered blocking, or nuisance, factors. Treatments are given Latin characters and are arranged in a matrix. The blocking factors are represented by the rows and columns.

A modified rocket design to explore the effect of four propellant mixtures on thrust is shown in Figure 11.17. This design is blocked on two factors: seeker and guidance system. It is recognized that slight variations may exists between individual components, and this design will eliminate seeker-to-seeker and guidance-to-guidance differences to get a truer understanding of the primary effect being considered.

Table 11.14 shows the general matrix for a three-level Latin Square design that illustrates how this results in n^2 design points. The same approach is used for designs with more levels. Modeling responses for experimental designs are explored in more depth in Chapter 4, but the fundamental model for Latin

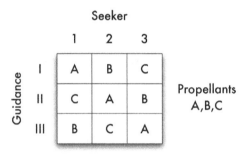

Figure 11.17 Latin Square rocket design focused on thrust.

TABLE 11.14 Three-Level Latin Square Matrix

DP	Row Blocking Factor	Column Blocking Factor	Treatment Factor
1	1	1	1
2	1	2	2
3	1	3	3
4	2	1	3
5	2	2	1
6	2	3	2
7	3	1	2
8	3	2	3
9	3	3	1

Square and related designs follows the form, where Y is the observation, η is the mean, R_i is the row effect, C_j is the column effect, T_k is the treatment effect, and ϵ_{ijk} is random error.

$$T_{ijk} = \eta + R_i + C_j + T_k + \epsilon_{ijk} \tag{11.15}$$

Graeco-Latin and Hyper-Graeco-Latin Square designs are extensions of the Latin Square method for studying more than one treatment simultaneously. Graeco-Latin squares, also introduced by Euler, are constructed by superimposing two Latin Square designs into one matrix and differentiating them by using Greek letters for one of the treatments. This design uses three blocking factors, instead of two. A Hyper-Graeco-Latin Square design considers more than three blocking variables. A Graeco-Latin Square design for a rocket system experiment conducted over three days, the new blocking factor, is shown in Figure 11.18.

Response Surface Method (RSM) designs are used to examine the relationship between a response variable and one or more factors, especially when there is *curvature* between a factor's low and high values, which has been assumed to not exist in designs so far. The full and fractional factorial designs described earlier were assumed to follow a straight line between values. RSM designs reveal the *shape* of a response surface, and are useful in finding *satisficing* or *optimal* process settings as well as weak points.

Figure 11.19, extracted from a Minitab® run of the full factorial design responses from the rocket design, clearly shows a nonlinear relationship between the number of fins and the choice of guidance system. Both plots show the greatest response from the increase in guidance system technology, which is consistent with the Pareto analysis shown in Figures 11.15 and 11.16. What is particularly revealing in both the *surface plot* on the left and the *contour plot* on the right, however, are the spikes in responses when the number of fins is at low and high points. This suggests that if, for example, the higher technology guidance system is not available or is too expensive, the rocket should achieve better results with a 5-fin design than

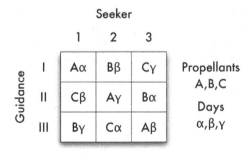

Figure 11.18 Graeco-Latin rocket design.

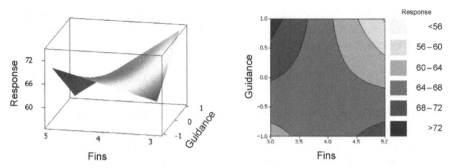

Figure 11.19 Response surface plots of the rocket design.

something less. Clearly, considering these interactions alone, a 4-fin design is least desirable using either guidance systems.

Other Design Techniques have been developed and are in use by solution designers. Though not as universally adopted as the techniques described above, these methods include modified factorial designs proposed by Taguchi in *System of Experimental Design* [32] and variations on Latin hypercubes, including nearly orthogonal designs by Cioppa [33].

11.5.5 Pareto Analysis

Pareto analysis further reduces the complexity of the solution space by critically examining factors that make up a solution and eliminating those that fail to make a meaningful contribution. It continues the drive to condense the problem to only the most essential factors.

The Pareto principle, also known as the *sparsity principle*, the *law of the vital few*, and the *80–20 rule*, was coined by quality management expert Joseph M. Juran in the 1951 edition of his *Quality Control Handbook* [34]. In a 1975 explanation, "The Non-Pareto Principle; Mea Culpa" [35], Juran traces how in the 1930s he came to understand that a small number of factors usually make the greatest contribution to the whole. He confesses that he extended Italian economist Vilfredo Pareto's concepts on the unequal distribution of wealth beyond their intended scope. In Le *Cour d'Economie politique* [36], Pareto noted that 80% of the wealth was owned by 20% of the people. Juran noticed that this ratio held true for many quality issues and used the term *Pareto principle* to describe his concept of the vital few and trivial many. Pareto analysis, regardless of the accuracy of the name, is now widely used to separate the *vital few* from the *trivial many* in many different applications.

Common Pareto analyses use a combination of bar and line charts to visually identify the vital few. Raw values are typically displayed in a bar chart sorted in decreasing order from left to right. A secondary y-axis shows a cumulative distribution using a line chart. Figure 11.20 uses this approach to display the results of the full factorial rocket design. A key to interpreting this kind of Pareto chart is to look for a *knee in the curve* of the cumulative distribution. This will indicate

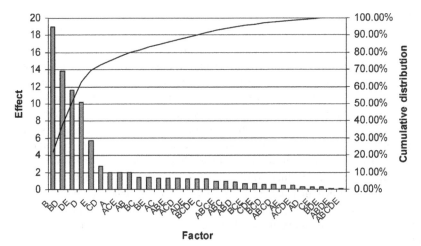

Factor

Figure 11.20 Pareto analysis of the rocket design experiment.

a point where the degree of contribution begins to level off, with factors on the steeper slope making a greater contribution. In Figure 11.20, this occurs at about the 70% point. If there is no distinct change in the slope of the line, a conservative approach is to consider factors that contribute 60% of the total. This reinforces Juran's confession that the 80–20 rule is more of a rule of thumb and should not be interpreted as an absolute metric. Table 11.15 summarizes the factors contributing to this point. Figure 11.20 is strong supporting evidence for screening out the trivial many factors in this solution design.

The Minitab® charts in Figures 11.15 and 11.16 show another method for conducting a Pareto analysis, with main and interaction effects plotted in decreasing order and a vertical line marking the boundary of statistical significance. In this case, the vital few are those that exceed the threshold for significance. Note that this figure reveals a conclusion that is identical to that of Figure 11.20: the five factors named in Table 11.15 are the most important vital few factors in this design.

Pareto analysis can be used throughout the systems decision process, whenever there is a large set of values to be prioritized. Other opportunities arise during brainstorming or stakeholder analysis, for example, to reduce the number of inputs to the vital few.

TABLE 11.15 The Vital Few Rocket Design Factors

Factor	Factor Name	Response	Percent of Total	Cumulative Percent
B	Seeker	18.938	21.97	21.97
BD	Seeker* guidance	13.812	16.02	38.00
DE	Guidance* thrust	11.563	13.42	51.41
D	Guidance	10.187	11.82	63.23
E	Thrust	5.688	6.60	69.83

11.6 SUMMARY

Successful solutions to complex problems are products of design, not chance. *Design* is a deliberate process for composing an object, architecture, or process. A *solution* is a process for solving a problem that results in feasible *alternatives* to present to a decision maker. *Solution design*, then, is a deliberate process for composing a set of feasible alternatives for consideration by a decision maker. As seen in Chapter 7, systems engineers build solution design teams specifically suited for the problems at hand. This chapter explores ways that those teams generate innovative ideas and refine them into alternatives that meet stakeholders' needs, wants, and desires. Solution designers further analyze alternatives through modeling and simulation, examined in Chapter 4, to provide greater understanding of how the system works over time. Systems engineers can then confidently provide decision makers with essential candidate solutions that require only the leader's experience and skill for a decision.

11.7 ILLUSTRATIVE EXAMPLE: SYSTEMS ENGINEERING CURRICULUM MANAGEMENT SYSTEM (CMS)—SOLUTION DESIGN

SOLUTION DESIGN

Robert Kewley, Ph.D. U.S. Military Academy

For the solution design phase of this problem, the system designers had to scan the environment for potential IT solutions for the problem identified in the previous chapter. On the basis of the unique constraints and challenges identified, they had to come up with a feasible subset of IT solutions that could be expected to solve the problem. This is truly a value-focused approach because they have defined a broad set of value-added functions that the system could support with limited understanding of the alternatives' abilities to achieve that functionality. They are asking the question, "What do we want the system to do?" as opposed to asking, "Which of these predefined alternatives helps us most?" With this approach, they are more likely to identify opportunities to improve curriculum management functions.

Once they had scanned the IT environment for potential software and development solutions, the team formulated the problem as a sequence of interrelated decisions. For this reason, they used Zwicky's morphological box, shown in Figure 11.21, to represent these alternatives. Each column of the box represents one of the development decisions for the design team. The options for each decision are represented in the rows.

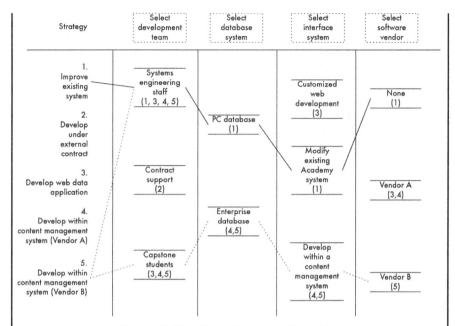

Figure 11.21   Alternative generation table.

Although this box gives a possibility of 54 ($3 \times 2 \times 3 \times 3$) alternatives, dependencies reduce this significantly. Other alternatives are not feasible or clearly inferior with no analysis required. The design team reduced the alternatives to a feasible set of five for which they would do additional analysis.

1. *Improve the Existing System.* Department IT personnel work with department leadership to develop a set of structured templates, storage folders, and processes to be applied to the current system of curriculum management. This alternative does not require significant IT development. Instead it seeks to identify, standardize, and enforce best practices employed in the current system.

2. *Develop the System Under External Contract.* The department hires an external contractor to develop an enterprise-wide system using database and interface solutions determined by the contractor and approved by the department. This outsourcing approach removes the development load from the department's IT staff, but requires significant external coordination with a contractor who may not understand department processes.

3. *Develop Web-Data Application.* Department IT personnel, supported when possible by capstone students (undergraduate students doing a two semester team research project), develop a Web-data application to handle

the curriculum management functions. They do this using Vendor A, Web services, and a service-oriented architecture to maximize usability of services and integration with existing systems. This alternative provides excellent opportunities for flexibility and customization, but it places a heavy burden on the department IT staff.

4. *Develop within a Content Management System (Vendor A).*   Department IT personnel, supported when possible by capstone students, performs development within the framework of a content management system from vendor A to integrate structured curriculum data into the portal. With this alternative, the content management system provides a significant portion of the required capabilities with limited or no development required. It does have some additional cost.

5. *Develop within a Content Management System (Vendor B).*   This alternative is distinguished from the previous one by a different vendor for the content management system. Vendor B's system has different functionality and cost. Given these alternatives, the design team must score each alternative against the value measures shown in the previous chapter in Figure 10.20.

Because this is a system concept decision, the team does not have sufficient design data to do modeling and analysis of the alternatives. For this decision, it is sufficient to research each alternative to subjectively assess the objectives in the values hierarchy.

11.8   EXERCISES

11.1. Identify and describe the seven main criticisms of traditional brainstorming sessions.

11.2. Delphi methods for idea generation seek to minimize some of the biasing effects of traditional brainstorming. Identify the three effects Delphi methods seek to counter and the three unique variations they use to do it. **Situation**. Assume that you are a member of a solution design team tasked with developing solution candidates for a new parking lot car-finder. The client wants to market a device that will allow shoppers to instantly find their cars in a crowded shopping mall parking lot. The client tells you that the device must be simple to use, very inexpensive to make but durable and reliable, and priced for a middle class mass market. Additionally, it must function in all weather conditions and be effective up to 100 m. It would be nice, he adds, if it could come in designer colors and styles to fit individual personalities. Finally, it must be completely secure, so that no legal action could reflect on the company if it is stolen or misused. Assume that one solution candidate considered material (plastic or metal), indicator (visual or electronic), and transmitter (none or radio frequency). Further assume that a trial takes place with shoppers executing the full factorial design. The average times for shoppers to find their cars for each design point are in the order 145, 133, 128, 118, 127, 110, 102, and 98 s.

11.3. Conduct a morphological analysis of the car-finder device that includes a two-dimensional matrix with at least four parameters and four variables.

11.4. Construct an ends-means chain to illustrate the objectives structure of the car-finder device.

11.5. Using the results of the morphological analysis, conduct a feasibility screening of the alternatives by the client's needs, wants, and desires.

11.6. Referring to the techniques explored in Chapter 4, identify the types of models appropriate for analyzing solution candidates that pass the feasibility screening test, and explain why they are appropriate.

11.7. Construct a full factorial experimental design matrix with all main and interaction effects, listing the factors in the order described above.

11.8. Calculate the main effects of changing each of the three factors.

11.9. Calculate the interaction effects of modifying material and indicator together and the three-way interactions of material, indicator, and transmitter.

11.10. Conduct a Pareto analysis of the car-finder experiment. Identify the vital few and the trivial many and explain how you reached that conclusion.

11.11. Reviewing your original morphological analysis, if only five alternatives passed a subsequent feasibility screening and each factor could be reduced to a single low and high value, how many design points would be required to conduct a full factorial design of these solution candidates?

11.12. Reducing the design in Question 11 to a Resolution III fractional factorial design would require what number of design points?

11.13. Fractional factorial designs run the risk of confounding effects. What combination of effects would be aliased, if any, in the design described in Question 12?

REFERENCES

1. *Webster's II New Riverside University Dictionary*. Boston, MA: Houghton Mifflin, 1988.

2. INCOSE SE Terms Glossary. Seattle: International Council on Systems Engineering, 1998.

3. Dalkey, NC. *The Delphi Method: An Experimental Study of Group Opinion*. Santa Monica, CA: United States Air Force Project RAND, 1969.

4. Osborn, AF. *How to Think Up*. New York: McGraw-Hill, 1942.

5. Osborn, AF. *Your Creative Power: How to Use Imagination*. New York: Charles Scribner's Sons, 1948.

6. Osborn, AF. *Wake Up Your Mind: 101 Ways to Develop Creativeness*. New York: Charles Scribner's Sons, 1952.

7. Osborn, AF. *Applied Imagination: Principles and Procedures of Creative Thinking*. New York: Charles Scribner's Sons, 1953.

8. Durfee, WK. Brainstorming Basics, ME 2011 Course Handout. Minneapolis, MN: University of Minnesota, 1999.

9. De Bono, E. *Serious Creativity: Using the Power of Lateral Thinking to Create New Ideas*. New York: HarperBusiness, 1992.

10. Isaksen, SG. A review of brainstorming research: Six critical issues for inquiry. Buffalo, NY: Monograph 302, Creative Problem Solving Group, 1998.

11. Taylor, DW, Berry, PC, Block, CH. Does group participation when using brainstorming facilitate or inhibit creative thinking? *Administrative Science Quarterly*, 1958;3:22–47.

12. Sage, AP, Armstrong, JE, Jr. *Introduction to Systems Engineering*. Wiley Series in Systems Engineering, New York: Wiley-Interscience, 2000.

13. Janis, IL. *Groupthink: Psychological Studies of Policy Decisions and Fiascoes*, 2nd ed. Boston, MA: Houghton Mifflin, 1983.

14. Dalkey, N, Helmer, O. An experimental application of the Delphi method to the use of experts. *Management Science*, 1963;9:458–467.

15. Girshick, M, Kaplan, A, Skogstad, A. The prediction of social and technological events. *Public Opinion Quarterly*, 1950;14:93–110.

16. Fowles, J, Fowles, R. *Handbook of Futures Research*. Westport, CT: Greenwood Press, 1978.

17. DeBono, E. *The Use of Lateral Thinking*. London, England: Jonathan Cape Ltd. Publishers, 1967.

18. DeBono, E. *Parallel Thinking*. Penguin Press: New York 1995.

19. Zwicky, F. *Discovery, Invention, Research Through the Morphological Approach*. New York: Macmillan, 1969.

20. Keeney, RL. *Value-Focused Thinking: A Path to Creative Decisionmaking*. Cambridge, MA: Harvard University Press, 1992.

21. Athey, TH. *Systematic Systems Approach: An Integrated Method for Solving Problems*. Boston, MA: Pearson Custom Publishing, 1982.

22. MacCrimmon, KR, Wagner, C. Supporting Problem Formulation and Alternative Generation in Managerial Decision Making. Presented at 24th Annual Hawaii International Conference on System Sciences, Honolulu, HI, 1991.

23. Sage, AP. *Systems Engineering*. New York: John Wiley & Sons, 1992.

24. Nutt, PC. A taxonomy of strategic decisions and tactics for uncovering alternatives. *European Journal of Operational Research*, 2001; 132: 505–527.

25. *Defense Acquisition Guidebook*. Fort Belvoir, VA: Defense Acquisition University, 2004.

26. Fisher, RA. *The Design of Experiments*, 9th ed. New York: Hafner Press, 1971.

27. Salsburg, D. *Lady Tasting Tea: How Statistics Revolutionized Science in the Twentieth Century*. New York: Henry Holt and Company, 2002.

28. Box, GE, Hunter, WG, Hunter, JS. *Statistics for Experimenters*. New York: John Wiley & Sons, 1978.

29. Law, AM, Kelton, WD. *Simulation Modeling and Analysis*. New York: McGraw-Hill, 2000.

30. Plackett, RL, Burman, JP. The design of optimum multi-factorial experiments. *Biometrika*, 1946;33:305–325.

31. Euler, L. Recherches sur une nouvelle espece de quarres magiques. *Verhandelingen/uitgegeven door het Zeeuwsch Genootschap der Wetenschappen te Vlissingen*. 1782;9:85–239.

32. Taguchi, G. *System of Experimental Design*, Vols. 1 and 2. White Plains, NY: UNIPUB/Kraus International Publications, 1987.

33. Cioppa, TM. Efficient nearly orthogonal and space-filling experimental designs for high-dimensional complex models. Ph.D. Dissertation. Monterey, CA: Operations Research Department, Naval Postgraduate School, 2002.

34. Juran, JM. *Quality Control Handbook*. New York: McGraw-Hill, 1951.

35. Juran, JM. The non-pareto principle; mea culpa. *Quality Progress*, 1975;8:8–9.

36. Pareto, V. *Le Cour d'Economie Politique*. London: Macmillan, 1896.

37. GroupSystems website, http://www.groupsystems.com. Accessed August 22, 2010.

Chapter **12**

Decision Making

MICHAEL J. KWINN, JR., Ph.D.
GREGORY S. PARNELL, Ph.D.
ROBERT A. DEES, M.S.

> Nothing is more difficult, and therefore more precious, than to be able to decide.
>
> —Napoleon Bonaparte

12.1 INTRODUCTION

To this point, we have learned how to work with our decision maker and our stakeholders to define our problem. We then developed candidate solutions to that problem. We now turn to the process of determining a recommendation and obtaining a decision. It is important to remember that when supporting a systems decision, which requires the level of detail of the process we describe in this book, systems engineers do not make the decision but rather they provide the necessary information to enable a logical, defensible decision by the decision maker. We will discuss how we obtain a decision later in this chapter.

In this chapter, we will first prepare to start the decision-making phase. Within this phase, we have three tasks that can be seen in the chapter's concept map shown in Figure 12.1. From the problem definition phase we have the problem statement, requirements, and value model. From the solution design phase, we have the candidate solutions and the life cycle cost model. We also may have some models and simulations that can be used in the decision making phase. The four tasks of the

Decision Making in Systems Engineering and Management, Second Edition
Edited by Gregory S. Parnell, Patrick J. Driscoll, Dale L. Henderson
Copyright © 2011 John Wiley & Sons, Inc.

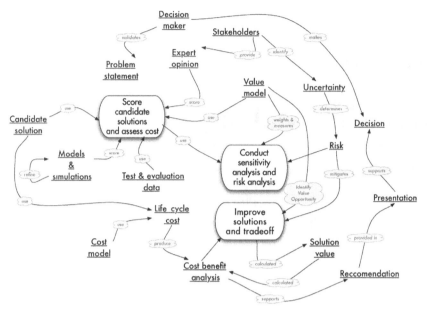

Figure 12.1   Concept map for Chapter 12.

decision-making phase are: score and cost the candidate solutions, conduct sensitivity and risk analyses, use value-focused thinking to improve solutions, and apply tradeoff analysis to compare value versus cost associated with candidate solutions. This chapter draws on material introduced by Parnell [1]. After performing these tasks, the systems engineer will present the recommended solution to the decision maker and obtain a decision. This will prepare the systems engineer for the implementation phase described in the next chapter.

Before the systems engineer begins scoring the candidate solutions, it is critical to review where we have been and ensure the process is still aligned with the problem.

12.2   PREPARING TO SCORE CANDIDATE SOLUTIONS

Like any task, before we begin the work, we have to ensure that we are prepared for success. In our case, we need to revisit some steps from previous phases of the process to ensure that we are still correctly aligned.

> Before we begin we have to review data from previous phases of the SDP to ensure the process is still aligned with the problem.

12.2.1   Revised Problem Statement

In the problem definition phase of the process, we developed a revised problem statement. Before we begin the decision-making phase of this process, we need to revisit the revised problem statement. Does the statement still capture the stakeholders' needs, wants, and desires? Does it still address the "real" problem? Is it still relevant? If the answer to any of these is no, then we need to do more work on properly structuring the problem statement. If we can still answer "yes" to each of these, we can continue.

12.2.2 Value Model

Also in the problem definition phase of this process, we developed the requirements and the value model. The requirements were used to narrow the alternatives to the potential solution candidates. The value model is very significant in the decision-making phase of the process. Though we will review the value measures and weights later in this chapter, we should pause now to review the entire value model. Is it still relevant? Does it capture all of the aspects of the problem, the functions, the requirements, and the values of the key stakeholders? If the answer to any of these is "no," we need to do some more work on the value model. If we can still answer "yes" to each of these, we can continue.

12.2.3 Candidate Solutions

We should also have our candidate solutions that form our solution set at this point in our process. The candidate solutions should have sufficient detail to allow us to score them on each value measure and assess their cost.

12.2.4 Life Cycle Cost Model

Using the principles and techniques in Chapter 5, we should have developed a life cycle cost model in the solution design phase to ensure that our candidate solutions were affordable and that they achieve any cost goals or requirements. We will use and possibly refine this model in this chapter to perform value versus cost tradeoffs.

12.2.5 Modeling and Simulation Results

When we did the solution design phase, our analysis determined the candidate solutions, and we may have accomplished some initial modeling and simulation. We should not discard these models and simulations. Some of the models and simulations we developed previously can be used in this phase to analyze candidate solutions and help improve the solutions.

12.2.6 Confirm Value Measure Ranges and Weights

Finally, this is a good time to reconfirm our ranges on the value measures and analyze their impact on the weights using the swing weight matrix for determining weights (see Chapter 10). We should have some more information from the modeling and simulation conducted to ensure the feasibility of our solutions. Do these results indicate that the ranges of our value measures may be different than we first assumed? If so, then we should make the adjustments now before we proceed to further evaluation.

So as we move forward, we have the revised problem statement, the requirements, the revised value model, the candidate solutions, a life cycle cost model, and possibly some models and simulations with associated outputs. In the steps below, we will begin to put them together to help form a recommendation for the decision maker.

12.3   FIVE SCORING METHODS

There are many methods for scoring the solutions against each value measure. Most of these fall into five categories: operations, testing, modeling, simulation, and expert opinion. We will discuss each of these, including their strengths and weaknesses.

> The five methods to score candidate solutions are operations, testing, modeling, simulation, and expert opinion.

12.3.1   Operations

The best data are usually obtained by using the system in the real operational environment. Unfortunately, operational data are usually only available for the baseline solution and candidate solutions that have been used for similar problems. It will usually be cost and schedule prohibitive to obtain operational data on all candidate solutions. For candidate solutions that require development, other scoring data will have to be used.

12.3.2   Testing

This is also known as development and operational testing.[1] This is essentially conducting testing using a prototype, a developmental, or a production system solution. Development and operational testing are key tasks in the system life cycle. We should use all available data to score our candidate solutions. Typically the development testing has less operational realism than operational testing.

In terms of accuracy and replicability, this is probably the best method to score solutions. It directly measures solutions against the value measures and reduces the number of assumptions required by the other methods.

The drawback of this method obviously is the costs associated with the scoring. Building prototypes, developmental, or production systems can be very costly and time consuming. That said, testing certainly has its important role in the design and evaluation process and is required prior to fielding any system. The other means of scoring solutions are less costly and are used in early system life cycle stages when test data are not available.

---

[1]Development testing is done by developers and operational testing is done by the users. Operational testing usually has more realistic environments. See Chapter 7 for discussion of the role of systems engineers and test engineers.

12.3.3 Modeling

Modeling usually refers to the development of mathematical models vice physical models (see Chapter 4). Mathematical models can prove to be very beneficial in the scoring of solutions. Because they are based on sound mathematical principles, they are accurate and can be easily replicated. Queuing models are important when determining service times for facilities layouts and other similar problems types. Often such problems lend themselves to a closed form solution using this type of model. If this type of problem becomes more complex, and especially if there is a stochastic nature of the problem, analysts use simulation to help determine the value measure scores. We turn to simulation now.

12.3.4 Simulation

Simulation is becoming more widely used as computing power increases and simulation packages make the building of these simulation models easier and quicker. Simulation is not limited to computer simulation, however. Simulation simply means using a representation of the solutions to determine its performance characteristics. Using this definition, simulation includes computer simulation and also physical representations of the candidate solutions (see Chapter 4).

Simulation is a very powerful tool for any systems engineer because it can be used to evaluate nearly any candidate solution. For example, a computer simulation could be built to assess the throughput of an assembly line layout. Also, a vehicle model could be manufactured similar to a candidate solution and placed in a wind tunnel or used in a physics-based virtual simulation to determine value measure scores required for the analysis.

Simulation is very useful because of its relative low cost when compared to developmental or operational testing. However, there are some significant limitations. One of the most significant is that the representation of a candidate solution may be biased by the analyst building the simulation. In other words, what is put into a simulation greatly affects what we get out of a simulation.

12.3.5 Expert Opinion

This is often considered the simplest means of obtaining the value measure scores for the candidate solutions. It is also considered the most questionable because it lends itself to more subjective analysis than the potentially more objective evaluation means described above. Certainly bias can creep into an analysis by relying too heavily on expert opinion for the value measure scores.

However, do not quickly dismiss this approach. It can be very valuable, depending on the time you have for the analysis and the level of fidelity required for the decision. For example, if the decision is required very quickly and all you have available is an expert on the subject, then this is a very sound approach to determining the value measure score. It is much quicker than building a model or a simulation and especially a prototype. It does add to the burden of sensitivity analysis, which we will discuss later in this chapter. Furthermore, experts often know operational data, development test results, and other related modeling and simulation results and use this knowledge in their assessment of the candidate solution scores.

12.3.6   Revisit Value Measures and Weights

A favorite saying of the second author is that "no value model ever survives first contact with the candidate solution scores!" After we have obtained the scores, we again return to the value measures to ensure that we can measure each of these and that the value model is weighted appropriately. Our revisiting the value measures and swing weights is critical to the success of the overall process to ensure they are correct and the weights accurately reflect the importance and impact of the variation of the value measures on the decision.

> We need to revisit the value measure ranges and update the value functions and swing weights if required.

Ensure that Each Candidate Solution Can Be Measured   We begin our review of the value measures to ensure that each candidate solution can be measured against each value measure. For consistency, we would like to use the same methodology to measure each value measure against each candidate solution. This is not always possible. An example would be if a candidate solution is so early in the development stage that it cannot be fully modeled or operationally tested, but all other candidate solutions can be operationally tested.

Adjust Value Model and Weights   If we have identified value measures that we cannot fully measure for each candidate solution, we have to change those value measures. This will change our value model. We should also take this time to ensure that we identify the potential ranges for each value measure. This is crucial to ensure that we have the correct weights on each value measure. In Chapter 10, we explained how to weight the value measures using the swing weight matrix. After confirming the weights, we are ready to begin our scoring of the candidate solutions.

12.4   SCORE CANDIDATE SOLUTIONS OR CANDIDATE COMPONENTS

In this section, we introduce two scoring approaches for assessing the value of competing feasible system solutions: candidate solutions and candidate components. Using the candidate solutions approach described in Section 12.4.1, we holistically score the candidate solutions for each value measure. Then we use the value functions developed in Chapter 10 to determine a value for each measure and then use Equation (10.1) to calculate the total candidate solution value. Using this approach, we calculate the value of the baseline, the candidate solutions, and the ideal solution. With a baseline, four candidate solutions, and n value measures to consider, this approach involves potentially obtaining a maximum of $5n$ scores. The baseline is a starting point for what we know. For existing systems, this could be the current system or systems used to perform the functions. In some cases, the problem is so new or technologically advanced that we many not have a baseline. The ideal solution score is the best score on each value measure. Notice that the candidate solutions approach does not evaluate all of the possible component combinations that could be made into a system.

The candidate components approach considers all feasible systems designs using the list of components to perform each system function. In this approach, we score each component using the values measures that are affected by the component. We assume that only one component contributes to each value measure's score. If we have four candidate components for each function, we would again obtain a maximum of $5n$ scores including the baseline. Next, we calculate the weighted value for the value measures affected by the component. Finally, we will use optimization (Chapter 4) to determine the candidate solution with the highest value (sum of the component value) subject to compatibility and cost constraints. The compatibility constraints may eliminate some of the component combinations that are not feasible. Section 12.4.3 describes this approach.

After obtaining scores for each candidate solution against each measure, we convert the scores to values so we can compare the solutions and develop a recommended solution decision. Some of the analysis can be done by hand, but there are software packages that can assist in more complicated analysis.

12.4.1 Software for Decision Analysis

As discussed in Chapter 1 and Chapter 9, the underlying mathematics of the SDP is multiobjective decision analysis. *OR/MS Today* has published four major surveys on decision analysis software [2]. Thirty-four of the packages trade off among multiple objectives, the focus of this chapter. Several of these packages allow the analyst to use some multiple objective decision analysis (MODA) techniques, but few can do all analyses described in this chapter. Logical Decisions™ exemplifies a typical MODA package with several built-in analysis techniques [3]. The disadvantages associated with these packages include their cost and the time needed to learn them.

Analysts can use spreadsheet models for MODA studies. Kirkwood's text offers Microsoft® Excel macros for converting scores to values [4]. Using Kirkwood's macros and spreadsheet add-ins (mathematical programming, Monte Carlo simulation, and decision trees), all the analysis described in this chapter can be performed.

Using spreadsheet models makes sense for several reasons. Spreadsheets are ubiquitous in that all computers today are configured with some office suite of software that contains a spreadsheet application. It has become the analysis environment of choice for most businesses and organizations. Consequently, clients may be more comfortable performing an analysis with spreadsheet models because they view them as less of a "black box" than more complex software. For the analyst building the model, having to construct each piece of the model builds a level of familiarity with the details and assumptions going into the model that can be useful in a decision briefing.

The main disadvantage of spreadsheet models is that analysts must design and program each model element—and this takes time. It is relatively straightforward to change a quantitative value model, such as by revising value functions, changing weights, or changing scores when the model is reasonably small. However, as a model size increases, making these updates in a spreadsheet becomes more demanding and time-consuming than similar updates using specially designed software packages. Additionally, most of the speciality software earn their value by being able to quickly produce key sensitivity analysis and summary graphics. This point will be illustrated in Section 12.6.1 for the case of using a Monte Carlo simulation for sensitivity analysis.

> There are several decision analysis software packages that simplify value models development and the analysis of candidate solutions.

12.4.2 Candidate Solution Scoring and Value Calculation

At this point, we have a value measure score for each measure for each solution. To assist in the further analysis and for proper documentation, we put this data into a table that provides scores for each of the measures for each candidate solution. This table is known as the *score matrix*, or *raw data matrix*. The raw data matrix of our rocket example is shown below in Table 12.1. For proper documentation, this table should include the dimensions for each measure and the source of the scores.

After completing this raw data matrix, we convert the raw data into the dimensionless value. This is accomplished by mapping the candidate solution's score for each measure against the value function for that measure. The resulting value is then recorded for each solution and measure in the *value matrix*. This value

TABLE 12.1 Raw Data Matrix

Candidate Solution	1.1.1 Speed of Platform (kph)	1.1.2 Percent Grade	1.2.1 Number of People	2.1.1 Thrust (lb)	2.2.1 Number of Payloads	3.1.1 Accuracy (m)	3.2.1 Range (km)
Baseline	30	20	4	1000	2	10	20
Global lightning	75	29	6	1546	5	2	100
Hot wired	66	56	3	2818	4	5	14
Star cluster	45	32	4	2993	3	4	55
Slow poke	30	42	2	1138	2	8	36
Ideal	90	60	0	3000	5	1	105

TABLE 12.2   Value Matrix

Candidate Solution	1.1.1 Speed of Platform (kph)	1.1.2 Percent Grade	1.2.1 Number of People	2.1.1 Thrust (lb)	2.2.1 Number of Payloads	3.1.1 Accuracy (m)	3.2.1 Range (km)
Baseline	14	10	40	0	10	0	19
Global lightning	77	19	5	27	100	90	94
Hot wired	65	87	60	91	60	72	13
Star cluster	35	26	40	100	20	80	52
Slow poke	13	51	90	7	10	36	33
Ideal	100	100	100	100	100	100	100

function might be discrete or continuous as described in Chapter 10. These conversions can be computed automatically using a variety of software programs such as macros in Excel or software such as Logical Decisions™. The value matrix for the rocket example is shown in Table 12.2.

The value functions convert raw data to value.

With the value matrix complete, we are ready to determine the total value for each solution.

Multiply Value by Weights and Obtain Overall Candidate Solution Values

MODA uses many mathematical equations to evaluate solutions. The simplest and most commonly used model is the additive value model introduced in Chapter 10. This model uses the following equation to calculate each candidate solution's value:

$$v(x) = \sum_{i=1}^{n} w_i v_i(x_i) \tag{12.1}$$

where $v(x)$ is the candidate solution's value, $i = 1$ to n is the number of the value measure, x_i is the candidate solution's score in the ith value measure, $v_i(x_i)$ is the single-dimensional value of the score of x_i, and w_i is the measure weight (normalized swing weight) of the ith value measure, so that all weights sum to one.

$$\sum_{i=1}^{n} w_i = 1 \tag{12.2}$$

We use the same equations to evaluate every candidate solution.

TABLE 12.3   Candidate Solution Value and Cost[a]

Candidate Solution	Cost	Candidate Solution Value
Baseline	30	10
Global lightning	85	76
Hot wired	96	62
Star cluster	48	55
Slow poke	148	26
Ideal	200	100

[a] The cost of the ideal is not known. For display purposes we assume a cost larger than the highest cost solution. An alternative would be a cost of 0.

Using the above equation, the weights for each measure developed in the swing weight matrix and the value matrix, the analyst can quickly determine the *total value* for each solution. The total value for each candidate solution in the rocket example is shown in Table 12.3 and plotted in Figure 12.5. We see that the candidate solution with the highest value is "Global lightning." Though this solution scored the highest, we still have a great deal of work until we can make a recommendation.

Again, many software packages can accomplish this quickly by linking all the data, the weights, and the value functions. For many large-scale problems, the use of these software packages can prove to be quite beneficial.

12.4.3   Candidate Components Scoring and System Optimization

In the candidate solutions scoring approach of the previous section, we holistically scored each candidate solution on each of the n value measures. Here, we extend our analysis of the rocket problem using the candidate components approach, applying system optimization in order to determine the best system solution. As discussed earlier, in this approach we assume that only one component affects each measure. This requirement can be relaxed, but this is beyond the scope of this book. Before performing component scoring, we have to align the individual value measures uniquely to one component. Notice, however, that this many-to-one assignment allows for several value measures to be aligned with each component.

Continuing our analysis of the rocket problem, consider the component raw data matrix shown in Figure 12.2. The five component types are mobility, logistics, rocket, number of payloads, and guidance. The first type has five candidate components while each of the rest has four candidate components. Scores are provided for each measure affected by the component. To illustrate an earlier point, notice how, for example, the component "Rocket A" affects two value measures (thrust and range), but each value measure is aligned with a single component. In Figure 12.2, this means that while we might have multiple value measure row entries for a component, each value measure will be assigned to only one component.

Type of component	Components	1.1.1 Speed of platform (kph)	1.1.2 % Grade	1.2.1 Number of people	2.1.1 Thrust (pounds)	2.2.1 Number of payloads	3.1.1 Accuracy (meters)	3.2.1 Range (kilometers)
Mobility	Vehicle A	30	0.2					
	Vehicle B	75.3	0.5					
	Vehicle C	66.2	0.55					
	Vehicle D	44.7	0.32					
	Vehicle E	29.6	0.42					
Logistics	Concept A			0				
	Concept B			3				
	Concept C			4				
	Concept D			6				
Rocket	Rocket A				1000			20
	Rocket B				1500			40
	Rocket C				2000			60
	Rocket D				3000			100
Number of payloads	Two					2		
	Three					3		
	Four					4		
	Five					5		
Guidance	Inertial						7.77	
	Global positioning system						1.97	
	Wire						4.65	
	Optical						3.5	

Figure 12.2   Component raw data matrix.

Using the same value functions developed in Chapter 10 that we did for the candidate solutions approach, we next convert the raw data score on each value measure to a value as shown in the component value matrix in Figure 12.3. We then calculate each component's total value using Equation (12.1), using only the value measures affected by the component. We also calculate an ideal component value by summing the weights of the value measures affected and multiplying

		1.1.1 Speed of platform (kph)	1.1.2 % Grade	1.2.1 Number of people	2.1.1 Thrust (pounds)	2.2.1 Number of payloads	3.1.1 Accuracy (meters)	3.2.1 Range (kilometers)	Component value	Ideal
Mobility	Vehicle A	14	10						3	
	Vehicle B	77	70						19	
	Vehicle C	65	87						16	24
	Vehicle D	35	26						8	
	Vehicle E	13	51						4	
Logistics	Concept A			100					6	
	Concept B			60					4	
	Concept C			40					2	6
	Concept D			5					0	
Rocket	Rocket A				0			19	3	
	Rocket B				25			37	8	
	Rocket C				50			56	14	26
	Rocket D				100			95	25	
Number of payloads	Two					10			2	
	Three					20			3	
	Four					60			10	17
	Five					100			17	
Guidance	Inertial						36		10	
	Global positioning system						90		24	27
	Wire						72		20	
	Optical						80		22	
										100

Figure 12.3   Component value matrix.

by 100. The ideal component value is a useful analytical tool for determining the *value gap* associated with any one component. During the value improvement task, we can compare the maximum component value with the ideal component value and decide if we need to continue to search or design for higher value component candidates. For example, the highest value vehicle component (B) has a value of 19 compared to the ideal vehicle component value of 24. If this difference was considered significant, it would motivate a search for a new component that could potentially close the value gap of 5.

Once we have calculated the component value for each component, we proceed with a system optimization. The particular system optimization technique we use is adapted from the project selection methodology proposed by Kirkwood [4]. There are many useful operations research books (e.g., Ragsdale [5]) and credible websites[2] that provide detailed explanations of how to use Excel Solver. Figure 12.4 shows the standard format for the system optimization table. The table is used to calculate the value and cost of any feasible system solution. We then use Excel Solver to identify the components that provide the highest solution value in light of the design constraints imposed on the system. For the rocket problem illustrated, these constraints are budget and weight limitations.

The left side of the table is used to calculate system value. The five types of components are shown on the left and repeated on the top of the table. For the component rocket problem, we assume that we will select only one component (or one level) from each of the component categories. A binary decision variable (0 or 1) is used for each of the component categories. The resulting optimal values identified by the Excel Solver then show whether the component is selected for the best solution (1) or not (0). For example, Vehicle B is identified for use as the mobility component in the optimal system solution, whereas the rest are not.

Each cell in the system total line contains an expression representing the sum of the binary variables for each component category. These become the left-hand side of the component constraints for the Excel Solver. The required line entries are numerical values limiting this sum. For the rocket problem, these indicate that only one component can be selected for each component category. These values become the right-hand-side values for the component constraints for the Excel Solver. The value column provides the component values calculated earlier in Figure 12.3. These act as objective function coefficients for each of the binary decision variables for creating the target cell objective function for the Excel Solver. The individual component values are included in the total system value being maximized only if their corresponding component is used in the design.

The right side of the table is used to calculate the system cost and the weight of payloads and guidance that must be launched by the rocket. The five-year research and development (R&D) and production costs are included for each component. The system cost is calculated in the system total row. Again, as in the case of component value, an individual component cost is only included in the total cost if the component is used in the system. Therefore, the system total cost cells shown

---

[2]Frontline Systems, Inc. (http://www.solver.com)(http://www.solver.com) accessed August 20, 2010.

		Mobility	Logistics	Rocket	Payloads	Guidance	Value	Cost Year 1	Cost Year 2	Cost Year 3	Cost Year 4	Cost Year 5	Production cost	Weight
Mobility	Vehicle A	0					3	1	1	1	1	1	1	
	Vehicle B	1					19	2	1	1	2	2	2	
	Vehicle C	0					16	2	3	3	3	3	3	
	Vehicle D	0					8	3	4	4	4	4	4	
	Vehicle E	0					4	4	5	5	5	5	5	
Logistics	Concept A		1				6	2	2	2	2	2	0	
	Concept B		0				4	3	3	3	3	3	0	
	Concept C		0				2	4	4	4	4	4	0	
	Concept D		0				0	6	6	6	6	6	0	
Rocket	Rocket A			0			3	3	3	6	6	6	3	4
	Rocket B			0			8	4	6	7	7	4	6	6
	Rocket C			0			14	7	8	8	8	7	9	8
	Rocket D			1			25	4	6	6	6	3	12	10
Payloads	Two				0		2	2	4	2	2	3	3	4
	Three				0		3	4	4	4	4	4	6	6
	Four				1		10	4	6	4	3	2	9	8
	Five				0		17	3	3	3	3	4	12	10
Guidance	Inertial					0	10	2	4	3	3	2	10	4
	Global positioning system					1	24	1	2	1	1	1	2	1
	Wire					0	20	3	6	3	3	3	3	1
	Optical					0	22	4	3	5	6	3	16	3
	System total	1	1	1	1	1		13	16	15	14	10	25	9
	Required	1	1	1	1	1		20	18	20	20	15	50	10

System value	84
Max value with above components	91

Estimated system cost	67
Total budget	92

Legend

Decison variable: 1 or 0
Input:
Calculation
Cell reference

Figure 12.4 System optimization table.

407

at the bottom of the table for each year contain an expression representing the sum product of the binary variables and the cost in that year. The total system production cost cell is similarly constructed. The required line for the cost columns are numerical entries listing the system annual budget limitations for R&D and production. The total payload and guidance weight is calculated in system total row. The maximum allowable weight is placed in the required row.

Four system totals are calculated in the bottom center of Figure 12.4. The system value cell contains an expression that is the sum product of the component binary variables and the component values. The "max value with above components" cell calculates the highest value that can be achieved with the component values in the table. The estimated system cost is the sum of the system total cost row on the right. The total budget is the sum of the five-year system costs and the system production cost in the required row on the right.

The system optimization using binary linear programming is performed using the following settings in Excel Solver.

- Maximize system value cell
- Decision variables: 21 component variables.
- Constraints
  - The decision variables are binary.
  - The number of components in the systems row on the left are set equal to the required row.
  - The system cost row is less than or equal to the required budget row.
  - The system payload and guidance weight is less than or equal to the maximum weigh.

Many additional constraints may arise in optimizing a system design. For example, Vehicle B is not compatible with logistics concept D, or perhaps Rocket A can only launch two payloads. Kirkwood [4] and Ragsdale [5] provide examples of additional constraints that may be appropriate for systems design constraints and many others.

Once we have determined the system value using component optimization we can plot a cost versus value plot to gain insights as to how each system solution lies in relation to each other solution, as illustrated in Figure 12.5. Since, when feasible, the optimization model will use the highest value components, the component optimization value is greater than Global Lightning. However, this high value was achieved by using more costly components.

The candidate components approach has advantages and disadvantages compared to the candidate solutions approach. The candidate solutions approach is simpler and quicker, requiring a systems engineer to develop three or four candidate solutions. Once the highest value candidate solution is identified, additional value-focused thinking is required to improve the solution. The candidate components approach requires more effort to obtain information on all of the components. Interestingly, the total number of scores generated by the two approaches is the same. The major advantage of the candidate solutions approach is that it provides assurance of identifying the highest value system solution possible given the components being considered and the constraint limitations within which all system solutions must lie.

Value-focused thinking can be used to change constraints and add better performing or less expensive components, if required by either the candidate components or candidate solutions approach. The systems engineer should select the best method to use based on availability of data and time to perform the analysis.

In the remaining sections of this chapter, we will use the candidate solutions approach. Recognize, however, that each of these sections would be similar had we chosen to use a candidate components approach.

12.5 CONDUCT SENSITIVITY ANALYSIS

When dealing with complicated decisions, such as the ones we are presenting in this book, systems engineers must be cognizant of the robustness of their analysis. The systems engineer should analyze the "sensitivity" of their candidate solution modeling assumptions and scores of the candidate solutions. The Pareto principle usually applies. Typically the decision will only be sensitive to 2 out of 10 factors. In the next section, we will discuss how the systems engineers should look at this sensitivity analysis and how it might affect their recommended solutions to a problem.

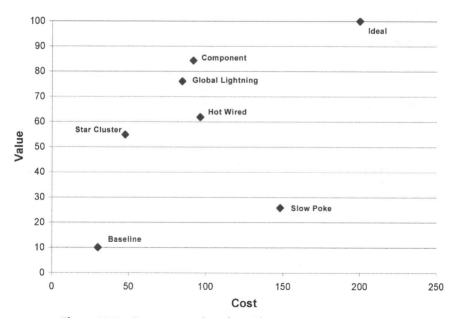

Figure 12.5 Cost versus value plot with component optimization.

> In order to tell the whole story, systems engineers must conduct sensitivity
> analysis to modeling assumptions and candidate system scoring uncertainty.

12.5.1   Analyzing Sensitivity on Weights

The purpose of sensitivity analysis is to see if a change in an assumption changes
the preferred solution. A parameter is sensitive if the decision maker's preferred
solution changes as the parameter is varied over a range of interest. The most
common sensitivity analysis is sensitivity to the assessment of weights. We plot
weight versus value for the range of interest for all solutions. If the solution
value lines do not intersect, we say the weight is not sensitive. If the lines
cross over, we consider the weights sensitive. The standard assumption for
analyzing weights sensitivity is to vary one weight and hold the other weights
in the same proportion so the weights still add to one. Since many factors are
involved in realistically large value models (weights, value functions, and scores),
typically less than 20% of the weights will be sensitive for realistic ranges of
interest.

There are several ways to do weights sensitivity. The first is to vary the weight
of each value measure from 0 to 1. This approach is shown in some MODA
books [4] for illustrative problems with only a few value measures. This approach
is not very useful for analytical purposes if there are a large number of value
measures and little disagreement about the weights assessment. The second way
when the weight assessments have not been controversial is to vary each of the
weights by ±0.1. This is a reasonable approach to determine if small changes in the
weights will change the preferred solution. However, sometimes weight assessment
is controversial and key stakeholders do not want to agree on one or more weight
assessments. The third way is to perform sensitivity analysis is to vary the weight
across the range of interest. This is typically the most useful and important weight
sensitivity analysis. If the preferred solution does not change across the range of
interest, then we do not need to spend time resolving the disagreement. If the
preferred solution changes across the range of interest, then we need to present this
to the key stakeholders and decision maker(s) for resolution.

Using the swing weight method (see top table in Figure 12.6), there are two
options for weights sensitivity analysis. The first option is to perform the sensitivity
analysis using the original swing weights. This approach has the advantage that
the analysis is done directly on the weight assessment judgment. The disadvantage
is that the measure weight variation depends on the other swing weights. Suppose
we vary a swing weight set at 85 out of 365 ($=$ measure weight of 0.23) from 0 to
100 (the highest swing weight), the measure weight will then vary from 0 to 0.26.
Applying the second approach varies the measure weight ±0.1, causing it to span
the interval 0.13 to 0.33 ($=$ swing weight assessment of 47 to 120). The advantage
of this approach is that is achieves the full range of ±0.1. The disadvantage is that
the swing from 100 to 120 may be unrealistic if there is agreement that the value
measure with 100 has the highest swing weight.

Swing Weight Matrix with swing weights link to calculation of value:

		Level of importance of the value measure								
		Mission critical			Mission enabling			Mission enhancing		
		Value Measure	Swing Weight	Measure Weight	Value Measure	Swing Weight	Measure Weight	Value Measure	Swing Weight	Measure Weight
Range of variation	Large capability gap	Accuracy	100	0.27	Range	50	0.14			
	Significant capability gap	Speed of Launch Platform	85	0.23	Thrust of Rocket	45	0.12	Percent Grade Platform can Traverse	5	0.01
	Small capability gap	Number of Different Payloads	60	0.16	Number of Operators	20	0.05			

Total 365

Data Table with cell referencing of Solution Value calculation cells:

Sensitivity Analysis on Speed of Platform

		0	25	50	85	100
Baseline	10					
Global Lightening	76					
Hot Wired	62					
Star Cluster	56					
Slow Poke	26					
Ideal	100					

Data Table completed by referencing Speed of Platform swing weight of 85:

Sensitivity Analysis on Speed of Platform

		0	25	50	85	100
Baseline	10	9	9	10	10	10
Global Lightening	76	76	76	76	76	76
Hot Wired	62	62	62	62	62	62
Star Cluster	56	62	60	58	56	55
Slow Poke	26	30	28	27	26	25
Ideal	100	100	100	100	100	100

Figure 12.6 Performing weight sensitivity analysis.

We recommend performing weight sensitivity analysis on the value measures that have been identified as being controversial during the weight assessment across the range of interest using the swing weights. Time permitting, we also recommend varying the weight ±0.1 on the remaining measure weights.

Next we consider how to perform weights sensitivity in Microsoft® Excel.

12.5.2 Sensitivity Analysis on Weights Using Excel

Weights sensitivity can easily be performed using the table function and graphical plots in Excel. We return to the Rocket Problem. Figure 12.6 repeats the swing weight matrix from Chapter 10, Figures 10.14 and 10.15. Suppose that stakeholders had difficulty assessing the swing weight for the "speed of the launch platform." Some stakeholders thought that the weight was too high and should be much less, and some thought it could be as highly weighted as distance from target. Suppose we decide to vary the swing weight currently set at 85 out of 365 (= measure weight of 0.23) from 0 to 100 (the highest swing weight). The measure weight would then vary from 0 to 0.26.

We can perform this analysis in six steps using Excel. First, we use the weights in the swing weight matrix in Figure 12.6 to calculate solution value in a worksheet of our spreadsheet. Second, we construct the table in the middle of Figure 12.6. Across

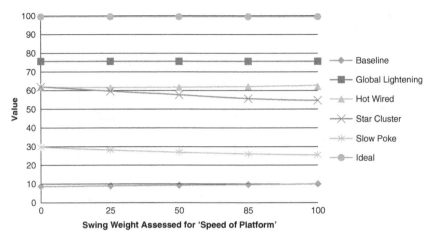

Figure 12.7   Swing weight sensitivity for Speed of Platform.

the top of the table we cell reference the swing weight name we are analyzing. In the second row, we put the swing weight range from 0 to 100 in several increments. Third, in the left-hand side of the table we cell reference the solution names and the cells used to calculate the solution value using the swing weight matrix assessments and the additive value model. Fourth, in Excel 2007 we select the "Data," "What If Analysis," and then "Data Table." Fifth, we cell reference the swing weight matrix cell with 85 in the Row Input cell in the Data Table and click on "OK." The Data Table generates the table in the bottom of Figure 12.6. Sixth, we plot the swing weight range versus solution value as shown in Figure 12.7.

Looking at Figure 12.7, we see that Global Lightning is always the highest value alternative regardless of the swing weight changing from 0 to 100. Therefore, we conclude that the preferred solution is not sensitive to the weight assessment of the speed of platform value measure. However, suppose we create a new case without Global Lightning and with an original swing weight assessment of 10 instead of 85. In this hypothetical case, the preferred solution would be Star Cluster for swing weight values below approximately 5 and Hot Wired for swing weight above 5. In this case, we would conclude that the preferred solution is sensitive to the weight assessment of the speed of platform value measure.

Swing weights are only one source of uncertainty that requires sensitivity analysis. A second source arises from our uncertainty about the solution scores on the value measures. We turn to that discussion now.

12.6   ANALYSES OF UNCERTAINTY AND RISK

In the previous section, we considered sensitivity analysis to our weight assessments. In this section we analyze the uncertainty about the scores of the candidate solutions. In the additive value model equation there are three elements: the weights, the value functions, and scores. To this point in this chapter we have assumed that the weights and value functions are known with certainty and the scores are deterministic. In many systems engineering and engineering management problems, new systems are being developed whose future performance (i.e., value measure scores) may be uncertain. In Chapter 3, risk is defined as a probabilistic event that, if it occurs, will cause unwanted change in the cost, schedule, or value return (e.g., technical performance) of an engineering system. In programmatic terms, performance uncertainty may be a major source of technical risk that could also provide schedule and cost risk.

In the rocket problem, the scores of some solutions on the value measures may be uncertain if operational or test data in operational environments is unavailable. In this chapter, probability distributions are used to assess our uncertainty about these scores and the events that can impact these scores. In the rocket problem, the impact of the uncertain scores could be a lower value than we would expect if we had used deterministic scores.

Uncertainties can impact one or more solution scores. The simplest case being considered is that each uncertainty impacts only one value measure score of one solution. We will call these *independent scoring uncertainties* and we can directly assess a distribution on each independent score. A more complex situation occurs when one uncertain event impacts the scores on the value measures of two or more solutions. We will call these *dependent scoring uncertainties* since the scores of the solutions depend on the outcome of an uncertain event.

An example of an independent uncertainty would be uncertainty concerning the accuracy of the inertial guidance used by Slow Poke (see Figure 11.5). Since Slow Poke is the only rocket using inertial guidance, this would be an independent uncertainty. In Section 12.6.1, we analyze the impact of independent uncertainties using Monte Carlo simulation. In Section 12.7.1, we analysis the impact of dependent uncertainties using decision trees [6]. An example of a dependent uncertainty would be a technical uncertainty with the effectiveness of fins. Since all solutions use fins (see Figure 11.5), a technical problem with fins could impact the range and accuracy of all the rockets.

In some analyses, we may need to consider alternative scenarios to capture our uncertainty about the future. When we consider multiple scenarios, we have some analysis options. We can consider all the scenarios and then develop one value model that captures the future planning space of the scenarios. This is the most common analysis approach. Alternatively, we could develop weights, value functions, and scores for each scenario. Clearly, this would require a lot of stakeholder time to develop multiple models and multiple scores. Since weights have a larger impact than value functions and scores, some studies have assessed different weights for each scenario and then displayed range in alternative value across all scenarios. One of the early studies using this approach was the Air Force 2025 study [7].

12.6.1   Risk Analysis—Conduct Monte Carlo Simulation on Measure Scores

When developing our raw data matrix, we identified a mean or expected score for each measure for each candidate solution. We did not identify a range of scores, but rather a single score that allowed us to calculate a total value for our candidate solutions. Oftentimes, this single score does not represent what we know—or, more precisely, *what we do not know*—of the measure.

In our rocket example, we developed scores for the thrust of the rocket and the accuracy of the rocket. However, uncertainty concerning technological performance of the rocket could cause the measure scores to vary, making a reasonably accurate final score unattainable until actual testing is completed in a later life cycle stage. For our continued analysis, we could use the mean or the mode associated with these measures, but as indicators of central tendency and common occurrence, they are ineffective as indicators of uncertainty.

This uncertainty in our measure scores gets propagated into the total value of each candidate solution through the value model. If this uncertainty is significant, it can affect the profile of the cost–value tradeoff. Using a single total value for each candidate solution masks this uncertainty. An appropriate sensitivity analysis would model the uncertainty present in these measure scores enabling us to estimate the extent of total value uncertainty.

Previously, our sensitivity analysis examined the impact that changes in a single value model element (weight) had on the total value for candidate solutions. Here, our interest is in assessing the impact of simultaneous uncertainty in the measure scores so that we can examine their combined effect on the uncertainty of total value. The most appropriate tool for this is Monte Carlo simulation.

There are many software packages which can be of great assistance in this analysis. Oracle® Crystal Ball [8] and Risk™ [9] are two of the more popular modeling environments for estimating the effects of uncertainty and risk in decision making. Both of these applications are completely integrated for use with Microsoft® Excel, which has the appeal of not having to leave the environment in which the value model was created.

Illustrating this type of sensitivity analysis with Crystal Ball, we use triangular probability distributions to model the uncertainty associated with each of the six shaded measures listed in Figure 12.8. The choice of distribution to use for modeling uncertainty depends on the information elicited from key stakeholders. Without

Candidate Solution	1.1.1 Speed of platform (kph)	1.1.2 % Grade	1.2.1 Number of people	2.1.1 Thrust (lbs)	2.2.1 Number of payloads	3.1.1 Accuracy (m)	3.2.1 Range (km)
Baseline	30.0	20.00%	6	800	2	10.00	20.0
Global lightning	75.3	28.62%	6	1546	5	1.97	99.5
Hot wired	66.2	55.63%	3	2818	4	4.65	14.2
Star cluster	44.7	32.46%	4	2993	3	3.50	55.4
Slow poker	29.6	42.33%	2	1138	2	7.77	36.5
Ideal	90.0	60.00%	0	3000	5	0.50	105.0

Figure 12.8 Six measures with triangular probability distributions.

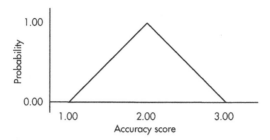

Figure 12.9   Triangular distribution for Global Lightning accuracy score.

a likeliest measure estimate, any value between some upper and lower measure score would occur with equal likelihood, motivating us to use a uniform distribution to model the uncertainty. Since this is not the case, the triangular distribution is more appropriate.

An example of the triangular distribution for Global Lightning's accuracy score is shown in Figure 12.9. The minimum accuracy of 1 meter and the maximum accuracy of 3 meters set the lower and upper limits of the distribution, respectively. The likeliest score being 2 m sets the peak location for the triangular distribution.

The triangular distribution is commonly used to model uncertainty in Monte Carlo simulations. It requires only three parameters: a lower and upper bound to set the limits of the distribution, along with a likeliest score to establish the peak location.

For this Monte Carlo simulation we used 1000 runs to produce data with which we could estimate the variability in total value for the candidate solutions. For each run, the software samples each of the six triangular distributions to obtain a random estimate of each measure score and then calculates the total value of the six candidate solutions. When the simulation is complete, the software calculates a probability distribution on the value of each candidate solution. The probability distribution results for all six candidate solutions are shown in Figure 12.10. Two of the solutions, baseline and ideal, had no uncertain measure scores. Therefore, there is probability equal to 1.0 for each of the values. Although one of the candidate solutions, Hot Wired, had uncertainty associated with its grade measure score, the combination of a small interval between lower and upper bounds in its triangular distribution (46 to 60) and the low measure weight in the value model (0.01) impose negligible variability in its total value. Thus there is very little uncertainty being propagated into this solution's value.

The other three solutions, Slow Poke, Star Cluster, and Global Lightning, contain significant amounts of uncertainty as indicated by the variability in their total value resulting from this Monte Carlo simulation. Figure 12.10 shows a very important result: Global Lightning has the highest value but the most uncertainty. However, even with this uncertainty, Global Lightning provides a higher value than every

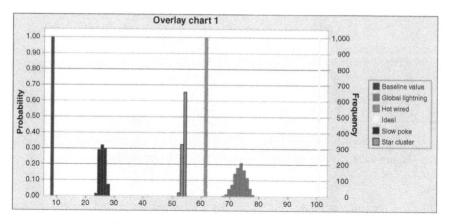

Figure 12.10   Monte Carlo simulation results.

solution except the ideal candidate solution. Its total value distribution does not overlap with any other, indicating that it *deterministically dominates* the other candidate solutions [6]. "Deterministic dominance" means that the worst outcome of the dominating solution is better than the best outcome of the dominated solution. This means that even in the face of measure uncertainty, Global Lightning can be expected to return the highest total value.

> When comparing the total value distributions from a Monte Carlo simulation sensitivity analysis, Distribution *A* deterministically dominates Distribution *B* if for every possible total value, the probability of getting a value that high is always better in *A* than in *B*.
>
> Monte Carlo simulation shows the effects of uncertainty about the scores on candidate solution values.

The next key question that the systems engineer should ask is which measure score uncertainty is making the most contribution to the variance observed in the Global Lightning value. Using Crystal Ball, we can answer this question using the Contribution to Variance Diagram shown in Figure 12.11. The measure score that creates the most uncertainty in Global Lightning value is the speed of the platform.

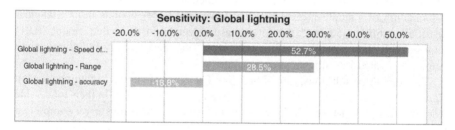

Figure 12.11   Global lightning contribution to variance.

It contributes about 53% of the variance. The systems engineer should conduct this sensitivity analysis on as many measure scores for the candidate solutions as required to ensure that the recommendation is robust for the known uncertainties. This provides the decision maker the confidence to make the system decision.

Although we have scored all of our candidate solutions and have examined sensitivity of their values to changes in weights and affects of uncertainty on the score, we are not ready to move to a recommendation yet. We first must try to develop even better solutions first using a process known as value-focused thinking [10].

12.7   USE VALUE-FOCUSED THINKING TO IMPROVE SOLUTIONS

Though we have scored all of our candidate solutions and have at least one that has scored the highest, it would be very rare indeed to have our highest value candidate solution be a perfect 100. That would mean that the candidate solution scored the highest possible for each measure. Though this is unlikely in practice, identifying an ideal solution is certainly our goal. After scoring our candidate solutions, we need to seek an even better solution.

Develop Better Solutions   To seek a better candidate solution, we return to our value model. The stacked bar chart in Excel provides an excellent means of doing this. Figure 12.12 illustrates the fundamental concepts involved with applying value-focused thinking during this phase of the SDP.

In the hypothetical situation shown, the baseline is being compared with three candidate solutions. Looking at the individual measure values, we ask a couple

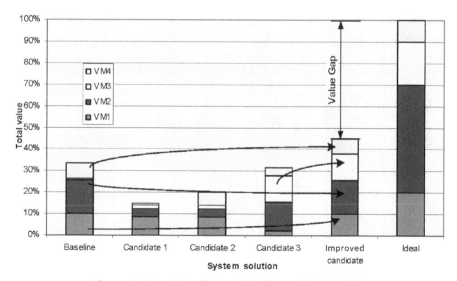

Figure 12.12   Value-focused thinking within the SDP.

of key questions. In what candidate solution do we come closest to achieving the stakeholder ideal scores? In applying value-focused thinking we first attempt to improve the best candidate solution by observing what is possible in the other candidate solutions. Assembling the maximum value measure scores into a single option produces an *improved* candidate solution. Is it possible to combine the known system elements from the different candidate solutions in this manner? What would this improved system solution look like? If it is not possible to combine the known elements generating the observed measure value levels in this manner, is there a new way of achieving a similar level of performance?

In a similar manner as described in Section 12.4.3, we notice that even this improved candidate solution falls short of the ideal levels as expressed by the stakeholders. Closing this gap will quite possibly require new design activities focusing on the candidate solution improvements that would need to be made in order to attain a better score for each measure.

It is also valuable to examine the individual measure value gaps when trying to improve candidate system solutions. Returning to the rocket design problem, Figure 12.13 shows the candidate solution stacked bar chart comparison without an improved solution. As a concept check, try assembling an improved candidate solution for this systems decision problem. By examining Figure 12.13 (and Table 12.1), we can see that to improve Global Lightning we need to improve the scores for grade, number of people, and thrust.

In order to attain an ideal score on the "% grade a platform can traverse" measure, we see that a 60% grade achieves an ideal score. What would the platform have to be to traverse a 60% grade? Can the existing platform be modified to achieve this? Does this require an entirely new platform be designed? Applying

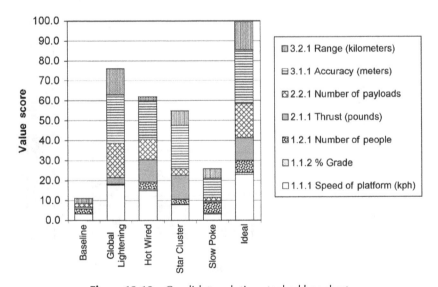

Figure 12.13   Candidate solution stacked bar chart.

value-focused thinking in this manner requires that we examine all the measures and determine what needs to be done to get closer to the ideal solution.

Once having identified the improved score for each measure, the project team works through the tradeoffs required to make any improved solution feasible. Armed with the new improved solutions, it is then time to rescore them.

Rescore the Improved Solutions When rescoring, the systems engineer can and should use scoring methodologies previously used in the process. Though we have now looked at each possible solution and analyzed the sensitivity of our process and measures, we still have to look at one other very important factor, risk, which we cannot completely eliminate. We discussed the various types of risks in Chapter 3. Here we examine how to mitigate some of the risks remaining from previous work.

12.7.1 Decision Analysis of Dependent Risks

Two major sources of uncertainty in systems development are technology development challenges and the potential actions of competitors or adversaries. Suppose for the rocket problem that two concerns are identified late in the solution design phase that could impact operational performance. The first concern is a technical concern and the second concern is a potential adversary threat.

Suppose that the engineers identify a technical concern with the new fin material that is planned to be used for both the Global Lightning and Star Cluster system solutions. This is a dependent uncertainty since the durability of the fins during flight has a direct impact on the range of the two candidate solutions. After working with the material and missile performance engineers, the systems engineer assesses the data shown in Table 12.4. If the fin material is durable, the range will achieve the original score. However, if there is some flight erosion of the fins, the range could decrease for both solutions.

Suppose that the intelligence agencies identify a potential future adversary threat that could result in degraded accuracy of guidance systems that use the Global Positioning System (GPS). Again, this is a dependent uncertainty since the accuracy of Global Lightning and Star Cluster will depend on the outcome of this event. After working with the navigation and missile performance engineers, the systems engineer assesses the data shown in Table 12.5.

We can use a decision tree to analyze the risk of dependent (and also independent) uncertainties. We use Precision Tree®, [11] a Microsoft® Excel add-in, to

TABLE 12.4 Fin Material Performance Uncertainty

		Range Score	
Fin Material	Probability	Global Lightning	Star Cluster
Durable	0.7	99.5	55.4
Erosion	0.3	80.0	40.0

TABLE 12.5   Global Positioning System (GPS) Performance Uncertainty

		Accuracy Score	
GPS Degrade	Probability	Global Lightning	Star Cluster
No Degrade	0.6	1.97	3.5
Degrade	0.4	4.0	7.0

perform the decision analysis. Decision trees are described in most decision analysis texts (e.g., reference 6), but they are typically used for single objective value and single objective utility. We use a decision tree with the multiple objective value model to determine the impact of the uncertainties on the preferences for Global Lightning and Star Cluster.

In Figure 12.14, the first node in the decision tree is a decision node, the second node is the Fin Material uncertainty, and the third node is the GPS Degrade uncertainty. Figure 12.15 shows the value calculations that are appended to the eight final branches of the decision tree. The value calculations are unchanged for the first five value measures. The value calculations for the last two value measures use the scores from Tables 12.4 and 12.5. The best decision is still Global Lightning, but the solution's expected value is now reduced from 76.3 to 71.9.

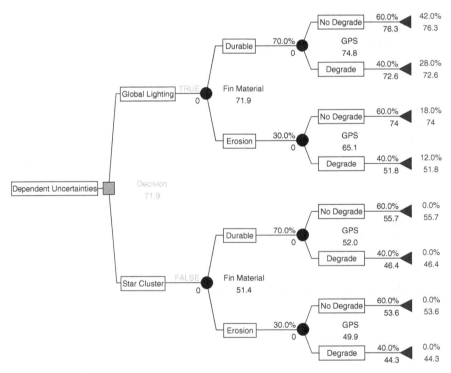

Figure 12.14   Decision tree for dependent uncertainty risk analysis.

	Speed of Launch Platform	Percent Grade Platform can Traverse	Number of Operators	Thrust of Rocket	Number of Different Payloads	Accuracy	Range	Solution Value
Global Lightning	77	19	5	27	100	90 No Degrade	96 Durable	76.3
Global Lightning	77	19	5	27	100	77 Degrade	96 Durable	72.6
Global Lightning	77	19	5	27	100	90 No Degrade	80 Erosion	74.0
Global Lightning	77	0	19	0	5	77 Degrade	80 Erosion	51.8
Star Cluster	34.6	26.2	40.0	99.7	20.0	80.0 No Degrade	55 Durable	55.7
Star Cluster	34.6	26.2	40.0	99.7	20.0	46 Degrade	55 Durable	46.4
Star Cluster	34.6	26.2	40.0	99.7	20.0	80.0 No Degrade	40 Erosion	53.6
Star Cluster	34.6	26.2	40.0	99.7	20.0	46 Degrade	40 Erosion	44.3
Weight	0.23	0.01	0.05	0.12	0.16	0.27	0.14	

Figure 12.15   Value matrix for decision tree.

Figure 12.16 shows the cumulative risk profiles for this situation. When we consider the two dependent uncertainties, we see a more complete picture of the risk of the two solutions. Since the cumulative risk profile of Global Lightning is down and to the right of Star Cluster, we conclude that Global Lightning stochastically dominates Star Cluster [6]. However, from the decision tree (and the cumulative risk profile) we see that there is a 12% probability that Global Lightning will have a value of 51.8 which is less than the original value of Star Cluster before we considered the two dependent uncertainties.

A decision maker would now be interested in knowing the impact of our assumptions about the two uncertainties. If we vary the probability that the Fin is durable from 50% to 100%, the expected value of Global Lightning ranges from 70 to 75 as shown in Figure 12.17. The decision trees, cumulative risk profiles, and sensitivity analysis are easily generated using Precision Tree®. Additional sensitivity analysis techniques—for example, two-way sensitivity—are also available in the within the software application.

This concludes our discussion of sensitivity analysis and uncertainty analysis that has focused on value. Next, we turn to cost analysis. Later we will consider value versus cost tradeoffs.

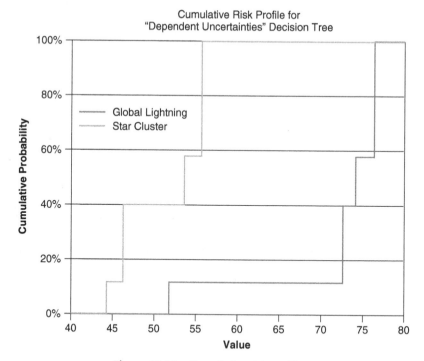

Figure 12.16   Cumulative risk profiles.

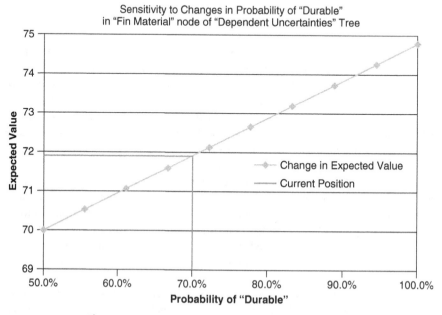

Figure 12.17   Sensitivity of dependent uncertainties.

12.8 CONDUCT COST ANALYSIS

In the Solution Design phase, after the alternatives were identified we began to develop the cost model by identifying the potential cost components and the life cycle costs using the principles and techniques described in Chapter 5. Cost analysts, system engineers, and various component engineers may have used the preliminary cost models to perform cost trades for system components or elements to improve the candidate solutions. The outputs of the Solution Design phase are the candidate solutions.

In this Decision Making phase, we continue our cost modeling and develop life cycle cost estimates for each of the candidate solutions. The systems engineer should know more about the candidate solutions at this point than when the cost components were initially developed. Typically we use two types of life cycle cost (LCC) approaches. Preparing budget estimates for a complete program typically entails preparing full LCC estimates. If detailed budget estimates are not needed, then delta LCC models are more appropriate. Delta LCC models need only estimate the solution cost deltas and not the total life cycle costs. Regardless of which approach is used, this phase is the time to draw on previous cost models and expand these models to provide additional coverage of a fuller range of costs. The key tools for cost analysts to identify the cost elements are the life cycle stages and work breakdown structures for each stage. Chapter 5 provides additional information about cost analysis tools and techniques such as production learning curve models that will be useful in later stages.

Once the systems engineer has ensured that the cost model is complete, the costs are computed for each candidate solution. Cost risks are just as important to consider as performance risks. Monte Carlo simulation can (and perhaps should) be conducted with the cost model to assess the potential cost uncertainty just as was accomplished with the value model when assessing value uncertainty. Chapter 5 describes this technique in detail.

The life cycle cost model can also be very useful in the Solution Implementation phase of the SDP. The system costs will need to be planned, executed, monitored, and controlled. The LCC model can provide a baseline for the initial plan. Cost monitoring and cost management are important implementation tasks. This will be discussed further in Chapter 13, "Solution Implementation."

12.9 CONDUCT COST/BENEFIT ANALYSIS

Armed with the cost data, it is time to plot the cost against the benefit, or value, of each candidate solution. This approach is also known as using cost as an independent variable. It highlights to the decision maker and the stakeholders the cost versus value tradeoffs. There are instances when the candidate solution with the highest value costs significantly more than the other candidate solutions that it may not be cost effective to select that solution for implementation.

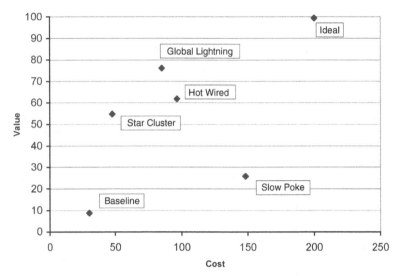

Figure 12.18   Candidate solution cost/benefit (value) plot.

Decision makers want to know the value provided for resources.

The objective is to show the decision maker the tradeoff between higher costs and higher values. This is best accomplished using a graphical representation. Figure 12.18 shows the cost/benefit chart for our rocket example. This plot is a great means to quickly convey to a decision maker the tradeoffs. From the graph, for example, the decision maker can see that he can choose the Star Cluster candidate solution which would have less value than the Global Lightning candidate solution but have a lower cost. On the other hand, the decision maker would never choose the Hot Wired or Slow Poke solutions because they have less value and higher costs than other available solutions. These two solutions are *dominated*. A dominated solution has the same value as, or a value lower than, that of another candidate solution but at greater cost than the candidate solution. Many times it is useful to put a value band (or cost band) on the chart to reflect the uncertainties.

12.10   DECISION-FOCUSED TRANSFORMATION (DFT)

When we developed the quantitative model in Chapter 10, we defined value measures based on preferences across wide, or global, ranges. This allows the systems design team to reap the benefits of identifying additional objectives during the Problem Definition phase and developing improved alternatives during the Solution Design phase [10]. As mentioned in Section 12.3.6, we should revisit the value measure ranges and associated swing weights. In practice, we often find that the candidate solutions do not span the entire global ranges specified on the value measures. The Decision-Focused Transformation [12] provides a methodology to revise value functions and swing weights prior to communication of analysis results without any additional consultation with stakeholders. When appropriate, the Decision-Focused Transformation offers the benefit of enhanced communication of analysis results by eliminating consideration of the common and unavailable value through rescaling of the value functions. Additionally, it revises our swing weights to reflect importance across these new value measure ranges. The transformation preserves the rank-ordering of the candidate solutions and the major differences of the nondominated candidate solutions. Through this, communication of results with stakeholders is enhanced, thereby increasing decision clarity.

12.10.1 Transformation Equations

The Decision-Focused Transformation is used as we prepare to present analysis to stakeholders. We have a set of alternatives, A, and their single-dimensional value, $v_i(x_{ij})$, on global value measures $i = 1, \ldots, n$. Additionally, we have measure weights, w_i, for all value measures. The magnitude of the value scale, S, may cover any range, but many use $0-1$, $0-10$, or $0-100$. We are able to calculate $v(x_j)$, or the global value of alternative $j \in A$ using the additive value model described in Section 10.4.4. Table 12.6 provides a summary of the distinctions, descriptions, and equations used in the Decision-Focused Transformation.

In our rocket problem, we perform the necessary calculations shown in Table 12.6 to find that the common value is 7.2%, the unavailable value is 9.3%, and the discriminatory value is 83.5%. In Table 12.7, we see that discriminatory value is the sum of the discriminatory power of the value measures. This means that tradeoffs only occur within 83.5% of our original value-focused decision space. Using the last equation in Table 12.6, we are able to easily calculate the transformed total decision value of our candidate solutions. We also provide the transformation of measure weights in Table 12.7 and single-dimensional value in Table 12.8 to illustrate the transformation of the value model.

In Table 12.8, we first note that we used the equation $v'(x_j) = \sum_{i=1}^{n} w_i' v_i'(x_{ij})$ from Table 12.6 to calculate total decision value. We could also use $v'(x_j) = \frac{1}{V_d}(v(x_j) - V_c)$ without having to calculate decision measure weights or single-dimensional decision value. In the transformed single-dimensional decision value, the Hypothetical Worst is made up of the worst performances on each value measure and the Hypothetical Best is made up of the best performances on each value measure. The Hypothetical Best alternative is similar in concept to the "improved candidate" shown in Figure 12.12. These two hypothetical alternatives now bound our decision trade-space. By looking for scores of 0.0 in the transformed score data, we see that the Baseline provides equal to the worst performance on Speed of Platform, Percent Grade, Thrust, Number of Payloads, and Distance from Target. On the other hand, decision value of 100.0 indicates that the Global Lightning is the best performer on Speed of Platform, Number of Payloads, Distance from Target, and Range. When graphed in a stacked-bar chart, the differences in this data become clearer.

TABLE 12.6 Decision-Focused Transformation Distinctions, Descriptions, and Equations

Notation	Distinction	Description	Equation
r_i	Value range Utilization	Amount of the single-dimensional value scale spanned by the set of alternatives on the ith value measure.	$r_i = \max_j v_i(x_{ij}) - \min_j v_i(x_{ij})$
d_i	Discriminatory power	Weighted value range utilization on the ith value measure. Relative ability of each value measure to distinguish between alternatives.	$d_i = w_i r_i$
V_c	Common value	Sum of the weighted minimum value achieved by all alternatives on all value measures. Common to all alternatives and does not provide distinction between them; also the value of a "hypothetical worst" alternative.	$V_c = \sum_{i=1}^n w_i[\min_j v_i(x_{ij})]$
V_u	Unavailable value	Difference between S, the magnitude of the value scale, and the sum of the weighted maximum value achieved by all alternatives on all value measures. Not achieved by any alternative and does not provide distinction between them; summation defines value of a "hypothetical best" alternative.	$V_u = S - \sum_{i=1}^n w_i[\max_j v_i(x_{ij})]$
V_d	Discriminatory value	Sum of the discriminatory power of the value measures given the alternatives' performance. Only value space where tradeoffs between alternatives occur.	$V_d = \sum_{i=1}^n d_i = S - V_c - V_u$
$v'_i(x_{ij})$	Decision value functions	Affine transformation of global single-dimensional value functions to a local scale defined by the alternatives in A.	$v'_i(x_{ij}) = \dfrac{v_i(x_{ij}) - \min_j v_i(x_{ij})}{\max_j v_i(x_{ij}) - \min_j v_i(x_{ij})}$
w'_i	Decision measure weights	Recalculated measure weights based on alternative scores.	$w'_i = \dfrac{d_i}{\sum_{i=1}^n d_i} = \dfrac{d_i}{V_d}$
$v'(x_j)$	Total Decision Value	Total value of alternative j in the local decision value model.	$v'(x_j) = \sum_{i=1}^n w'_i v'_i(x_{ij}) = \dfrac{1}{V_d}(v(x_j) - V_c)$

TABLE 12.7 Transformation to Decision Weights

DFT Element	Speed of Platform (kph)	Percent Grade (%)	Number of Operators (#)	Thrust (lb)	Number of Payloads (#)	Distance from Target (m)	Range (km)
$\max_j v_i(x_{ij})$	77.0	88.0	90.0	99.7	100.0	90.0	96.7
$\min_j v_i(x_{ij})$	14.0	1.4	5.0	0.0	10.0	0.0	13.5
r_i	63.0	86.6	85.0	99.7	90.0	90.0	83.2
w_i	0.23	0.01	0.06	0.12	0.17	0.27	0.14
d_i	14.67	1.19	4.66	12.29	14.79	24.66	11.39
w_i'	0.18	1.01	0.06	0.15	0.18	0.29	0.14

Discriminatory Value $V_d = 83.5$

12.10.2 Visual Demonstration of Decision-Focused Transformation

With the transformation complete at the single-dimensional value function and measure weight level, we can see the effect of the decision-focused transformation on our rocket problem in Figure 12.19. In boxes 1 and 2, we display the original stacked bar chart and cost/benefit chart similar to Figures 12.13 and 12.18. In this new stacked bar chart, we also show hypothetical best/worst alternatives as defined by the best/worst performance of our candidate solutions on each single-dimensional value measure. In box 3, we display the decision-focused stacked-bar chart. The hypothetical worst now has a value of 0.0 and the hypothetical best has a value of 100.0. The transformation removed common and unavailable value. In box 4, we display the updated cost/benefit chart after the decision-focused transformation.

12.10.3 Cost/Benefit Analysis and Removal of Candidate Solutions

With the candidate solutions under consideration, we might initially conclude that our value model is relatively well-scaled in that the discriminatory value is large at 83.5% and the effect of the transformation does not drastically enhance our ability to communicate the differences between our alternatives. The decision-focused transformation is still useful when the decision is made to reduce the candidate solution set based on dominance. In our rocket problem, the Hot Wired and Slow Poke are dominated. Although an analyst does not make the decision to remove solutions, typically dominated solutions (Hot Wired and Slow Poke) are eliminated during final presentations to the decision maker. When DFT is implemented in software, we can remove these alternatives in real-time and update the transformed model. If the dominated candidate solutions are either the best or worst sole performer on any value measure, then our decision trade space will be further narrowed beyond the original transformation. As shown in Figure 12.20, we can consider the differences between the three remaining alternatives (Baseline, Star Cluster, and Global Lightning).

TABLE 12.8 Transformation to Decision Value

$v'_i(x_{ij})$	Speed of Platform (kph)	Percent Grade (%)	Number of Operators (#)	Thrust (lb)	Number of Payloads (#)	Distance from Target (m)	Range (km)	Total Decision Value ($v'(x_j)$)
Hypothetical worst	0.0	0.0	0.0	0.0	0.0	0.0	0.0	**0.0**
Hypothetical best	100.0	100.0	100.0	100.0	100.0	100.0	100.0	**100.0**
Baseline	0.0	0.0	41.2	0.0	0.0	0.0	7.8	**3.4**
Global lightning	100.0	11.5	0.0	27.5	100.0	100.0	100.0	**82.6**
Hot wired	80.0	100.0	64.7	91.6	55.6	77.8	0.0	**65.2**
Star cluster	33.3	19.2	41.2	100.0	11.1	85.2	49.9	**57.0**
Slow poke	0.0	51.3	100.0	7.2	0.0	35.6	27.1	**21.5**

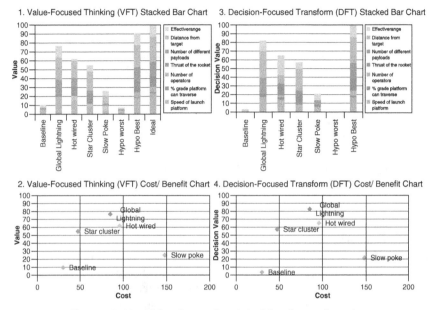

Figure 12.19   Value-focused and decision-focused results.

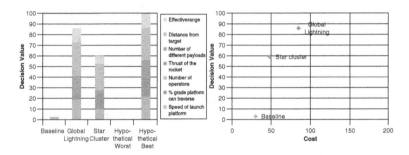

Figure 12.20   Decision-focused results after removal of dominated solutions.

With the dominated solutions (Hot Wired and Slow Poke) removed, we see that the Baseline is the worst performer on all measures other than the number of operators value measure. The absence of a bar indicates that a candidate solution is the worst performer in the remaining set. Since Star Cluster provides gains over the Baseline on several value measures for a relatively small increase in cost, it might be reasonable to believe that the decision maker would decide to remove the Baseline from consideration. If so, we can easily update the Decision-Focused model without the Baseline as shown in Figure 12.21. This second update leaves us with only two remaining solutions, the Star Cluster and Global Lighting.

Any time that we remove additional candidate solutions, we perform another distinct iteration of the Decision-Focused Transformation based on the common,

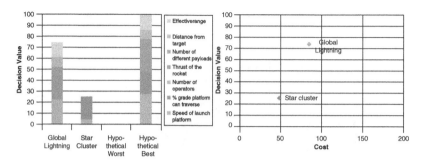

Figure 12.21 Decision-Focused results after removal of dominated and Baseline solutions.

unavailable, and discriminatory value of the reduced solution set. Now that we have only two remaining alternatives in our Decision-Focused model, we are able to see the tradeoffs clearly. Global Lightning costs more, but provides decision value on Speed of Launch Platform, Number of Different Payloads, Accuracy (Distance from Target), and Effective Range value measures. Star Cluster is less expensive and does provide decision value on Grade Platform Can Traverse, Number of Operators, and Thrust of the Rocket value measures. Although Star Cluster performs better on Grade Platform can Traverse, the difference in decision value is minor. Based on the small variation in performance between the two alternatives, the Grade Platform can Traverse measure has been reduced significantly in weight to the point where it nearly drops from our analysis; this measure does not have much discriminatory power. Discriminatory power, and the resulting decision measure weights, vividly display how measure weights depend on the importance of swinging across the established range of the value measures [12].

Additionally, we also observe that Global Lightning now has a decision value of 74.7 and Star Cluster now has a decision value of 25.3. With only two alternatives remaining, each alternative contributes to the hypothetical best for differing value measures. Their value scores now sum to 100.0, and the value scores are only based on the relative strengths between the two solutions. At a deeper theoretical level, this highlights the concept that the numeric value depends on the size of the decision trade-space. When we have a large value space initially, we see less differentiation between solutions. If we narrowly define our preferences based on a smaller set of candidate solutions, the model will display greater differentiation.

In Figure 12.21, the decision maker is presented with the simplest possible form of results and is able to consider the tradeoffs between the final two solutions with greater clarity. In the process of applying Value-Focused Thinking to build a qualitative and then quantitative model, we retain the benefits of generating additional alternatives and uncovering hidden objectives along with other benefits [1, 10]. By applying the Decision-Focused Transformation with a fixed set of solutions, we facilitate communication and clarity concerning the value and cost tradeoffs as the set is reduced, thus reinforcing commitment to action. In sum, we use Value-Focused Thinking to understand and model the complexity inherent in the situation and the Decision-Focused Transformation to simplify the understanding of tradeoffs as we move to decide.

Since Decision-Focused Transformation is a tool designed to enhance communication, we should consider the circumstances under which we would use the transformation and when it may not be useful. Dees et al. [12] prescribe that the transformation is most useful when both common and unavailable value are large (resulting in small discriminatory value) and when multiple nondominated alternatives exist. Additionally, we have shown above that the transformation is useful as we reduce the set of nondominated solutions. However, the analyst using Decision-Focused Transformation must communicate the size of the reduced decision trade-space each time the transformation is applied. Common value and unavailable value do not provide any distinction between our alternatives and cloud our ability to understand tradeoffs, but understanding their magnitudes is important. Figure 12.22 displays the magnitude of the common, unavailable, and discriminatory value after each iteration in which we removed alternatives.

As shown in Figure 12.22, we had tradeoffs in 83.5% of our value trade-space in our original Value-Focused model. When we removed the dominated alternatives, the discriminatory value reduced slightly to 79%. On the top end, we see more (12.9%) unavailable value because Hot Wired is the sole best performer on Percent Grade and Slow Poke is the sole best performer on Number of Operators.

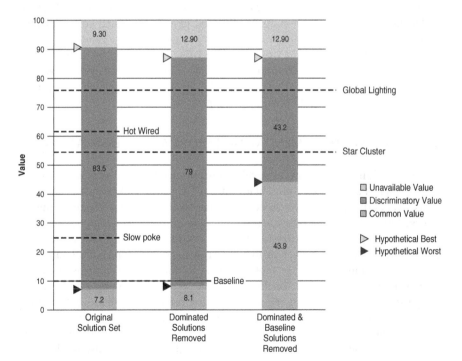

Figure 12.22 Common, Discriminatory, and Unavailable Value as alternatives are removed.

On the bottom, we see more (8.1%) common value as Hot Wired is the sole worst performer on Range and Slow Poke is the sole worst performer on Speed of Platform. In Figure 12.20, we showed that removal of the dominated alternatives along with a second iteration of the Decision-Focused Transformation produces a clearer picture of the tradeoffs than the original Value-Focused model.

Also in Figure 12.22, we show that the discriminatory value reduces to 43.2% when we further eliminate the Baseline and perform a final iteration of the Decision-Focused Transformation with only Global Lightning and Star Cluster. Unavailable value remains the same at 12.9%, but common value significantly increases to 43.9% as the Baseline is the sole worst remaining performer on Speed of Platform, Percent Grade, Thrust, Number of Payloads, Distance from Target, and Range. As we reduced the decision trade-space to 43.2% of the original size through the transformation and removal of alternatives, we must ensure that the decision maker understands we are comparing differences between solutions in a narrower band. In Figure 12.21, we offered the simplest possible tradeoffs between the final two candidate solutions in 43.2% of our original trade-space.

Low/high common value either indicates that the alternatives are all poor/good or that they are very different/similar. Similarly, low/high unavailable value indicates either that the alternatives are all good/poor or that expectations may be low/high [12]. The Decision-Focused Transformation is most useful in complex problems when there is both high common and unavailable value which imply a small discriminatory value, or decision trade-space. As discriminatory value becomes smaller, the alternatives have more similar total global value, and communicating the size of the reduced decision trade-space becomes more important.

This completes the systems engineer's analysis, but not the work. Before the systems engineer can move to the next phase in the systems decision process or the overall life cycle process, it is time to tie all the work together in a written and/or oral presentation and obtain a decision.

12.11   PREPARE RECOMMENDATION REPORT AND PRESENTATION

After all this painstaking work to develop a recommendation, it is now time to put it all together in a report or a presentation for the decision maker and stakeholders. This is a very critical step in the process. Outstanding analytical work can be quickly dismissed by a decision maker when the presentation is overly complicated or too simplistic. The perceived professionalism of a written report or oral presentation can convince a decision maker of the validity of a recommendation, confuse the decision maker into inaction, or motivate them to find a better systems engineer.

There is no one set order for developing the report or making a presentation. Though some decision makers request a decision briefing and then want a follow-up report that includes the decision and implementation plan, most decision makers request a final decision briefing accompanied by a detailed report. The implementation plan is then developed separately. We assume that this latter case is addressing what follows. While we do make several specific suggestions concerning the format of a presentation to increase its usefulness to the decision maker, our primary objective here is to provide general guidance on important factors that give a report and presentation the greatest chance of success.

12.11.1 Develop Report

Organizations will often establish a standard format for written technical reports. Standardizing the content and format lessens the burden on an analyst who has to prepare the report and makes it easier for decision makers to locate specific items of interest in the report. Regardless of the existence or absence of a specific format, there are some basic principles in the development of a technical report for a decision maker. Reports should include an executive summary, main body, and appendices (as necessary). These are in order in the report and in the order of detail.

The key to a successful technical report is a clear, concise executive summary.

Executive Summary The executive summary is designed to provide a brief overview of the content of the report. It provides the decision maker with enough supporting facts to make a decision without having to read the entire report. It should include the objective of the report (often to obtain a decision), the most compelling evidence to support a decision, and a quick overview of the methodology used. The best executive summaries can be crafted to fit on a single page. It should very rarely be longer than 10% of the overall length of the main body or over five pages, whichever is less. Additional details are provided in the body of the technical report.

Main Body The main body of the report is designed to be a much more detailed explanation of the study. Here the systems engineer must tell the story of what the analysis means to the decision maker and key stakeholders. This is a technical report. The writing should be very concise and restricted to the important parts of the analysis which support the recommendation. It should be organized to allow the decision maker and key stakeholders to follow the analysis from the initial problem statement until the recommended decision. Rarely should the systems engineer include steps not taken. For example, if the analysis did not lend itself to operational testing, the analyst should not include a paragraph on operational testing even if the organization's standard report format calls for such a paragraph. The only exception would be if the absence of this step has a significant impact on the recommendation (e.g., in risk mitigation).

The main body should be detailed enough for understanding the analysis conducted and how it supports the decision, but should refrain from being so detailed that the analysis obscures the recommendation.

Appendices The appendices of the report should include detailed formulations of models, simulation code, and data. These are rarely of interest to the decision maker, unless he or she is extremely technical or there are questions in the analysis, but are very useful to other analysts or stakeholders. A decision maker may ask other analysts to comment on your report, in which case, these appendices are very important.

The final crucial part of any written work is the documentation and references to any support received or researched in the analysis. Proper documentation provides two things. First, it provides credibility to the work as it shows support by previous respected work. Second, and most important, it supports the integrity of the analyst. Nothing destroys an analysis as quickly as questionable documentation and even experienced systems engineers cannot easily recover from integrity problems.

12.11.2   Develop Presentation

The single most important consideration when developing a presentation is understanding what the decision maker needs in the presentation in order to make a decision. Written reports are commonly tailored to the type of problem being addressed and accepted report format of an organization. Oral presentations must be tailored to the decision maker. They must include the detail required to make a decision and to capture and hold the interest of the decision maker throughout the presentation. There are some general guidelines to follow when constructing a presentation used to obtain a decision.

The most successful presentations stay on message and stay within time limits.

Opening   The opening should set the tone for the remainder of the presentation. The presenter should immediately state the purpose of the presentation to focus the expectations of the decision maker. In this case, the purpose is to obtain a decision. Immediately following the purpose, the presenter should provide the decision maker with enough background on the problem to frame and focus their attention on the topic being presented. Although the topic may be fresh in the mind of the presenter, the decision maker may have just left a situation involving a topic entirely different than the one at hand. The presenter should explain why the problem and the current presentation are important to the decision maker.

The final part of the opening is the recommendation. This is known as "the Bottom-Line Up Front" (BLUF). This provides the decision maker with both a good idea of where the presentation is heading and the recommended solution decision. Knowing the final recommendation helps the decision maker focus on the questions critical to the decision that he or she will make.

Presentation of Analysis   The presenter should start the description of the analysis from an accepted point of common knowledge. This might be a summary of the previous meeting or even going back to the original problem statement. This allows the decision maker to feel knowledgeable and comfortable at the start of the discussion of the analysis.

From there, the briefing should take the decision maker through the process at a detail required to maintain his or her interest and understanding until the recommendation is reached. Some decision makers are very detail-oriented and want the formulations and the data. Some want only highlights. In the absence of prior knowledge, the presenter should present limited details and have backup information ready to address specific questions.

The presentation should logically flow from the start point until an ultimate conclusion. This keeps the decision maker knowledgeable and comfortable. A knowledgeable and comfortable decision maker will be much more likely to support an analysis and make a decision at the end of a presentation than a decision maker who is overwhelmed with information and confused. This decision maker is more likely to put off a decision rather than make a wrong decision.

A good presenter will know exactly how much time the decision maker has available and will keep the briefing shorter than the time allotted. This allows more time for questions, and a busy decision maker will appreciate the extra time. Do not assume that the decision maker will allocate extra time for this presentation because he or she may leave prior to making a decision. Always have a one chart summary if the decision maker has to shorten the time.

Always have a one chart summary in case the decision maker has to shorten the time.

Concluding the Presentation   After presenting a concise and convincing argument, the presenter should restate the recommendation and ask for a decision. When a presenter states at the start of the presentation that the purpose is to obtain a decision, the decision maker will be prepared for this request. The decision maker might want to put off the decision; and, if so, the presenter should politely ask when a decision might be forthcoming. Though some decision makers do not like to be pressed, when the timing of the decision is critical (e.g., in the progress of a manufacturing or development process), it is worth the effort to press the issue.

Whether the decision maker makes a decision or not, the presenter should continue with the future actions required based on the decision or lack thereof. Since the decision may significantly change the information prepared in advance of the presentation, the presenter should be prepared to adjust the plans as necessary.

Some final thoughts on briefings:

- *Do Not Read the Slides*. Nothing detracts from a presentation and infuriates an audience as quickly. Summarize the slide or the chart.
- *Have Simple Slides and Quick Thoughts*. A slide or concept in a presentation that tries to convey too much information often loses the audience and conveys little.
- *Transition the Decision Maker to Focus on the Problem Topic Very Early in the Presentation*. Yours is not the only problem on their mind.
- *Be Careful with the Use of Pointers*. These can often distract the audience from the presentation, especially if the presenter is nervous!

- *Keep Text Font Size Consistent Throughout the Presentation*. Using larger font sizes has an effect similar to that of capitalizing letters in e-mail: giving the impression of yelling.
- *Dress Professionally*. The presenter should always be more formal than the decision maker, but not overly so.
- *Speak Professionally*. Do not use quaint or colloquial phrases or try to be too funny. The briefing is designed to obtain a decision, not audition for stand-up.
- *Stay on Message*. Do not introduce tangential material that is not essential to the decision.
- *End in Control of the Presentation in a Way That Lets the Decision Maker Know Where the Project Is at, Where It Is Going, When Deliverables Should Be Expected, and What Actions Are Required of the Decision Maker in Order to Make the Project a Success*. It is a parting shot to reframe the presentation content before you lose the decision maker's attention.

Using a Storyline Approach   One straightforward method for organizing information and presenting it effectively using is called the *storyline* method. There are two principles invoked when creating a presentation using this approach: horizontal integration of the story and vertical integration of support. Conceptualizing each presentation slide as a single page of a book, the area at the top of the slide typically reserved for a slide title is used to "tell the story" of the presentation content from start to finish using a single sentence on each slide. Done correctly, the decision maker should be able to read across the top of every slide and understand the main messages that the system team wants to convey. This effect is known as achieving *horizontal integration* in the sense that if all the slides were laid out on a table in order, the presentation storyline could be read by reading horizontally across the slides.

The main body of each slide is then used to present key evidence (e.g., text, graphics, mathematics, simulation results, etc.) supporting the storyline sentence present in the slide title area. This is known as achieving *vertical integration* of the presentation material. It is "vertical" in the sense that the typical audience member will logically look to the title area first, encountering the storyline statement, and then "drill down" into the supporting evidence below the statement to understand the logical basis for the statement. Figure 12.23 illustrates a comparison between the storyline approach and a typical default presentation format that uses simple labels as slide titles in Microsoft® PowerPoint.

One attractive feature of this method is that it forces a presenter to clearly address the salient points needing to be made, the logic connecting these points, and the key elements of convincing evidence that the statement is factually based in its claim. This frees the presenter to add value during the presentation by providing the audience with insights and reasoning that complement what they are seeing instead of reading the content of the slides to the audience, which is considered bad practice.

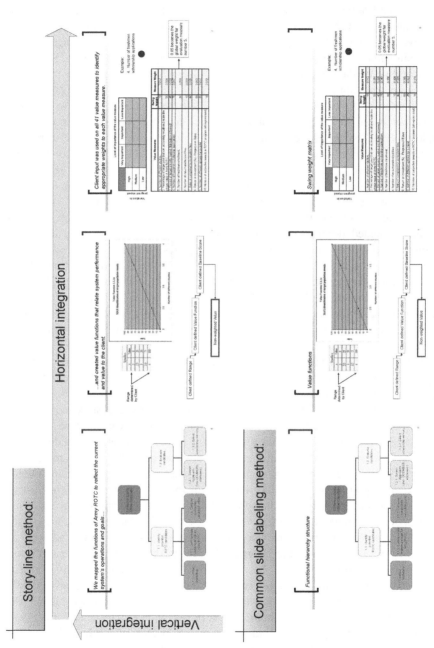

Figure 12.23 Two methods for organizing and presenting information in slideshows [13].

The storyline method delivers two additional benefits that add to its appeal for presentations supporting systems decision making. First, it is not uncommon for decision briefings to be circulated widely throughout an organization after the presentation concludes. Vertical and horizontal integration helps prevent individuals who were not present for the presentation from misinterpreting the message because the main points are clearly present along with their supporting evidence. Similarly, the storyline method enables slide handouts to function as stand-alone references for the presentation at a later date.

Secondly, slides created using this method tell the intended story. The resulting presentation can serve as a logical template for creating a technical report on the topic as well. Using the presentation in this fashion requires each slide to be first placed on a single page in a document. Next, one elaborates on the title line statement and describes to the reader the evidence providing vertical integration support to the statement. Any graphics or images required to make the important points clear are retained as figures and charts in the technical report. Any slide images that contain purely textual information will eventually be deleted, replaced by the expanded description crafted in the report body. Adding any necessary references and section organization nearly completes the report.

Lastly, a storyline approach is very helpful to "story board" the flow of the presentation prior to implementing it in software. One way of doing this when classroom or conference facilities are available is as follows. Estimating as a rule of thumb that every slide will consume approximately a minute of presentation time on average, draw an empty box (placeholder slide) for each minute of the presentation. Below each slide, block out and identify groups of slides that will contain the general content of the presentation as the team intends it to unfold. These contents consist of, but are not limited to: title slide/team identification, agenda, current project timeline, bottom-line up-front (optional but encouraged), problem background and description, methodology, modeling, results and analysis, conclusions, recommendations, areas for further improvement, updated timeline, references (optional). The logical organization of these placeholder slides aligns with the horizontal integration of the slides when the presentation is complete.

Next, identify the content of each slide (in general terms, not specific detail) needed to support the storyline. The idea here is to see the presentation from a single, macroscopic perspective in the hope that by doing so any gaps in logic, analysis, content, and so on, will be revealed. Finally, by examining the information the team actually possesses to support the storyline, the team's workflow can be adjusted as necessary to fill-in any missing information prior to the presentation being given.

Presentation software is not the only choice for conducting effective presentations. Very successful briefings can be conducted using butcher charts, simple paper slides, or even chalk. The key is that the presentation is professional and it is concise. A decision maker will appreciate a presentation much more if it conveys a simple message than if the words come flying in from the side and there are explosions and movies. Many experts suggest that slides or charts should include no more than three ideas and have fewer than four very short lines per slide, chart, or board space.

12.12   PREPARE FOR SOLUTION IMPLEMENTATION

In this chapter, we have worked through the decision making phase of the system decision process. We started with the important items developed in the previous phases of the process, the revised problem statement, the requirements, the value model, some cost models, the candidate solutions and some previous results of the initial analysis from the modeling and simulation and testing efforts. After reviewing these important elements of our analysis to ensure they are still relevant, we were ready to proceed.

We identified the data we were going to use to complete the raw data matrix. We then used the value model to convert the raw data to the values and obtained the value for each candidate solution. After evaluating the risks, the sensitivity of our analysis and the costs of our candidate solutions, and developing improved solutions, we were ready to develop a recommendation.

After developing a recommendation, we prepared written reports and oral presentations in order to obtain a decision from the decision maker. We now have to determine what we are going to do with the solution decision. Depending on where in the system life cycle the process has been employed, the systems engineer must now develop a plan to implement the solution decision. This leads to the next and final step in the systems decision process: solution implementation. The systems engineer will find that the easy part of the process has been done. The implementation is the difficult part of process.

12.13   ILLUSTRATIVE EXAMPLE: SYSTEMS ENGINEERING CURRICULUM MANAGEMENT SYSTEM (CMS)—DECISION MAKING

Robert Kewley, Ph.D. U.S. Military Academy

Decision Making

Once they had alternative solutions from the solution design phase of the system concept decision, the design team had to score those solutions against the values developed during the problem definition phase in order to come up with a recommended decision to present to the department leadership, the customer, for funding. In order to score each candidate solution, the design team used research from the solution design phase to evaluate each objective using the measures of effectiveness. The raw data matrix in Figure 12.24 shows the results of this subjective assessment.

The constructive scores had the following values:

−1 Worse than current system

  0 Same as current system

	Objectives	Measures of effectiveness	Solutions				
			Improve existing system	Contractor development	Web data development	Content management (A)	Content management (B)
Value matrix	Maximize content sharing	Constructive scale comparison to current system	1	2	2	3	3
	Enforce standards	Constructive scale comparison to current system	1	2	2	2	2
	Maximize availability	Constructive scale comparison to current system	0	0	0	-1	-1
	Maximize collaboration	Constructive scale comparison to current system	1	2	1	3	3
	Maximize usability	Constructive scale comparison to current system	1	1	1	3	2
	Maximize flexibility	Constructive scale comparison to current system	-1	0	1	3	2
	Maximize security	Constructive scale comparison to current system	0	2	2	2	2
	Minimize dev. and maint. time	Months to achieve course-level functionality	1	6	4	2	3
	Minimize dev. cost	Total development cost in dollars	$0	$250,000	$1,000	$2,000	$0
	Integrate	Constructive scale comparison to current system	0	1	1	2	2

Figure 12.24   Raw data matrix for CMS system concept decision.

+**1** Marginal improvement to current system

+**2** Some improvement to current system

+**3** Significant improvement to current system

In order to determine the total value of each candidate solution, these raw data scores had to be converted to value scores (on a scale of 0 to 10) using the value functions developed in the value modeling step of the problem definition phase. The results of this transformation are shown in Figure 12.25.

In order to get a total value for each alternative, the design team calculated the value of each candidate solution using the additive value model [Equation (9.1)]. Figure 12.26 shows a graph of these results.

This analysis shows that the content management system for Vendor A is the best solution for providing IT support to the curriculum management functions in the Department of Systems Engineering. A closer look at the measures of effectiveness shows that the usability and collaboration capabilities of content management systems give them significantly more value than other forms of development. This is primarily due to built-in capabilities for file management, discussion, survey response, and e-mail. Furthermore, Vendor A provides an advantage over Vendor B with respect to instructor ease of use, flexibility, and integration. The design team assessed Vendor A's product to provide more drag-and-drop functionality, more user customization features, shorter development time to tweak or add features, and a better capability to integrate with other Academy-wide IT systems.

	Objectives	Measures of effectiveness	Solutions				
			Improve existing system	Contractor development	Web data development	Content management (A)	Content management (B)
Raw Data Matrix	Maximize content sharing	Constructive scale comparison to current system	3	6	6	10	10
	Enforce standards	Constructive scale comparison to current system	3	6	6	6	6
	Maximize availability	Constructive scale comparison to current system	4	4	4	0	0
	Maximize collaboration	Constructive scale comparison to current system	3	6	3	10	10
	Maximize usability	Constructive scale comparison to current system	5	5	5	10	7.5
	Maximize flexibility	Constructive scale comparison to current system	0	2.5	5	10	7.5
	Maximize security	Constructive scale comparison to current system	4	9	9	9	9
	Minimize dev and maint time	Months to achieve course-level functionality	8.3	0.0	3.3	6.7	5.0
	Minimize dev cost	Total development cost in dollars	10	0	10	10	10
	Integrate	Constructive scale comparison to current system	1	3	3	6	6

Figure 12.25 Value matrix for CMS system concept decision.

Because usability is such a significant factor in the total value, the design team performed sensitivity analysis to see how the scores might change if this factor were not weighted so heavily. This analysis calculated the resulting total value if the importance of usability in swing weighting were medium or low, instead of high. The resulting swing weighting values would be 75 and 20, instead of 100. Figure 12.27 shows the results of this analysis. The ordering of the alternatives did not change as collaboration importance ranged from very important to important to less important. This gives increased confidence in the recommendation. Similar analysis was done for other measures of effectiveness.

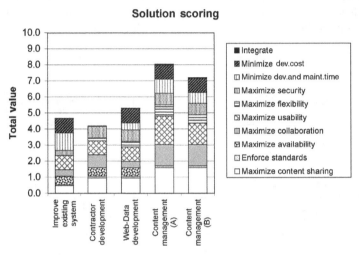

Figure 12.26 Solution scoring for CMS system concept decision.

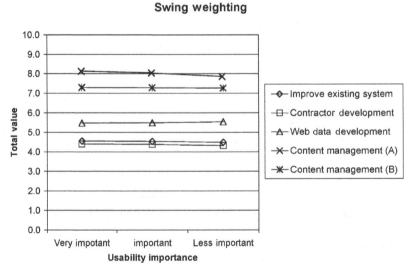

Figure 12.27 Sensitivity analysis for importance of usability.

Based on the scoring results, the design team recommended that the department head authorize the purchase of content management software, database software, and development tools from Vendor A so that they could begin development of the proposed system. As the project progressed, they would attempt to integrate capstone students into the development process. This recommendation is based on the fact that content management tools support both information and collaboration with features like file management, discussion, survey response, and e-mail. Vendor A's features provide an advantage over Vendor B with respect to instructor ease of use, flexibility, and integration. The department head, who had been a part of the process from the problem definition phase, agreed with the recommendation and approved the system concept, the software purchase, and the development timeline.

12.14 EXERCISES

12.1. What important things brought forward from the previous phases in the SDP should a systems engineer review prior to the start of the decision-making step? What should the systems engineer ensure about them prior to moving forward?

Situation. Assume you are the member of a systems design team tasked with designing baggage handling system for the new Baghdad International Airport.

12.2. One measure your team identifies is "Handling Time" or the time to process a bag from the plane to the passenger in the terminal. Describe a method to evaluate this measure for each of the four different ways of scoring alternatives. Identify the strengths and weaknesses of each method.

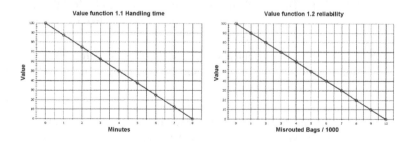

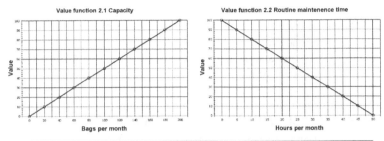

	Handling time (minutes)	Reliability (misrouted bags/1000)	Capacity (Bags/hrs)	Routine maintenance time (Hrs/Month)
High Tech	2.1	3	190	35
Low Tech	5.3	9	160	45
Mixed	6.2	7	100	25
Manual	7.7	2	75	5

Situation. Your team developed the value curves shown below in previous steps of the process. You have now ducted the scoring of all the alternatives and completed the raw data matrix shown below.

12.3. Given the value curves and the raw data matrix above, develop the value matrix for these candidate solutions.

12.4. Given the weights for each measure in the table below, calculate the value for each candidate solution.

Measure	Weight
Handling time	0.37
Reliability	0.28
Capacity	0.22
Routine maintenance time	0.13

Situation. The Hi-Tech candidate solution calls for the use of very advanced technologies. Some of these technologies are not yet commercially viable (still in development.)

12.5. Is the above statement about the technologies of the Hi Tech candidate solution an example of risk or uncertainty? How would a systems engineer address this in the analysis?

Situation. During the scoring of the candidate solutions, the handling time for the Hi Tech solution and the capacity of the manual solution, though shown above in the raw data matrix as deterministic, were really stochastic.

12.6. Is the above statement about the variability of the measure scores an example of risk or uncertainty? If the handling time for the Hi-Tech solution and the capacity for the manual solution both followed triangle distributions ([1.1, 2.1, 3.1] and [60, 75, 90], respectively), how should a systems engineer address this variability?

12.7. Given the sensitivity chart shown below, how should a systems engineer present the sensitivity of the top-level function "Manage Baggage"? (*Note*: The values in this chart do not completely correspond to the previous numbers in this example as the entire value model is not presented.)

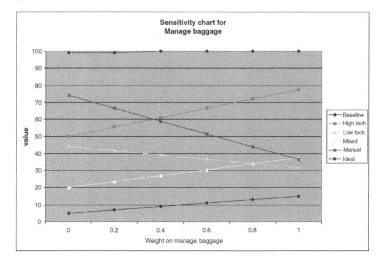

12.8. Use Decision-Focused Transformation to better communicate the value tradeoffs of the nondominated candidate solutions. Is DFT helpful for this problem? Why or why not?

12.9. Given the costs shown below for each candidate solution and the values determined in Exercise 12.4, above, develop the cost versus value graph. Identify the dominated solutions determined in the graph.

Candidate Solution	Cost (millions)
Ideal	$150
High tech	$120
Low tech	$95
Mixed	$75
Manual	$45
Baseline	$15

12.10. Based on all the analysis conducted in the previous exercises for this chapter and especially the cost versus value graph, make a recommendation to a decision maker and justify the recommendation in one or two paragraphs.

12.11. Prepare a one-page executive summary for the decision maker on your analysis of this baggage handling system. Also, develop a top-level outline of the slides you will prepare for the decision briefing that you will present to the decision maker on the analysis conducted above. Assume you only have 20 minutes and you have not presented anything to this decision maker since the approval of your revised problem statement.

REFERENCES

1. Parnell, G. Value-focused thinking. *Methods for Conducting Military Operational Analysis*. Washington, DC: Military Operations Research Society, 2007.

2. Maxwell, D. Decision Analysis: Aiding Insight VII: Decision Analysis Software. In: *OR/MS Today*; Marietta, Georgia, Lionheart Publishing, 2004.

3. Logical Decisions. 2010. Available at http://www.logicaldecisions.com. Accessed August 20, 2010.

4. Kirkwood, C. *Strategic Decision Making: Multiobjective Decision Analysis with Spreadsheets*; Pacific Grove, CA: Duxbury Press, 1997.

5. Ragsdale, C. *Spreadsheet Modeling and Decision Analysis*, 4th ed. Thomson South-Western, 2004.

6. Clemen, R, Reilly, T. *Making Hard Decisions with Decision Tools Suite*. Pacific Grove, CA; Duxbury Press, 2004.

7. Parnell, G, Jackson, J, Burk, R, Lehmkuhl, L, Engelbrecht, J. R&D concept decision analysis: Using alternate futures for sensitivity analysis. *Journal of Multi-Criteria Decision Analysis*, 1999; 8: 119–127.

8. Oracle® Crystal Ball. Denver, CO: Decisioneering, Inc, 2005. Available at. http://www.decisioneering.com. Accessed April 10, 2010.

9. @Risk. Palisade Corporation. Available at http://www.palisade.com, Accessed April 10, 2010.

10. Keeney, R. *Value-Focused Thinking: A Path to Creative Decisionmaking*. Cambridge, MA: Harvard University Press, 1992.

11. Precision Tree by Palidade, http://www.palisade.com/precisiontree/. Accessed August 20, 2010.

12. Dees, R, Dabkowski, M, Parnell, G. Decision-focused transformation of additive value models to improve communication, *Decision Analysis*, 2010; 7: 172–184.

13. Balthazar, T, Dubois, P, Heacock, J, Stoinoff, C, Driscoll, P. A systems perspective on Army ROTC. Department of Systems Engineering Capstone Conference, U.S. Military Academy, West Point, 2007.

Chapter 13

Solution Implementation

KENNETH W. MCDONALD, Ph.D.
DANIEL J. MCCARTHY, Ph.D.

> The focus of a project manager during a solution implementation is to do everything possible to insure the system delivers the value expected, on-time, and within cost.
> —Mr. Jack Clemons, Lockheed-Martin Corp., 2008

13.1 INTRODUCTION

Once a decision is made, we focus our attention to implementing the chosen solution design. Simply deciding to implement the selected solution does not imply that the solution will be successfully implemented. The systems engineer hopes to encounter "blue skies and smooth sailing" as indicated by the blue color depicted in the systems decision process (SDP) when implementing the solution. However, solution implementation may be the most difficult and frustrating phase of the SDP if sufficient attention is not given to detailed planning [1]. Even the best of solution designs, if poorly implemented, can fail to meet the needs of our client. Successfully implementing a solution depends on the emphasis and consideration given to the eventual solution implementation during the three phases that precede it: problem definition, solution design, and decision making. Planning for implementation must begin in defining the problem in phase one, continue throughout the design of candidate solutions, and be a consideration in the decision making phase of the SDP.

Decision Making in Systems Engineering and Management, Second Edition
Edited by Gregory S. Parnell, Patrick J. Driscoll, Dale L. Henderson
Copyright © 2011 John Wiley & Sons, Inc.

> Implementing a successful solution depends on the emphasis and consideration given to the eventual solution implementation during the three phases that precede it.

The phases of the SDP are highly interdependent as are the stages of a system life cycle. Decisions we make in one phase of the SDP inevitably impact the decisions we make in other phases of the process. How we define our problem certainly impacts our solution design space. Similarly, the solution design selected during the Decision-Making phase of the SDP will shape the plan we develop in the Solution Implementation phase. The same interdependence is true between life cycle stages. For example, the decisions (e.g., design choices) made in the Design and Develop the System life cycle stage will impact the decisions and latitude available to the project manager (PM) responsible for manufacturing the system in the Produce the System life cycle stage. For this reason, it is imperative that neither the designer nor the PM carry out his or her role independently of the other. Both should work toward the same set of objectives in trying to ensure that the system solution chosen during the Decision-Making phase is met with success. In this sense, making decisions and implementing them are essentially indistinguishable, except in terms of their order [1].

A concept map of the solution implementation phase is shown in Figure 13.1. The map depicts the interrelationship of the activities and tasks as well as the inputs and outputs within the solution implementation phase. The tasks and relationships are described in greater detail in this chapter.

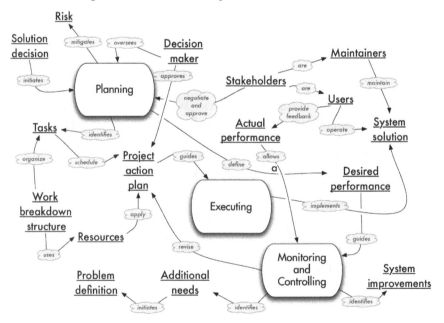

Figure 13.1 Concept map for Chapter 13.

This chapter focuses on the activities and elements that contribute to a successful implementation of the system solution. We begin by discussing the solution implementation phase and its relationship to specific stages of the system life cycle introduced in Chapter 3. Next, we recognize the implementation of a solution as a project. We then describe each of the critical solution implementation tasks and discuss various project management tools and techniques available for use in completing these tasks. Next, we describe critical solution implementation tasks as applied through selected system life cycle stages. Lastly, we provide a case study as an illustrative example to highlight many of the concepts covered in the chapter and some of the key challenges one might face in the Solution Implementation phase of the SDP.

13.2 SOLUTION IMPLEMENTATION PHASE

The Solution Implementation phase is the fourth phase of the SDP. It is probably one of the more difficult phases to accomplish successfully because the activities of this phase focus on turning the client's expectations for the system into reality. Entry into this phase may induce a false sense of completion as moving from the decision making phase to the solution implementation phase is often viewed as moving to completion. However, as mentioned earlier in the text, there always exists the possibility of returning to an earlier phase of the SDP based on evolving project conditions. Engaging the systems decision process is an iterative and cyclical in its progression as is intentionally depicted in the circular nature of the SDP diagram. Additionally, the Solution Implementation phase can be needed in any one of the seven life cycle stages listed in Table 13.1.

This phase of the SDP requires a substantial amount of coordinated effort and emphasis. The first step toward success is conceptualizing the action of "implementing a solution" as a project. By doing so, the complete arsenal of project management principles are available to help plan, execute, monitor, and control the conversion of the system solution into reality.

A *project* is a temporary endeavor undertaken to create a unique product, service, or result [1]. As a temporary endeavor, a project has a definite beginning and end, but this does not mean a project is short in duration. On the contrary, many projects may last for years, making the concept of "temporary" a flexible term. Here, conceptualizing the Solution Implementation phase as a project means that the project end is defined as producing a result in concert with the intended outcome of Solution Implementation phase. Properly moving a project through to completion requires expertise in project management capabilities. Project management encompasses the knowledge, skills, tools and techniques applied to activities in order to meet the project objectives [2]. Project management applies and integrates the project management processes: initiating, planning, executing, monitoring, controlling, and closing.

A broad, detailed exposition of project management in the context of the Solution Implementation phase exceeds the scope of this chapter. Indeed, entire texts are

TABLE 13.1   Life Cycle Stages

Life Cycle Stage	Purpose
Establish system need	• Define the problem • Identify stakeholder needs • Identify preliminary requirements
Develop system concept	• Refine system requirements • Explore concepts, examine technology readiness, and assess risks • Perform capability versus cost tradeoffs • Select concept
Design and develop the system	• Develop preliminary and final designs • Build development system(s) for test and evaluation • Test for performance, integration, robustness, effectiveness, etc. • Assess risk, reliability, maintainability, supportability, life cycle cost, etc.
Produce the system	• Acquire long-lead-time components • Develop production plan and schedule • Perform low-rate initial production (LRIP) • Perform full-rate production (FRP) • Monitor and test production items for conformance to specifications
Deploy the system	• Identify deployment locations • Provide training for installation, maintenance, and operation • Transport to chosen locations • Plan and execute logistical support
Operate the system	• Operate system to satisfy user needs • Gather and analyze data on system performance • Provide sustained system capability through maintenance, updates, or planned spiral developed enhancements
Retire the system	• Store, archive, or dispose of system

devoted to project management and in many cases full texts are devoted to particular tools and techniques used by project managers. As such, the focus of this chapter is to highlight select portions of project management processes in sufficient detail to enable a systems engineer or engineering manager to successfully complete the Solution Implementation phase.

There is more than one way to manage a project to successful completion, and one could argue that project management is more of an art than a science. Successful project managers (PM) tailor the five project management processes to fit the characteristics of a particular project. Initiating defines and authorizes the project; planning defines scope, objectives, and the course of action; executing integrates people and resources to execute the project management plan; monitoring and controlling track progress and identify shortcomings requiring action; and closing is the formal acceptance of the product, service or outcome which brings the project to an end [1].

Figure 13.2 depicts a model that illustrates how these different project management processes relate to one another. Viewing the model from left to right, the model begins with the initiating process. The initiating process develops a project charter and a scope statement. These two items establish the breadth of the project and the objectives that need to be accomplished by the project end. From the initiating process, the model flows into the planning process which requires a project team to collect and consolidate information from many sources to identify, structure, and finalize the project scope, cost and schedule. These three items along with a plan as to how, when, why, and where the available people, tools, and resources are going to be used define a complete project management plan.

The executing process that naturally follows consists of the activities used to complete the work identified in the project management plan. The planning process and the executing process are iterative, allowing the PM to reassess the plan as new information arises or the project scope requires adjusting. The monitoring and controlling processes observe project execution so that any potential problems and challenges to successful completion may be identified as early as possible. Once challenges/issues are identified, corrective action is taken to avoid schedule slippage, cost overruns, and other detrimental effects imposed by deviations from the plan. Monitoring and controlling must be performed frequently enough to allow the PM sufficient cognizance of the health of the project so that any corrective action required may be taken prior to events having an adverse impact on the project's cost, schedule, or performance. For systems decision problems, the performance element of the project plan is comprised of the total system value returned by properly implementing system functions of the qualitative and quantitative value

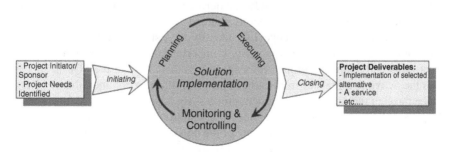

Figure 13.2   The project management process.

models. The closing process employs techniques to formally terminate the activities of project management plan and verify project completion [1].

13.3   THE INITIATING PROCESS

Initiating defines and authorizes the project by creating a project charter and a scope of work statement. The project charter is a document that provides authorization for the project. Using the project charter and the project statement of work, a preliminary project scope statement is developed. The project scope statement defines what needs to be accomplished and generally includes the following items [1, 3]:

1. Project and product objectives (clearly defined and achievable)
2. Project assumptions
3. Project constraints
4. Project requirements and deliverables
5. Project boundaries
6. Solution requirements and characteristics
7. Initial project organization
8. Initial project risks (risk register update)
9. Initial work breakdown structure (WBS)
10. Schedule milestones
11. Project configuration management requirements
12. Cost estimate
13. Solution acceptance criteria

Obviously, each project scope statement will vary and not all of the listed items are necessary for every Solution Implementation. The project scope statement may also be redefined as the situation dictates. However, changing the project scope requires the PM to review each of the subsequent processes as well, since the project scope statement is the basis from which subsequent processes unfold. If a project is large or complex enough, it may need to be decomposed into phases in order to be properly managed. This should again appear in the scope statement. Finally, a feasibility analysis is conducted during this process in order to assess the practicality of completing the project as planned.

As noted in the project scope list, identifying risks that might threaten the solution is vital to success. This activity's importance cannot be overemphasized. Identifying and analyzing project risks specific to this phase of the SDP in concert with the active risk register from earlier phases continues a common thread of vigilance through the project management process. By addressing risks at this point in the project specific to cost, schedule, logistics, liability, laws, regulations, and so on, the PM achieves a greater understanding of the impact of uncertainty going forward.

As discussed in Chapters 11 and 12, life cycle cost analysis continues into the Solution Implementation phase. An updated cost estimate that now includes many new elements specific to this phase will become part of the project scope statement. The PM uses this cost estimate to develop a realistic projected budget sufficient to carry the overall effort through to completion. As described further in Section 13.6, this budget serves as a primary indicator for tracking progress in comparison to schedule and performance.

13.4   PLANNING

The *planning* process is critical to setting the conditions for overall success of the implementation phase. Inadequate planning is a primary contributor to projects failing to achieve their schedule, cost and performance objectives. The planning process lays out the course of action to attain the scope and objectives of the project. There are several techniques and approaches to assist the systems engineer in this planning effort. The first step in the planning process is analyzing the preliminary project scope statement and project management processes in order to develop a *project management plan*. The project management plan includes the activities needed to identify, define, combine, unify and coordinate the various processes to successfully accomplish the project. The project management plan uses all the necessary subordinate plans and integrates them into a cohesive effort to accomplish the project. The subordinate plans include but are not limited to [1–3]:

1. Project scope management plan
2. Schedule management plan
3. Cost management plan
4. Quality management plan
5. Process improvement plan
6. Staffing management plan
7. Communication management plan
8. Document control management plan
9. Risk management plan
10. Procurement management plan

Each one of these subordinate plans includes a number of project management techniques which allow the specific plans to be implemented and monitored. For example, the project scope management plan is used to ensure all required work necessary to complete the project successfully is identified. It is just as important to identify what is not included in the project. The work breakdown structure (WBS) is one of the more effective techniques to help in this scoping process.

Similar to the logic of describing the functional structure of a system, the WBS is a hierarchical representation of all the tasks that must be accomplished in order to successfully complete a project. Four rules are used when developing a WBS. First, each task that is broken down to a lower level must have at least two subtasks. Second, if it is difficult to determine how long a task will take or who will do the task, it most likely requires further decomposition. Third, any task or activity that consumes resources or takes time should be included in the WBS. Fourth, the time needed to complete an activity at any level of the hierarchy should be the sum of the task times on branches below it. Properly completed, the WBS ultimately serves as the basis for identifying and assigning appropriate task responsibilities.

The WBS defines the exact nature of the tasks required to complete the project. While the hierarchical structure is certainly helpful, the WBS is not limited to one particular format but takes a number of different forms. A WBS can appear as a tree diagram (Figure 13.3) with level one tasks directly below the overall project objective followed by level two tasks [3]. In the case of the rocket example, Figure 13.3 illustrates a classic WBS.

The overall objective of this project is production of a mobile rocket. Under the level one task #5—plan production training—the two subtasks include identify test requirements and identify test location. This logical breakdown allows planners working the project scope management plan to accurately identify the requirements associated with every task supporting the production of the mobile rocket objective. Although this example may seem simplistic, the WBS tool is highly effective in supporting the project scope management plan. Other effective tools and techniques identified during the planning process include, but are not limited to, those shown in Table 13.2 [1–7].

The linear responsibility chart is an excellent technique that is used in conjunction with the WBS. When complex tasks are broken down to basic tasks, a linear responsibility chart takes those tasks and assigns personnel and organizations responsibility for each one. A linear responsibility chart shows the critical interfaces between tasks and organizations/individuals and highlights areas that require special management attention [3]. Such a chart is illustrated in Figure 13.4, which takes our rocket WBS in Figure 13.3 and assigns a number of tasks to different individuals and teams.

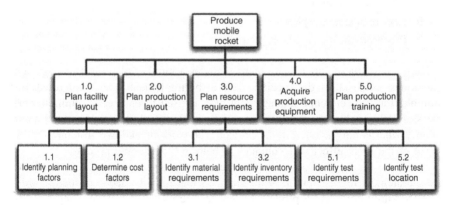

Figure 13.3   Work breakdown structure (WBS) for the rocket problem.

TABLE 13.2   Tools and Techniques Supporting the Planning Process

- Linear responsibility charts
- Scheduling
  - Schedule milestone list
  - Activity sequencing
  - Activity resource estimating
  - Activity duration estimating
- Project configuration management requirements
- Order of magnitude cost estimate
- Resource allocation
  - Resource loading
  - Resource leveling
  - Constrained resource scheduling
- Staffing management plan
- Earned value analysis
- Schedule baseline
- Cost baseline
- Quality baseline
  - Quality assurance
  - Quality control
- Risk
  - Risk identification
  - Qualitative risk analysis
  - Quantitative risk analysis
  - Risk response
- Critical path method
- Value engineering
- Stakeholder communication plan
- Document control register
- Change control system

For clarification and understanding, a number of implied tasks are not listed on the WBS, including a project plan and budget. On the linear responsibility chart, the associated managers and teams are added to illustrate the complexity of the overall organization and the interfaces between departments. For example, establishing the project plan is the responsibility of the lead planner and his planning team. Moving from left to right in the chart, certain relationships between particular individuals and teams are evident. Obviously, the plan is exceptionally important because it is the foundation for executing the overall project. Therefore, formal approval is most likely to occur all the way through the chain of command from the project manager, through the program manager, to senior VP for programs.

Additionally, the linear responsibility chart also shows the relationships between the lead planner and his team and the other departments. For planning purposes, it is important to consult each department because they have valuable information that the planner uses. At a minimum, the lead planner needs to consult each department manager. Therefore a "3" is used to identify a mandatory consultation requirement on the part of the lead planner. You could argue that the lead planner has a mandatory requirement to consult with the team as well but that is not necessarily how the organization runs. In this case, it is assumed that the department managers are the "gatekeepers" to their departments and that the lead planner needs to consult with them versus going directly to the department team.

	Senior Vice President for Programs	Program Manager	Project Manager	Land Planner	Planning Team	Engineering Manager	Engineering Team	Operations Manager	Operations Team	Logistics Manager	Logistics Team	Lead Budget Analyst	Budget Team	Resource Manager	Resource Team
Establish Project Plan	5	6	2.4	1.2	1	3	4	3	4	3	4	3	4	3	4
Establish Project Budget	5	6	2.4	4	4	3	4	3	4	3	4	1.2	3	3	4
Plan Facility Layout	5	6	1.2	3	4	1.2	1	3	4	3	4	3	4	3	4
- Identify Planning Factors		5	6	6.2	3	3	1	3	4	3	4	3	4	3	4
- Determine Cost Factors		6	6	3	3	3	3	3	3	3	3	1.2	3	3	3
Plan Production Layout		6	6.2	3	4	1.2	3	3	3	4	4	4	4	3	3
Plan Resource Requirements		5	6.2	3	3	4	4	3	3	4	4	4	4	1.2	3
- Identify Material Requirements		5	6	3	4	4	4	3	3	4	4	4	4	1.2	3
- Identify Investory Requirements		5	6	3	4	4	4	3	3	4	4	4	4	1.2	3
Acquire Production Equipment		6	6.2	3	4					3	3	3	3	1.2	3
Plan Production Training		5	6.2	3	4	3	3	3	3			3	4	1.2	3
- Identify Test Requirements		5	6	3	4	3	3	3	3			3	4	1.2	3
- Identify Test Location		5	6	3	4	3	3	3	3			3	4	1.2	3

1 Responsible   4 Consultation Possible

2 Supervision   5 Must be notified

3 Consultation Mandatory 6 Formal Approval

Figure 13.4   Example linear responsibility chart.

Tremendous effort must go into the planning process to ensure success of the project. Many projects fail, regardless of size, when left to planners or individuals who have limited practical experience. Therefore, a successful plan must include experienced PMs on the planning team. Their expert advice brings a level of practical experience which equates to time and money savings when the final project management plan moves to execution. If possible, a PM should be identified to lead the project during the planning phase. It is not necessarily imperative that a PM be brought in this early, but it should be viewed as a "best practice" effort to improve the overall quality and efficiency of the project.

Figure 13.2 illustrates the planning process as an iterative process which includes the two processes: (a) executing and (b) monitoring and controlling. As execution begins, there are inevitable changes that occur, such as information updates, challenges with resource allocation, scheduling delays, value engineering, and so on. These changes are identified during either the executing or the monitoring and controlling processes. Once identified, the change or information is fed back into the planning phase to allow the project management plan to be updated accordingly. For the most part these changes are unpredictable. Therefore close integration of the planning, execution, and monitoring and controlling processes is essential for success. The PM must constantly seek this type of feedback in order to adjust the plan and execute accordingly.

13.5   EXECUTING

The *executing* process requires the PM team to perform a myriad of actions to put the project management plan into action. As discussed, the project management plan is made up of specific management plans that employ various techniques and tools to execute the overall management plan. The PM team must orchestrate the integration, sequencing, and interfacing of these plans. Additionally, the PM team must track the deliverables (products, results, and/or services) from each of the subordinate plans. The communications plan becomes exceptionally important here as well because distributing accurate information keeps the PM team informed of the project's progress and status.

Equally important is managing the expectations of stakeholders. The involvement of stakeholders during this process is troublesome at times. The PM should be aware of the type and amount of information that is passed along to stakeholders. Most projects do not proceed smoothly at all times because of natural occurring and frequently uncontrollable variation in project components affecting scheduling milestones. However, over the long haul these same projects are successful. Exposing stakeholders to a complete view of this variation may cause unwarranted celebration (upside variation), concern (downside variation), or over-reaction. Communicating information to stakeholders can be problematic if not properly conducted by the PM. Additional tasks during the Executing process include but are not limited to [1]:

1. Perform activates to accomplish project objectives.
2. Expend effort and spend funds to accomplish the project objectives.
3. Staff, train, and manage the project team members assigned to the project.
4. Obtain quotation, bids, offers or proposals as appropriate.
5. Obtain, manage, and use resources including materials, tools, equipment, and facilities.
6. Implement the planned methods and standards.
7. Create, control, verify, and validate project deliverables.
8. Manage risks and implement risk response activities.
9. Manage sellers.
10. Adapt approved changes into the project's scope, plans and environment.
11. Establish and manage project communication channels, both external and internal to the project management team.
12. Collect project data and report cost, schedule, technical and quality progress, and status information to facilitate forecasting.
13. Collect and document lessons learned and implement approved process improvement activities.

It is vital to ensure information that affects the project management plan be updated as quickly as possible to ensure appropriate corrective/improvement action can be implemented in a timely manner. Most project deliverables take the form of tangible items such as roads, buildings, systems, reports, devices, and so on. However, intangible deliverables such as professional training, information, professional image enhancement, and security, among others, can also be provided [1].

13.6 MONITORING AND CONTROLLING

The *monitoring and controlling* process serves the purpose to monitor the other project processes so that effective control measures can be directed to keep the project performing correctly, on time, and below cost [1]. One of the first steps to proper monitoring is identifying those essential elements requiring control, which for typical systems projects are performance, time, and cost. The PM must establish clear boundaries for control and identify the level of importance for each category. It is safe to say that the boundaries and level of importance are not the same for each project and are driven by the project's overall scope statement and stakeholder input [3, 8, 9]. Continuous monitoring in each of the subordinate plans (Section 13.4) allows the PM team to keep current with the changing dynamics of the project and to register the project's health through the prism of performance, time and cost.

The linear responsibility chart shown in Figure 13.4 underscores the need for good monitoring. Figure 13.5 demonstrates how continuous monitoring of a project motivates confidence for a PM that the project is on-track or that it requires action to restore it to this status. It is a three-dimensional "snapshot" of the project status at time τ_1 in comparison to the cost, schedule, and value estimates promised by the chosen solution at the decision gate just prior to Implementation. The dot inside the box represents the project plan's ideal state of these three project elements by time τ_1. The box imposed around this ideal location represent the acceptable levels of variation for these elements at time τ_1 under conditions in which "normal" (planned for) variation in these three elements occurs. If the current estimates for these three planning elements locate the project state within this box, then for all practical purposes the project is on-track at time τ_1 to deliver the total system value (typically represented by functionality or performance) on time and under or at cost by the end of project.

The dashed line in the figure illustrates the hypothetical development path that the project proceeded along up to time τ_1. Note that in this situation the initial cost outlay for the project at the start of the implementation phase was greater than zero. Unfortunately for the PM, the current state of the project at time τ_1, shown by the dot below the box, indicates that the project implementation has issues that must be addressed. While the project is a bit ahead of schedule ($S(\tau_1) < S^*(\tau_1)$, and cost is less than or equal to the planned estimate ($C(\tau_1) \leq C^*(\tau_1)$, the system is not achieving the planned for value return ($V(\tau_1) < V^*(\tau_1)$) at this time. Thus, some corrective or controlling action is required.

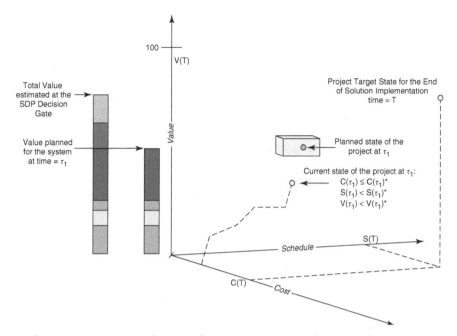

Figure 13.5 Conceptualization of project management during implementation.

An example of a corrective action taken when the schedule is behind is called "crashing the schedule." Crashing the schedule is a technique used in the critical path method to bring a project back on schedule if a particular task is going over the schedule time allowed. This technique requires placing money and/or resources (personnel and/or equipment) against the task in order to reduce its duration. The obvious outcome of using this technique is increasing the cost of resources which increases the overall project cost. The PM team understands the importance of these monitoring and controlling techniques and will ensure analysis of the results provide an accurate picture of the project's status.

Another method for monitoring a project is earned value. *Earned value* (EV) analysis is a commonly used method for measuring the overall performance of a project [1, 3]. It involves analysis of the actual work accomplished as compared to the projected budget and actual expenditures. Figure 13.6 illustrates an EV graph representing the facility layout portion of the rocket project in Figure 13.3. In this case we again see that the EV is lagging behind the budgeted amount and the actual expenditures. This leads the PM to conclude the project is behind schedule. The money spent on the project to-date exceeds the value accumulated by the system for the work performed. An earned value graph presents the PM with an effective means of monitoring critical project elements in a way that clearly highlights the links existing between cost, schedule, and value (performance).

There is more information in the EV chart shown in Figure 13.6 than simply a summary of goals being achieved or not. The stair-stepping pattern of the EV line

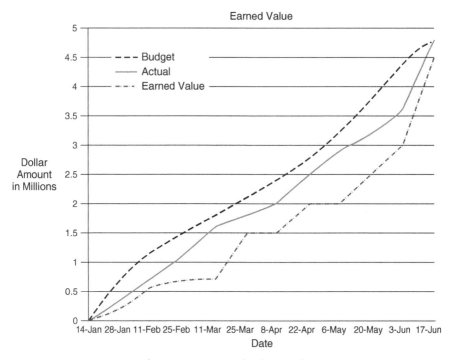

Figure 13.6 Earned value graph.

indicates surging by the contractor. While there are several reasons this pattern can occur, it often indicates challenges with the contractor. At the 11 February date, the contractor is stagnant (EV does not increase). The PM in this case responds with an increase in actual budget expenditure. More than likely, there is a possible cash flow problem with the contractor, and he required more money to get the work done. Even though the EV is below what the PM would want, there are situations where paying the contractor ahead of the EV is the best course of action for project success. This occurs most often when a good working relationship is established between the contractor and PM. Working with the contractor to keep maintaining progress instead of, for example, dismissing him from the project and pursuing litigation is probably the best course of action for the PM in this case.

The data past 11 March shows that the contractor responded with an upsurge in the EV. However, on 25 March, the contractor slows down again and EV begins to plateau. In this instance, the PM reacted differently and lowers the payment to the contractor in order to get him to respond. You will notice also that near the end of the facility layout portion of the project, there is great gain in EV for the overall project. This is a very typical pattern for a systems implementation project. Engineers have a tendency to estimate a concave shape over the project, indicating optimism in terms of how quickly value (functionality, performance) can be developed. In reality, the EV curve assumes a convex shape because of re-work imposed by test failures, delayed schedules and response surges in activity, and cost-conserving measures put in-place to mitigate the threat of running out of budget before the expected (or required) value has been delivered.

Overall, the monitoring and controlling process is concerned with but not limited to the following list of activities [1]:

1. Comparing actual project performance against the project management plan
2. Assessing performance to determine whether any corrective or preventive actions are indicate and then recommending those actions as necessary
3. Analyzing, tracking, and monitoring project risks to make sure the risks are identified, their status is reported, and the appropriate risk response plans are being executed
4. Maintaining an accurate, timely information base concerning the project's products and their associated documentation through project completion
5. Providing information to support status reporting, progress measurement and forecasting
6. Providing forecasts to update current cost and current schedule information
7. Monitoring implementation of approved changes when and as they occur
8. Updating the Project Management Plan accordingly
9. Updating stakeholders and their status with regards to the project as appropriate
10. Conducting PM team meetings to develop an updating rhythm for analysis, tracking, and monitoring
11. Conducting proper change control methods
12. Approving/rejecting change requests
13. Updating project management plan
14. Updating project scope statement
15. Approving corrective actions
16. Approving preventive actions
17. Monitoring deliverables

13.7 CLOSING

The *closing* process involves all the necessary administrative and contractual closing procedures to ensure proper project closeout. This process includes all the subordinate project plans as well as any phases (for complex/large projects) that are associated with the project. The administrative closing procedures are those procedures and project relationships dealing with different aspects of the project [1]. A major portion of the administrative work required here includes documentation archiving (see Section 13.4). This process is exceptionally important for possible future legal inquiries. Many contracts start out with good intentions on the part of both parties, but more often than not, there are lawsuits brought against a company by a contractor for numerous reasons. A good document control plan and register assists the PM team respond to litigation. The contract closure procedure includes those activities needed to complete contractual obligations. One aspect of this process that is important is ensuring proper contractor documentation completion. For example, near the end of most projects, the daily requirements of progress reports sometimes get overlooked, especially if a contractor is near 100% paid. If a PM fails to hold the contractor to full compliance for contractual requirements such as progress reporting, the PM is liable. It is imperative that the PM team ensure all aspects of close-out procedures are adhered to and followed.

13.8   IMPLEMENTATION DURING LIFE CYCLE STAGES

After careful analysis of the implementation phase, it is important to understand its role during all stages of a systems life cycle. The life cycle is a dynamic living model that requires thorough understanding of the SDP and how it is used in every stage of the life cycle. As such, the implementation phase of the SDP takes different approaches to meet the particular needs as dictated by the solution and the life cycle stage. The guidelines presented in the previous sections are still relevant here as well but are adjusted to meet the unique circumstances presented by the situation. The following sections will focus on the implementation phase as it applies to the Produce the System, Deploy the System, and Operate the System stages of the system life cycle.

13.8.1   Implementation in "Produce the System"

The fourth life cycle stage and the first to be discussed in this section is "Produce the System." One inherent and necessary objective of the Design and Develop stage is to create the design such that the system and all of its elements are produced effectively and efficiently. For most physical systems, the primary objective of the production stage is to turn the system solution into reality. During this time, the PM team handles any design changes justified by requirements or by market demands [6]. Inspection and testing of the product occurs in this stage. The PM team is required to validate that the product meets the specifications identified in the requirements. Project management techniques as discussed earlier can be integrated with the SDP and systems engineering procedures and practices to organize and implement a production or manufacturing requirement. This can be an exceedingly complex set of activities requiring an excellent PM team [10].

Planning for "Produce the System"   To achieve success during the produce the system life cycle stage, detailed planning must precede the execution of production. Planning the implementation of the system solution helps ensure that the system solution does get implemented in such a way that the expected performance is realized [11]. For the producing the system stage to successfully occur, it is supported by specific planning actions such as those discussed in Section 13.4. The purpose of planning is not to eliminate uncertainties, but rather to prepare as much as possible for anticipated events and to adjust when unexpected events occur. By preparing a plan, we establish a common baseline from which to coordinate activities. Detailed planning is essential for the producing the system stage because it minimizes the risk inherent in projects.

The systems engineer must realize that although we are focused on producing the system, the majority of planning for deploying and operating the system also occurs in this stage. That is to say that the planning of one stage is performed in relation to other life cycle stages. For example, how the system solution is deployed is considered when planning how the system solution is produced. Likewise, how the system solution operates is considered when planning how the system solution is deployed. Planning must occur not in a vacuum but concurrently and with the end state in mind.

Executing for "Produce the System" When planning the production of the system solution is complete, the next step is to execute the production plan. Resources are required to execute the plan as well as support the deployment and operation of the system solution. Execution includes resource management for implementation of the system solution. Attention is now given to defining resource requirements for production to begin. Resources include everything needed to accomplish each task during production: people, money, supplies, inventory, equipment, facilities, infrastructure, external services, and technology, to name a few. Spreadsheets and cost-estimating models are techniques that help organize and layout resource requirements (amount and time).

As an example, Microsoft® Project has the ability to display all of this information. This enables the PM to identify a task, its duration, start and finish dates, resources, resource requirements, and when those resources are required. The products of this stage are successfully executed if resources are available. As this stage is carried out, assessment and control of the stage must occur. During the execution of this stage, the key tasks of monitoring and controlling are critical to ensure that the system will function as expected and executed as planned.

Monitoring and Controlling for "Produce the System" The SDP includes measures and methods that allow for monitoring and controlling system solution performance during all life cycle stages. A feature of Microsoft® Project is that it may be used to measure progress in terms of time, budget, and project performance. Since Microsoft® Project does not monitor progress in terms of system performance, other methods (simulations and testing) are used to assess system performance. The project action plan allows the PM team to compare actual and planned task durations, resource usage, and expenditures at any level of activity. The project action plan is used as a control document by measuring comparisons. These comparisons dictate what project performance information is monitored. This gives the PM team the ability to control the project and take corrective action if the project is not proceeding according to plan [5]. The elements that are controlled during the produce the system stage include but are not limited to:

1. Cost
2. Schedule
3. Risk
4. System performance requirements
5. Design changes
6. Production and manufacturing process
7. Quality
8. Reliability
9. Safety of product and personnel

Additionally, there may be others defined by the PM and/or stakeholders.

13.8.2   Implementation in "Deploy the System"

At the end of the produce the system stage, the system solution enters the fifth life cycle stage: Deploy the System. The PM team must receive prior approval from the decision maker that the system is ready to proceed to the next life cycle stage. Upon approval, the manufactured system solution becomes the deployment system, which delivers fully operational system solutions to users.

Planning for "Deploy the System"   The purpose of the Deploy the System stage is to transfer the system solution from the development facility to the operational location and to establish full operational capability. Distribution facilities, marketing, and sales organizations are required to support the implementation of this stage. Use of these and other resources are planned in great detail. In this section, the planning elements and methods for successful deployment of the system are addressed.

Deploying the system solution is a process that must also be planned for in order to meet stakeholder objectives. Elements of the development process requiring planning include but are not limited to:

1. Geographical distribution
2. Deployment schedule
3. The type and number of system components at each location
4. Logistical support
5. Type training required (e.g., installation, maintenance, and operation)
6. Resource requirements to support the required training
7. Testing

Marketing elements are keys in determining the best locations to deploy the system solution. The strategy for deployment and support requirements are identified. Acceptance testing or a full operational testing occurs prior to moving the system solution into the next life cycle stage. Acceptance testing often results in minor adjustments to system operation [12]. This testing is conducted with stakeholders present. Stakeholders ensure that the system solution continues to operate as intended according to desired preferences. Risk is always present, and it is possible that the events and activities inherent in the deployment process may affect the system's operational performance. As such, a risk management plan is developed and contains elements as outlined earlier.

An operational test or demonstration is another type of test which is very useful in communicating the operability of the system. Before testing is complete, operational testing in the environment and under conditions in which the system solution operates are performed. Another critical activity in this stage that requires planning is training. Training produces trained installers, operators, and maintainers to support the operations of the system solution. Training resources are planned and phased into the project as required. The process planning methods described during the Produce the System stage are applied in this stage as well.

Executing for "Deploy the System" Deploying a system solution requires excellent documentation concerning how to install the system in its operational environment. In some cases, special analysis and testing is carried out so as to assure field operability. Training is also designed and delivered in formal well-developed programs. The required training is scheduled, not only in this stage but also in the Operate the System stage. Subsequently, training specialists are part of the team of personnel required to execute the project.

Upon completion of planning for deployment, the plan is carried out. Resources are present to execute the deployment process. Many products are late to market simply because of being starved for resources. When this occurs, the costs can be enormous. To execute the deployment plan correctly, timing and resources are critically important. The right kind and the right amount of resources are required and on time when needed. When an increase or adjustment in resources is needed, senior leaders must prepare to respond quickly. Some examples of resource requirements in this stage are listed but not limited to:

1. Competent trainees
2. Users to perform operational testing
3. Tentative user locations
4. Transportation assets
5. Training equipment

Monitoring and Controlling for "Deploy the System" The need to exert proper control over deploying the system mandates the necessity for monitoring and controlling the proper activities and elements during this stage. According to Table 13.1, we are verifying that the system solution meets system performance measures. In order for the system solution to perform during full operational capacity as intended, it is monitored, assessed, controlled. The details of the deployment plan identify additional elements to control commensurate to the unique deployment requirements. As the system performs outside of its intended functions, corrective action is taken to bring performance in conformity with stakeholder preferences. The fundamental items controlled are time, cost, and performance, which were discussed earlier in the chapter. It is prudent and necessary to perform testing of the system functions prior to the deployment process, during the deployment process, and upon arrival at user location if possible. These efforts ensure that the system operates as intended upon reaching the end user.

13.8.3 Implementation in "Operate the System"

A system is not considered successful until it is successfully implemented and is turned over to the user. A full-scale operation generally occurs in the sixth life cycle stage: Operate the System. This stage begins as users receive the first operational systems. The objective of the operation stage is to fulfill the stakeholder's need. The stakeholders' needs are fulfilled when the system solution is realized. Planning, monitoring and assessment and control during this stage is critical to ensure that the system solution is successfully implemented into full-scale operation. It is hoped that the events and activities preceding this stage were conducted thoroughly.

Planning for "Operate the System"   In the Operate the System stage, the system has attained full operational capability. This means that the system solution operated and maintained in conformance with user requirements. This also includes satisfying the user, gathering data on system performance, sustaining and maintaining operability, adding enhancements to the system solution, and identifying improvements for future implementation. During this long period in which the system is operational in the field, emphasis should be placed on the continuous measurement of the system's performance.

The planning performed in this stage centers on operating and maintaining the system and identifying system improvements. Preliminary planning for retiring the system also occurs here. Since emphasis during this stage is on system performance, data are gathered to assess system performance. Measurement procedures range from simple manual data sheets to automated sensors that record operational status continuously [2]. Some companies try to maintain contact with consumers through hot lines, reports of satisfaction, and online usage monitoring. To support these methods, it is important that procedures on how to install and sustain a performance measurements program be explicitly defined. Emphasis should be placed on maintaining the system solution to ensure that it continues to function in accordance to its operational requirements. Maintainability is the ability of the system solution to be retained in or restored to a performance level when prescribed maintenance is performed [5].

Executing for "Operate the System"   The system operators and maintainers execute the functions of operating and maintaining the system solution. The resources are adequately planned and acquired when needed. These individuals are involved in the collection of data to assess whether the system functions in accordance with its intended design. Trained individuals are assigned to perform periodic evaluations of the system as it operates in its natural environment and to perform the necessary maintenance needed to sustain system performance. Some examples of resource requirements include, but are not limited to, the following:

1. Data collection methods
2. Data collection equipment
3. Personnel resources

Monitoring and Controlling for "Operate the System" Once a system is operating, it is controlled; that is, its operation is regulated so that it continues to meet expectations [11]. Continual operational evaluation and testing of the system is a method used to identify what is controlled. System audits are also performed. This method is used after the recommended alternative is implemented to see how the actual system performs, whereas an operational evaluation can occur prior to operational implementation and during actual implementation. System improvement relies on the identification of deviations between the actual operation of the system and what is termed as normal or standard. After these deviations are pinpointed, their causes are identified in order to correct malfunctions. Feedback is another valuable tool. Users and maintainers provide feedback about what they like and do not like, which is used during refinement to make changes in the design, leading to upgrades of the system [13].

A set of specific evaluation test requirements and tests are evolved from the objectives and needs determined in the final requirements specifications. Each objective and critical evaluation component is measured by at least one evaluation test instrument. If it is determined that the resulting system product can no longer meet stakeholder needs, the problem enters phase one of the SDP Problem Definition and repeats the procedure set forth in this text.

Solution Implementation can often be one of the most difficult and time consuming challenges faced by the systems engineer. Even the best of solutions, if not properly implemented, can fail to meet the needs of the stakeholders. History is full of engineering projects that have come in late, over budget, or that failed to perform as intended. Thankfully, there are many project management tools and techniques that are available to assist the systems engineer during the Solution Implementation phase of the SDP. These tools and techniques can be used in any of the lifecycle stages of a system and are often tailored to fit the needs of the specific lifecycle stage as described in this chapter. However, as has been said project management is more of an art than a science. Many times, successful solution implementation comes down to the ability of the systems engineer to work with the many stakeholders involved in the Solution Implementation phase of the SDP. The importance of these people skills and the involvement of leaders early and often are highlighted in the case study that follows. The case study describes the implementation of an automated system for college applications in the early years of information systems.

CASE STUDY: DESIGN AND IMPLEMENTATION OF THE PROSPECTIVE STUDENT INFORMATION SYSTEM

Bobbie L. Foote, Professor Emeritus, University of Oklahoma

The Beginning

In 1967 the University of Oklahoma decided to venture into new computer technology to consider automating their student admissions system. The University at this time was making a commitment to use commercial computing power instead of continuing full-scale development of their own computer, the OSAGE, which was designed and built from conception by University faculty. They had realized that they did not have the resources to develop the myriad software applications necessary to run the University information needs based on the one of a kind operating system for the OSAGE. The University had bought an IBM 650 computer by 1957 and after 1962 had purchased an IBM 1620 and had started developing the software and operating specialists to run a punch card operation.

A young professor in Industrial Engineering was selected to design and implement the new system. A team of administrators was formed to advise him. A bright young graduate student who was an employee of a local air conditioning manufacturing firm and had an undergraduate degree in Industrial Engineering from the University of Oklahoma was hired to gather information and write prototype software to show feasibility.

In the early planning stages, a standard stakeholder analysis was performed. A survey was sent out to determine potential users. These users were interviewed as to what information they wanted. This led to a listing of actions and decisions that they needed to make. A standard flowchart was developed to look at timing.

This was an admissions system. So, the first day of enrollment in the summer or fall served as one of the drop-dead time goals. To be admitted, the student had to supply high school transcripts, a copy of a medical exam, records of immunizations, religious preferences, housing preferences, and their interest in various academic programs. There were deadlines for these submissions. If a student failed to submit the required information, the University manually went through the file and noted omissions and sent out a letter with boxes marked to indicate the information shortage. Some students went through several cycles. Students who showed late interest were a problem as the University needed to expedite their record analysis and communicate decisions quickly. Appeals had to be handled.

Another problem was the exponential growth of the applications as more and more people realized the value of a college education. This led to a big problem with applications for a scholarship as the donors and other supervisors had deadlines to make decisions. Scholarship applications required extra information such as letters of recommendation and student essays. This put a real burden on the manual system.

As a part of the routine information gathering, histories of other processing applications were sought. It was at this point an unpleasant realization occurred. Over 90 percent of the past projects had either failed or never even started. An investigation of these failures ensued.

The Pause

At this point a series of meetings with upper level managers including the Vice President (later Provost) for Academics was initiated. A series of surveys were sent out to get information about these failures. The returns indicated a variety of reasons: the system did not meet their needs, low budget, hence a requirement that people who used the system kick in from their resources, awkward means of interacting with the system, inability to modify the systems once flaws were found.

During this time the student research assistant created a system to handle the requirements as understood. This system consisted of: a set of decisions required by each user, the time deadlines, and the information required by the decisions, a set of actions required that were to be automated such as letters of acceptance, letters asking for more information, and letters reminding students of deadlines; and most importantly a system flow diagram that pinpointed interactions among future users of the system. Student letters asking for information were separated into two groups: (a) one that required a form answer and (b) more complex issues that required a personal answer. This list was reviewed by University software developers and feasibility was assessed. IBM had an interest and donated time to help assess feasibility.

The above study highlighted the amount of cooperation required from University functions and their clerical personnel to use the system successfully. Everyone realized that a training package had to be developed before the system could be activated.

The Implementation Plan

The chairman of the IE department and the study manager had a series of meetings to address the problem of implementation. The need for training was determined to be an easy problem to design and manage. The major problem was the financial issue and the perceived quality issues that plagued previous projects. A series of personal interviews with users and system designers revealed a further problem: the selective memory of users. After a new system was begun users would frequently say that they had unmet needs that had been communicated to designers but ignored in the final product. Users also had little concept of the changes in operating protocol that would be needed.

The following solution emerged. If you wanted help, you had to pay, but the Provost would match the dollars from the managers' budget from his own. The final part of the implementation protocol was to circulate the plan to the managers (users) and then meet personally with them. The provost required one of two responses: The plan suits my need or it fails and here is what is missing. Time deadlines for responses were given. A new plan was devised and the same process was carried out. When no gaps were found by users, then there was a final document circulated which required the signature of the user. They either signed saying that the system met their needs or the system failed. If the system failed, they had to give reasons. Those users with problems met with the study manager and the provost to discuss their problems. In front of the Provost, managers were candid and cooperative. After this meeting, every user who had not signed, signed. The head of the computer center took responsibility for developing and testing the software in time for implementation just after Christmas.

The Outcome

The resulting system was programmed and proved to be a rousing success. The offload on users and their clerical staff was huge. The growth of the University was enabled and further data processing projects grew rapidly. PSIP proved to be the basis of the complex information system that encompasses the University today.

ILLUSTRATIVE EXAMPLE: SYSTEMS ENGINEERING CURRICULUM MANAGEMENT SYSTEM (CMS)—IMPLEMENTATION

Robert Kewley, Ph.D. U.S. Military Academy

Planning For Action

The CMS project encompasses the design, development, deployment, training, use, maintenance, improvement, and retirement of the CMS system for the Department of Systems Engineering. The department's CIO has oversight of this project, and development will take place using a combination of internal IT staff and capstone students. The CIO, with approval from the department head, will schedule a phased development and deployment of the system to support the teaching calendar and ABET accreditation requirements. The goal is to have all components of the system deployed within 13 months in order to support ABET data collection beginning in August 2007. Deployment during the 2006–2007 academic year will allow the department faculty to learn the system, use it, and provide feedback to the development team as to how it can better support their needs. The development team created a phased project plan using project management software. Their first step was to load all of the functions from functional analysis into the project plan. The next step was to break those functions into phases so that development could proceed in accordance with the academic calendar and ABET requirements. In addition, some functions naturally precede others. For example, the Develop Program function relies on the use of course data. Therefore, it made sense to sequence the Develop Course and Execute Course functions before the Develop Program function. The Integrate Department Academic Operations function was least critical to ABET assessment, so it was sequenced last. Figure 13.7 shows a breakdown of the project by phases. Once the development team had developed a project plan, they took a detailed look at the Phase I activities and developed a work breakdown structure using project management software. They first had to look at each function and identify any additional development tasks needed in order to support implementation of each function. For example, the function Develop Course Strategy required the addition of two supporting tasks, Develop Course-Level Data Tables and Develop Course-Level Portal. Once all tasks were added, they had to be scheduled in order to meet Phase I deployment timelines—to include an acceptance test at the end of Phase I on 1 August 2006. In addition, some tasks required other tasks to be completed before they could be started. In order to implement the task Develop Interface for Course Objectives, the Develop Course-Level Data Tables task had to be completed so that the objectives could be stored in the database.

EXECUTION

In order to begin execution of the project plan, the team identified available development resources from the department's internal IT staff and assigned development engineers to the identified tasks. A portion of the resulting work breakdown structure with tasks assigned to different developers is shown in Fig. 13.8.

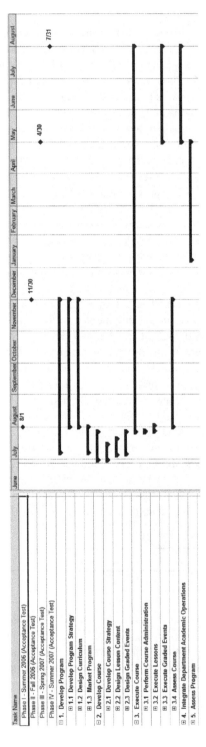

Figure 13.7 CMS project plan by phase.

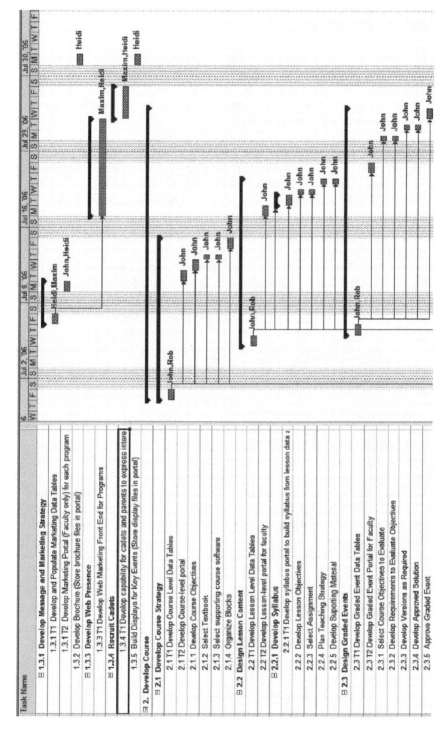

Figure 13.8 Detailed work breakdown structure for Phase I tasks.

ASSESSMENT AND CONTROL

The purpose of monitoring is to ensure that the system enables execution of the supported functions in accordance with the system design goals and expectations of the stakeholders. In order to monitor the performance of the CMS across all phases of the system life cycle, the development team incorporated three different techniques.

Acceptance Testing. At the end of each phase of the development process, the program directors will do acceptance testing of the system to ensure it meets the design requirements specified in the needs analysis. Deficiencies will be noted and corrected. If serious enough, these deficiencies could delay deployment of the system until its performance is acceptable.

Online Feedback. The development team will incorporate an online feedback capability for users of the CMS system. At any time, users of the system, including cadets, instructors, and program directors, can follow a link to a free text feedback page that allows them to provide feedback about how the system meets their needs. This feedback will be stored in the system database and reviewed monthly by the development team. Based on that feedback, they can change and update the CMS as required.

Focus Groups. At the end of each semester, the development team will conduct a focus group with CMS users in order to get a more structured assessmente of how the system is performing in accordance with stakeholder needs. The results of these focus groups will be integrated into the development and maintenance plan.

As the department continues to develop and use the CMS, the assessment feedback will guide decisions about maintaining or upgrading the system. They will also be able to assess adequacy of user training. At some point during the life cycle, information from the assessment process will indicate that the system is beyond its useful life. This will lead to retirement of the CMS and possible development of a new system to perform the curriculum management function.

13.9 EXERCISES

13.1. Describe the Solution Implementation phase.

13.2. During the Solution Implementation phase, what signifies the completion of the project? Explain your answer.

13.3. Draw and describe the project management process.

13.4. Why is the Initiating Process so important?

13.5. There are several techniques used to assist in the project manager in developing a project plan. List one method and explain how it works.

13.6. Why is project management considered more of an art than a science?

13.7. Define the tasks within the Solution Implementation phase. Which task is the most important? Explain your answer.

13.8. What is a work breakdown structure? Why is it important to a PM?

13.9. What is linear responsibility chart? Why is it important to a PM?

13.10. Why is the Closing Process so important?

13.11. Explain the relationship of the Solution Implementation phase and the project life cycle stages.

13.12. The case study reveals a common problem among users and stakeholders: unmet user needs. What do you recommend to ensure that user and stakeholder needs are met?

13.13. Why is the Solution Implementation phase considered the most difficult phase of the SDP? Explain your answer.

13.14. Reviewing the case study, what do you think was the primary reason for solution implementation success? Why?

REFERENCES

1. Project Management Institute. *A Guide to the Project Management Body of Knowledge*, 3rd ed. Newtown Square, PA: Project Management Institute, 2004.

2. Kerzner, H. *Project Management: A Systems Approach to Planning, Scheduling, and Controlling*, 9th ed., Hoboken, NJ: John Wiley & Sons, 2006.

3. Meredith, JR, Mantel, SJ. *Project Management*. New York: John Wiley & Sons, 2006.

4. van Gigch, JP. *System Design Modeling and Metamodeling*. New York: Plenum Press, 1991.

5. Mantel, SJ, Jr., Meredith, JR., Shafer, SM, Sutton, MM. *Project Management in Practice*. Hoboken, NJ: John Wiley & Sons, 2001.

6. Forsberg, K, Mooz, H, Cotterman, H. *Visualizing Project Management*, 2nd ed. New York: John Wiley & Sons, 2000.

7. Smith, PG, Reinertsen, DG. *Developing Products in Half the Time*, 2nd ed. New York: John Wiley & Sons, 1998.

8. Fisk, ER, Rapp, RR. Engineering Construction Inspection. Hoboken, NJ: John Wiley & Sons, 2004.

9. Palmer, D. *Maintenance Planning and Scheduling Handbook*, 2nd ed. New York: McGraw-Hill, 2006.

10. Eisner, H. *Essentials of Project and Systems Engineering Management*, 2nd ed. New York: John Wiley & Sons, 2002.

11. Athey, TH. *Systematic Systems Approach: An Integrated Method for Solving Problems*. Boston, MA: Pearson Custom Publishing, 1982.

12. Sage, AP, Armstrong, JE, Jr. *Introduction to Systems Engineering*. New York: John Wiley & Sons, 2000.

13. Buede, DM. *The Engineering Design of Systems: Models and Methods*. New York: John Wiley & Sons, 2000.

14. Blanchard, BS, Fabrycky, WJ. *Systems Engineering and Analysis*, 3rd ed. Upper Saddle River, NJ: Prentice-Hall, 1998.

Chapter **14**

Summary

GREGORY S. PARNELL, Ph.D.

The system votes last.
—Jack Clemons, Senior VP Engineering (Ret), Lockheed Martin Corporation

Systems are developed to perform critical corporate and public functions. Effective and efficient systems provide great benefits to corporations (and their stockholders) and governments (and their citizens). Ineffective systems can lead to deaths (the two space shuttle tragedies), cause financial ruin of a corporation (Enron), or drastically impact the safety of a region (Hurricane Katrina) or a nation (9/11). Systems engineering and systems decision making are tools to help us deal with complex, interconnected systems designed to provide products and services in dynamic environments in the face of natural hazards and adaptive adversaries. As the above quote reminds us, there are consequences for using poor processes and making decisions that are not supported by timely, fact-based data. Systems decisions are hard enough if decision makers are provided the essential data to support timely, logical, defendable decisions. Without the essential data, systems decisions are problematic.

Systems thinking, systems engineering, and systems decision making are powerful tools to help decision makers make fact-based decisions that involve key stakeholders and consider the future operating environment of the system. In this chapter, we summarize the major insights that we hope the reader has learned from this book. In total, the book is organized into three parts: systems thinking, systems

Decision Making in Systems Engineering and Management, Second Edition
Edited by Gregory S. Parnell, Patrick J. Driscoll, Dale L. Henderson
Copyright © 2011 John Wiley & Sons, Inc.

engineering, and the systems decision process. Each part introduces the reader to fundamental ideas in such a way as to logically construct a cohesive understanding of how to completely support decision making in the context of systems engineering and management. In the sections that follow, we summarize the themes of the three parts and key insights for successful systems engineering, concluding with some thoughts on future systems engineering challenges.

14.1 SYSTEMS THINKING—KEY TO SYSTEMS DECISION MAKING

Part I addressed the philosophy and concepts associated with systems thinking. In essence, systems thinking resists the instinctive response to decompose a systems decision problem into its smallest parts without first understanding the system as a whole. Possessing such a holistic understanding unveils the structure of a system while clearly providing a framework to explore systems interactions occurring at the seam between the system boundary and other entities in the environment. This is a critical capability for systems engineering teams because it is widely acknowledged that today's complex, interacting systems can only be understood by considering the dynamic behavior of the system with its environment.

In Chapter 1, we introduced key terminology and concepts that were subsequently developed in the book. In Chapter 2, we introduced the philosophy, language, and techniques used to engage with systems thinking. In Chapter 3, we described the structural concept of a system life cycle, illustrating the benefits of leveraging this temporal organization to perform a host of key activities, including risk management. In Chapter 4, we described the essential role of modeling and simulation in systems analysis. In Chapter 5, we presented the concepts and techniques for developing system life cycle cost analysis. Five major themes were developed in Part I and reinforced in the later chapters of the book.

14.1.1 Systems Thinking Reveals Dynamic Behavior

Systems do not exist in isolation. Systems thinking begins by carefully identifying the boundary along with a list of stakeholders who may be impacted by the decision. They exist, operate, and compete within an ever-changing environment that not only evolves in its state condition across a host of factors, but also alters the type, intensity, and purpose of interactions with the system under study. Understanding the current and future operating environment is essential to identifying critical functions, objectives, interfaces, constraints, and requirements that drive systems decision making. Systems engineers and systems managers are educated and trained to identify stakeholders, their vested interests, and any potential influences these stakeholders might affect on the system under study.

14.1.2 The System Life Cycle Must Be Considered

Major decisions affecting the well-being of systems are made during each stage of the system life cycle. While these decisions oftentimes occur during discrete time intervals, they are nonetheless interdependent. Decisions made early in the system life cycle can dramatically impact future system performance, life cycle costs, and the likelihood of specific risk events transpiring. Therefore, in supporting systems decision-making, systems teams must extend the impact horizon of decisions made in the near term to consider and predict as accurately as possible what these decisions imply for the future. This is part of proper due diligence associated with helping key stakeholders decide between candidate solutions in which a tradeoff must be considered.

14.1.3 Modeling and Simulation—Important Tools

For most modern systems, it is simply not practical to expect systems teams to construct physical prototypes of candidate solutions to provide decision makers with sufficient and substantial information concerning critical dynamic behavior. Enter modeling and simulation as the widely accepted modern day testbed for these candidate solutions. Early in the system life cycle, modeling and simulation are the essential tools to develop system concepts and help the systems designers and engineers understand system performance issues in the operating environment. Models and simulations also have many uses throughout the system life cycle to support all major systems decisions.

14.1.4 The System Life Cycle is a Key Risk Management Tool

The fundamental purpose of system analysis, development, and testing is to manage risks that have the potential to threaten the value return that a system promises. The system life cycle is a stage-and-gate process. Risk assessment should be performed at each decision gate. The risks carried forward from the previous stage and the risks expected in the next stage must be assessed at the start of each stage and managed during the next stage. The system should not enter the next stage until the key risks have been assessed and managed. Systems engineers and engineering managers have an important role in risk management identification, assessment, communication, and (identification and assessment) throughout the system life cycle.

14.1.5 Life Cycle Costing Is an Important Tool for Systems Engineering

Decision makers, clients, system owners and other stakeholders need to understand the system resource requirements so they can adequately plan and budget for the system. In addition, systems decisions must be informed by a complete understanding of the resource implications for stakeholders in each stage of the system life cycle. Life cycle costing provides useful techniques to estimate systems costs in each stage and provide key decision information required for successful system realization.

14.2 SYSTEMS ENGINEERS PLAY A CRITICAL ROLE IN THE SYSTEM LIFE CYCLE

Part II transitioned from fundamental systems thinking concepts to the practice of professional systems engineering. Systems engineering is an interdisciplinary approach and means to enable the realization of successful systems. It focuses on defining customer needs and required functionality early in the development cycle, documenting requirements, then proceeding with design synthesis and system validation while considering the complete problem. In Chapter 6, we provided an introduction to systems engineering. In Chapter 7, we described what professional systems engineers do. In Chapter 8, we introduced system reliability and operational suitability techniques used by systems engineers to design, develop, and monitor system operation. Four major themes were developed in Part II.

14.2.1 Systems Engineers Lead Interdisciplinary Teams to Obtain System Solutions that Create Value for Decision Makers and Stakeholders

Systems engineers must develop and lead interdisciplinary teams in each stage of the system life cycle. Systems engineers must consider the complete problem in major systems decisions in each stage of the system life cycle. The problems change in each stage, but an interdisciplinary approach to systems thinking and problem definition is always required due to the complexity of the system and the diversity of stakeholders. Systems engineers need to understand the system component engineering disciplines, work effectively with many disciplines, and know when to bring interdisciplinary teams together to solve requirements, design, test, and operational problems. Systems engineers need to focus on identifying opportunities to create value for consumers, customers, and users.

14.2.2 Systems Engineers Convert Stakeholder Needs to System Functions and Requirements

Systems engineers convert stakeholder needs to technical statements that engineering designers can use to develop the system design. Systems engineers work with future clients, system owners, systems users, and consumers of system products and services to determine the functions the system must perform and requirements the system must meet to provide products and services in the future environment. Requirements analysis and trade studies are key tools to ensure the system concepts, designs, products, and services will be affordable.

14.2.3 Systems Engineers Define Value and Manage System Effectiveness

Systems are created to provide value to stakeholders. Systems engineers define performance effectiveness measures to guide design synthesis, system validation, and test systems to solve the defined problem. Systems engineers use availability, reliability, effectiveness, and maintainability modeling to define and assess system performance. The basis for system design and validation is usually an iterative sequence of functional analysis, requirements analysis, modeling, simulation, development, test, production, and evaluation. Once the system design is validated, the systems engineer must continue to work on the successful deployment and operation of the system.

14.2.4 Systems Engineers Have Key Roles Throughout the System Life Cycle

In each stage of the system life cycle, systems engineers play important roles. Chapter 7 defined these roles in some detail. The key role of systems engineering is to assemble and lead teams to identify and resolve systems issues during the life cycle. On many programs they are systems thinker and the technical "honest broker" for the program manager. For major decisions, they analyze technical tradeoffs and present a "system perspective" recommendation to the project manager.

14.3 A SYSTEMS DECISION PROCESS IS REQUIRED FOR COMPLEX SYSTEMS DECISIONS

Part III of the text developed and described the systems decision process. For complex, interconnected, and dynamic systems involving multiple stakeholders, systems decisions may require extensive analysis, justification, reviews, and approvals by other organizations or agencies besides the program management organization. For these systems, a systems decision process may be required. In Chapter 9, we introduced our four-phase systems decision process. We also discussed when to use and when not to use the process. In Chapter 10, we introduced the three problem definition tasks: research and stakeholder analysis, functional and requirements analyses, and value modeling. In Chapter 11, we described the three solution design tasks: idea generation, alternative generation and enhancement; and cost analysis. In Chapter 12, we described the three decision making tasks: solution scoring, sensitivity analysis, and value-focused thinking. In Chapter 13, we described the three solution implementation tasks: planning, executing, and monitoring and controlling.

We summarize each of the major themes for the four phases of the systems decision process.

14.3.1 Problem Definition Is the Key to Systems Decisions

Problem definition is the most important phase of our process. The initial problem is never the final problem. Research and stakeholder analysis are key first steps for defining the problem from the perspective of stakeholders from many disciplines. The key stakeholders are clients, owners, users, and consumers. The key techniques of stakeholder analysis are interviews, focus groups, and surveys. Each has advantages and disadvantages. The second set of tasks include functional and requirements analyses. Systems engineers must identify what functions a system must perform. As we progress through the system life cycle, additional information (requirements, constraints, interfaces, etc.) will be developed. The third task is value modeling. Systems engineers must define the objectives and the value measures that will guide solution development to create value for consumers of system products and services. These three tasks are critical for the systems decisions in each life cycle stage.

14.3.2 If We Want Better Decisions, We Need Better System Solution Designs

The key to architecture, system-of-systems, and systems design is to create candidate solutions that provide great value for clients, owners, users, and consumers. Creative design engineers are the key to better design solutions. The creativity techniques introduced in Chapter 11 can help the designers. Modeling and simulation can help ensure that alternatives are feasible, effective, and efficient. We need to screen the infeasible alternatives and then use value-focused thinking to help identify opportunities to create better candidate solutions. We need to develop cost models during solution design to ensure that the candidate solutions are affordable.

14.3.3 We Need to Identify the Best Value for the Resources

When we complete the solution design phase, we have several candidate solutions. The purpose of the decision-making phase is to provide the essential information that the decision maker needs to make a timely, sound, and defensible decision. Each of the candidate solutions should be scored using the most appropriate method (operational data, development data, test data, modeling and simulation data, or expert opinion). The systems engineer should identify the most sensitive assumptions of the analysis and identify and analyze key uncertainties and risks affecting success. Finally, the systems engineer should use insights from scoring and sensitivity analysis to encourage the design team to create higher value solutions. We assess the system resources with life cycle cost models. When time runs out (as it always does!), we present the nondominated solutions and let the decision makers choose which system solution provides the best value for the resources. Our oral presentation and written reports must be clear, concise, and cogent.

14.3.4 Solution Implementation Requires Planning, Executing, and Monitoring and Controlling

The implementation of the solution can be very frustrating. The key stakeholders must support the implementation of the system solution. A sound decision can fail to achieve the planned value due to poor implementation. The keys to solution implementation are previous stakeholder involvement, good task identification, clear task assignments, development of performance measures, feasible monitoring procedures, clear assessment plan, and effective control measures. An important part of solution implementation is risk identification, assessment, communication, and management.

14.4 SYSTEMS ENGINEERING WILL BECOME MORE CHALLENGING

As systems become more complex, interconnected, and dynamic, it is difficult to imagine that the task of systems engineers will become easier. As more stakeholders become interested in the consequences of systems decisions, more stakeholders will become interested in systems decisions that will affect them. We believe that increasing complexity, increasing interconnectedness, increasing security challenges, and increasing stakeholder interest will make the systems engineer's job interesting and challenging for the foreseeable future.

Appendix A

SDP Trade Space Concepts

Each of the $n = 1, 2, \ldots N$ competing solutions created during an SDP application are evaluated against a set of $m = 1, 2, \ldots M$ value measures whose numerical estimate of value, $VM_{nm} = f_m(x_{nm}) \in [0, 100]$, results from translating data estimates (aka: scores) x_{nm} into a common unit of value via stakeholder assigned value functions f_m. Value functions can be discrete or continuous over their domains and easily accommodate both objective and subjective data estimates. These value functions are, in turn, aggregated into a value return estimate, V_n, for each of the n competing feasible solution alternatives, using the additive value model most commonly seen in multiobjective decision analysis [1]: $V_n = \sum_1^M w_m \cdot f_m(x_{nm})$, where $0 < w_m \leq 1$ and $\sum_1^M w_m = 1$. Infeasible alternatives are eliminated from consideration prior to this point because they fail to meet mandatory requirements (screening criteria) identified by stakeholders and accepted by the decision maker. Efforts to establish and maintain independence between value measures begin during the construction of the qualitative value model and continue throughout the SDP as data estimates and new information become available.

The individual weights w_m used as multipliers for the m value functions are normalized swing weights s_m such that $w_m = s_m / \sum_{m=1}^M s_m$, $0 < s_m \leq UB$, where UB is an arbitrary upper bound [2]. These swing weights reflect a decision maker's assessment of relative importance of each of the m value measures combined with an estimate of the impact of value measure range swings ([3],[4]) on the decision. This latter estimate enhances the benefits afforded by weighting beyond simple preference ranking, especially for those value measures in which a small change in numerical value has a very large impact on the decision [5].

Decision Making in Systems Engineering and Management, Second Edition
Edited by Gregory S. Parnell, Patrick J. Driscoll, Dale L. Henderson
Copyright © 2011 John Wiley & Sons, Inc.

This quantitative value model, while similar to the normative approach in utility models [6] from the standpoint of using a function to convert dissimilar units into a common unit of measure, differs from utility modeling in that these value functions are not required to be supported by certain equivalence relationships, are not assessed with lotteries, and are not affected by stakeholder risk attitude [7]. Thus, when properly constructed a value model simply translates and aggregates key system performance estimates into a stakeholder-weighted, value return estimate for each solution. This translational layer of value functions also distinguishes the SDP quantitative value model structure from data envelopment analysis (DEA) models that effectively employ ratios to eliminate disparate data units for aggregating measures [8].

Cost enters consideration neither as a value measure nor as screening criteria in the construction of tradeoff models. Estimates of cost for each feasible solution alternative, C_n, are developed in separate models incorporating cost estimating relationships, time value of money, system life cycle, reliability, and other considerations (see Chapter 5). While typically defined as life cycle monetary costs, C_n can be more generically defined in terms of resource expenditures such as time or effort, elements of risk such as loss of life or compromise of critical information, and so on, thereby affording a good deal of flexibility in applications.

Definition 1. The Cartesian plot of each solution's life cycle cost estimate C_n versus value return estimate V_n defines a deterministic *decision trade space*, $D \subset \mathbb{R}^2$, in which each solution $A_n = (C_n, V_n) \in D$ strives to achieve a single, stakeholder expressed ideal, $A_{\text{ideal}} = (C_{\text{ideal}}, V_{\text{ideal}}) \in D$.

The ideal values in each dimension are characterized by an upper bound on total value: $V_{\text{ideal}} = V_{UB}$, and a lower bound on total life cycle cost: $C_{\text{ideal}} = C_{LB}$. In most applications, the chosen ideals are set to 100 and 0, respectively, predominantly because for ease of understanding on the part of stakeholders. The existence of a stakeholder ideal establishes a partial preference ordering on D.

Definition 2. For any two solutions $A_k, A_n \in D$, with $A_k = (C_k, V_k)$ and $A_n = (C_n, V_n), k \neq n, A_k$ is preferred to A_n, denoted by $A_k \prec A_n$, whenever $C_k \leq C_n$ and $V_k \geq V_n$ (unless equality holds for both, which yields indifference).

The preference ordering underlying both axes in the decision trade space motivates the principle that a rational decision maker should not desire lower value for higher cost. This rationality, coupled with the properties of trade space efficiency and dominance defined in what follows, supports the construction of a *choice set* C^* of solutions from which a rational decision maker should select a solution.

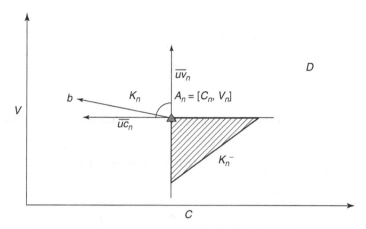

Figure A.1 The cone K_n and polar cone K_n^- of alternative A_n.

Let D be the deterministic decision trade space created via the SDP with preference ordering on C_n and V_n as defined. Let $A_n = (C_n, V_n) \in D, n = 1, \ldots N$ be a set of solutions with $A_{\text{ideal}} = (C_{LB}, V_{UB}) \prec A_n$ for all n, so that A_{ideal} is logically preferred over A_n. For each A_n, define two linearly independent *generating unit vectors*: $\vec{uc}_n = [-1, 0]$ and $\vec{uv}_n = [0, 1]$, originating from translated origins centered on each A_n. Figure A.1 illustrates these elements with respect to a single solution A_n.

Definition 3. The *cone* $K_n \subseteq D$ of A_n is the set of all vectors (points) $b \in D$ such that $b = A_n + \lambda_1 \vec{uc}_n + \lambda_2 \vec{uv}_n$, with $\lambda_1, \lambda_2 \geq 0$.

Since all points in D correspond to solutions, Definition 3 states that any solution lying in the cone K_n of A_n is a nonnegative linear combination of the generating unit vectors $\vec{uc}_n$ and $\vec{uv}_n$. Figure A.2 shows Solution 4 as being contained in the cone of Solution 5 as $A_4 = [C_4, V_4] = A_5 + \lambda_1 \vec{uc}_5 + \lambda_2 \vec{uv}_5$ with $\lambda_i \geq 0, i = 1, 2$.

Definition 4. Given a cone $K_n \subseteq D$ of A_n, the *polar cone* of $A_n, K_n^- \subseteq D$ is defined as the set of all points $l \in D$ such that $l = A_n - \lambda_1 \vec{uc}_n - \lambda_2 \vec{uv}_n$ with $\lambda_i \geq 0, i = 1, 2$.

Equivalently, the polar cone can be defined using the inner product $\langle l, b \rangle = \|l\| \|b\| \cos \theta$ as $K_n^- = \{l \in D : \langle l, b \rangle \leq 0 \forall b \in K_n\}$. Thus, the polar cone accounts for all points in D whose vectors extending from the translated origin of solution A_n make an angle $90 \leq \theta \leq 180$ degrees with any vector corresponding to points within or on the boundary of the cone K_n (See [9], or [10]).

Figure A.3 shows a set of solutions with superimposed generating unit vectors. Given the preference ordering on D, a solution A_k lying in the cone K_n of another

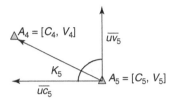

Figure A.2   Solution 4 is contained in the cone K_5 of Solution 5: A_4 strictly dominates A_5.

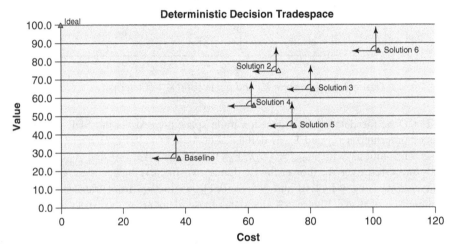

Figure A.3   Solutions with their cone generating unit vectors.

solution A_n has $C_k \leq C_n$ and $V_k \geq V_n$. It follows that there is a need to define equivalence between solutions in this trade space.

Definition 5. A solution $A_n \in D$ *dominates* solution $A_k \in D, k \neq n$, (equivalently, A_k is trade space inferior to A_n) if and only if $A_n \in K_k$ with $\lambda_1 > 0$ or $\lambda_2 > 0$, but not both.

Definition 6. A solution $A_n \in D$ *strictly dominates* solution $A_k \in D, k \neq n$, (equivalently, A_k is strictly trade space inferior to A_n) if and only if $A_n \in K_k$ with both $\lambda_1 > 0$ and $\lambda_2 > 0$.

Remark. Let $A_k, A_n \in D, k \neq n$. A_n dominates A_k if its value is greater ($V_n > V_k$) and its cost is at least a small ($C_n \leq C_k$), or its cost is less ($C_n < C_k$) and its value is at least as great ($V_n \geq V_k$). If both inequalities are strict, then A_n strictly dominates A_k.

Corollary 6.1. A solution A_n *dominates* A_k if $A_n \neq A_k$ and A_k lies in the polar cone of A_n.

Corollary 6.2. A solution A_n *strictly dominates* A_k if $A_k \in K_n^-$ but does not lie on its boundary.

Finally, we define the notion of trade space efficiency that is fundamental to decision support via the SDP.

Definition 7. A solution $A_n \in D$ is *trade space efficient* if no other solution $A_k, k \neq n$ lies in its cone except the ideal, A_{ideal}.

Definition 8. The *choice set*, C^*, is the set of all trade space efficient solutions.

This concept of trade space efficiency depends solely on the determination of dominance among solutions. In a systems setting, it is sometimes the case that the existing (baseline) system is efficient for returning value for cost. However, degraded value return over time due to maturing life cycle stages, changes in the environment, and other factors initiate a decision problem in which improved value return for potentially greater cost is sought. The goal of value modeling in decision support is to identify a trade space efficient set of solutions from which the decision maker should pick.

Remark. The choice set $C^* \in D$ constructed by quantitative modeling via the SDP is a trade space efficient (nondominated) set of solutions.

Trade space efficiency is for most value modeling applications not equivalent to the formal economic notion of Pareto efficiency [11] underscoring data envelopment analysis (DEA) models. This characterization of choice set membership via cones is a much simpler means of identifying the subset of trade space efficient solutions a related optimization-dependent *value free efficiency* calculation affords [12] for weighted aggregate data, largely due to the existence of A_{ideal} which is absent in a DEA setting.

The trade space cost modeling as described in this book will always conclude with a nonempty choice set of trade space efficient solutions. Once an appropriate uncertainty analysis is performed on the choice set solutions by the systems engineering team using Monte Carlo simulation, the decision maker should either select from these solutions, apply decision-focused thinking, or use value-focused thinking to develop creative solutions not previously identified (see Chapter 12).

REFERENCES

1. Kirkwood, C.W. *Strategic Decision Making: Multiobjective Decision Analysis with Spreadsheets*, Belmont, CA: Duxbury Press, 1997.
2. Dillon-Merrill, R., G.S. Parnell, Buckshaw, D. Hensley, W.R. Caswell, D. "Avoiding common pitfalls in decision support frameworks for Department of Defense analyses," *Military Operations Research Journal*, 2008; 13(2): 19–31.
3. Parnell, G.S., Trainor, T.E. "Using the swing weight matrix to weight multiple objectives," to appear in *Systems Engineering*, Manuscript 2009.
4. Belton, V., Stewart, T.J. *Multiple Criteria Decision Analysis: An Integrated Approach*, Boston: Kluwer 2002.
5. Ewing, P., Tarantino, W. Parnell, G.S. "Use of decision analysis in the Army base realignment and closure (BRAC) 2005 military value analysis," *Decision Analysis*, 2006; 3 (1): 33–49.
6. Bell, D.E., Raiffa, H. Tversky A. (Eds.). *Decision Making: Descriptive, Normative, and Prescriptive interactions*, New York; Cambridge University Press, 1988.
7. Howard, R.A., J.E. Matheson (Eds.). *Readings on the Principles and Applications of Decision Analysis*, Menlo Park, CA: Strategic Decisions Group, 1984.
8. Cooper, W.W., Seiford, L.M. Tone, K. *Data Envelopment Analysis*, New York: Springer Science + Business, 2007.
9. Curtis, C.W. *Linear Algebra: An Introductory Approach*, New York: Springer-Verlag, 1984.
10. Rockafellar, R.T. *Convex Analysis*, Princeton, NJ: Princeton University Press, 1996.
11. Lotov, A.V., Bushenkov, V.A. Kamenev, G.K. *Interactive Decision Maps*, Applied Optimization, Volume 89, New York: Klewer, 2004.
12. Yun, Y.B., Nakayama, H. Tantino, T. "On efficiency of data envelopment analysis," in *Research and Practice in Multiple Criteria Decision Making*, Lecture Notes in Economics and Mathematical Systems, Y.Y. Haimes and R.R. Steuer (Eds.), New York: Springer, 1998.

Index

WILEY SERIES IN SYSTEMS ENGINEERING AND MANAGEMENT

Andrew P. Sage, Editor

YACOV Y. HAIMES
Risk Modeling, Assessment, and Management, Third Edition

DENNIS M. BUEDE
The Engineering Design of Systems: Models and Methods, Second Edition

ANDREW P. SAGE and JAMES E. ARMSTRONG, Jr.
Introduction to Systems Engineering

WILLIAM B. ROUSE
Essential Challenges of Strategic Management

YEFIM FASSER and DONALD BRETTNER
Management for Quality in High-Technology Enterprises

THOMAS B. SHERIDAN
Humans and Automation: System Design and Research Issues

ALEXANDER KOSSIAKOFF and WILLIAM N. SWEET
Systems Engineering Principles and Practice

HAROLD R. BOOHER
Handbook of Human Systems Integration

JEFFREY T. POLLOCK and RALPH HODGSON
Adaptive Information: Improving Business Through Semantic Interoperability, Grid Computing, and Enterprise Integration

ALAN L. PORTER and SCOTT W. CUNNINGHAM
Tech Mining: Exploiting New Technologies for Competitive Advantage

REX BROWN
Rational Choice and Judgment: Decision Analysis for the Decider

WILLIAM B. ROUSE and KENNETH R. BOFF (editors)
Organizational Simulation

HOWARD EISNER
Managing Complex Systems: Thinking Outside the Box

STEVE BELL
Lean Enterprise Systems: Using IT for Continuous Improvement

J. JERRY KAUFMAN and ROY WOODHEAD
Stimulating Innovation in Products and Services: With Function Analysis and Mapping

WILLIAM B. ROUSE
Enterprise Tranformation: Understanding and Enabling Fundamental Change

JOHN E. GIBSON, WILLIAM T. SCHERER, and WILLAM F. GIBSON
How to Do Systems Analysis

WILLIAM F. CHRISTOPHER
Holistic Management: Managing What Matters for Company Success

WILLIAM B. ROUSE
People and Organizations: Explorations of Human-Centered Design

MO JAMSHIDI
System of Systems Engineering: Innovations for the Twenty-First Century

ANDREW P. SAGE and WILLIAM B. ROUSE
Handbook of Systems Engineering and Management, Second Edition

JOHN R. CLYMER
Simulation-Based Engineering of Complex Systems, Second Edition

KRAG BROTBY
Information Security Governance: A Practical Development and Implementation Approach

JULIAN TALBOT and MILES JAKEMAN
Security Risk Management Body of Knowledge

SCOTT JACKSON
Architecting Resilient Systems: Accident Avoidance and Survival and Recovery from Disruptions

JAMES A. GEORGE and JAMES A. RODGER
Smart Data: Enterprise Performance Optimization Strategy

YORAM KOREN
The Global Manufacturing Revolution: Product-Process-Business Integration and Reconfigurable Systems

AVNER ENGEL
Verification, Validation, and Testing of Engineered Systems

WILLIAM B. ROUSE (editor)
The Economics of Human Systems Integration: Valuation of Investments in People's Training and Education, Safety and Health, and Work Productivity

ALEXANDER KOSSIAKOFF, WILLIAM N. SWEET, SAM SEYMOUR, and STEVEN M. BIEMER
Systems Engineering Principles and Practice, Second Edition

GREGORY S. PARNELL, PATRICK J. DRISCOLL, and DALE L. HENDERSON (editors)
Decision Making in Systems Engineering and Management, Second Edition